The kinematics of mixing: stretching, chaos, and transport

J. M. OTTINO
Department of Chemical Engineering,
University of Massachusetts

CAMBRIDGE
UNIVERSITY PRESS

PUBLISHED BY THE PRESS SYNDICATE OF THE UNIVERSITY OF CAMBRIDGE
The Pitt Building, Trumpington Street, Cambridge CB2 1RP, United Kingdom

CAMBRIDGE UNIVERSITY PRESS
The Edinburgh Building, Cambridge CB2 2RU, United Kingdom
40 West 20th Street, New York, NY 10011–4211, USA
10 Stamford Road, Oakleigh, Melbourne 3166, Australia

First published 1989
Reprinted 1997

Printed in the United Kingdom at the University Press, Cambridge

A catalogue record for this book is available from the British Library

Library of Congress Cataloguing in Publication data

Ottino, J. M.
The kinematics of mixing : stretching, chaos, and transport / J. M.
Ottino.
p. cm.
Bibliography : p.
Includes index.
ISBN 0 521 36335 7.   ISBN 0 521 36878 2 (paperback)
1. Mixing. I. Title.
TP156.M5087   1989
660.2'84292–dc19
88-30253

ISBN 0 521 36335 7 hardback
ISBN 0 521 36878 2 paperback

Contents

Preface

The objective of this book is to present a unified treatment of the mixing of fluids from a kinematical viewpoint. The aim is to provide a conceptually clear basis from which to launch analysis and to facilitate the understanding of the numerous mixing problems encountered in nature and technology.

Presently, the study of fluid mixing has very little scientific basis; processes and phenomena are analyzed on a case-by-case basis without any attempt to discover generality. For example, the analysis of mixing and 'stirring' of contaminants and tracers in two-dimensional geophysical flows such as in oceans; the mixing in shear flows and wakes relevant to aeronautics and combustion; the mixing of fluids under the Stokes's regime generally encountered in the 'blending' of viscous liquids such as polymers; and the mixing of diffusing and reacting fluids encountered in various types of chemical reactors share little in common with each other, except possibly the nearly universal recognition among researchers that they are very difficult problems.[1]

There are, however, real similarities among the various problems and the possible benefits from an overall attack on the problem of mixing using a general viewpoint are substantial.

The point of view adopted here is that from a kinematical viewpoint fluid mixing is the efficient stretching and folding of material lines and surfaces. Such a problem corresponds to the solution of the dynamical system

$$d\mathbf{x}/dt = \mathbf{v}(\mathbf{x}, t),$$

where the right hand side is the Eulerian velocity field (a solution of the Navier–Stokes equations, for example) and the initial condition corresponds to the initial configuration of the line or surface placed or fed into the flow ($\mathbf{x}$ represents the location of the initial condition $\mathbf{x} = \mathbf{X}$). Seen in this light, the problem can be formulated by merging the kinematical foundations of fluid mechanics (Chapters 2 and 3) and the theory of dynamical systems (Chapters 5 and 6). The approach adopted here is to analyze simple protypical flows to enhance intuition and to extract conclusions of general validity. *Mixing is stretching and folding and stretching and folding is the fingerprint of chaos.* Relatively simple flows can act as prototypes of real problems and provide a yardstick of reasonable expectations for the completeness of analyses of more complex flows. Undoubtedly, I expect that such a program would facilitate the analysis of mixing problems in chemical, mechanical, and aeronautical engineering, physics, geophysics, oceanography, etc.

The plan of the book is the following: Chapter 1 is a visual summary to motivate the rest of the presentation. In Chapter 2 I have highlighted, whenever possible, the relationship between dynamical systems and kinematics as well as the usefulness of studying fluids dynamics starting with the concepts of *motion* and *flow*.[2] Mixing should be embedded in a kinematical foundation. However, I have avoided references to curvilinear co-ordinates and differential geometry in Chapters 2 and 4, even though it could have made the presentation of some topics more satisfying but the entire presentation slightly uneven and considerably more lengthy. The chapter on fluid dynamics (Chapter 3) is brief and conventional and stresses conceptual points needed in the rest of the work. The dynamical systems presentation (Chapters 5 and 6) includes a list of topics which I have found useful in mixing studies and should not be regarded as a balanced introduction to the subject. In this regard, the reader should note that most of the references to dissipative systems were avoided in spite of the rather transparent connection with fluid flows.

A few words of caution are necessary. Mixing is intimately related to flow visualization and the material presented here indicates the price one has to pay to understand the inner workings of deterministic unsteady (albeit generally periodic in this work) two-dimensional flows and three-dimensional flows in general. However, we should note that the geometrical theory used in the analysis will not carry over when **v** itself is chaotic. Though mixing is still dependent upon the kinematics, the basic theory for analysis would be considerably different. Also, even though many of the examples presented here pertain to what is sometimes called 'Lagrangian turbulence', the reader might find a disconcerting absence of references to conventional (or Eulerian) turbulence. In this regard I have decided to let the reader establish possible connections rather than present some feeble ones.

I give full citation to articles, books, and in a few cases, if an idea is unpublished, conferences. When only a last name and a date is given, particularly in the case of problems or examples, and the name does not appear in the bibliography, it serves to indicate the source of the problem or idea. It is important to note also that some sections of the book can be regarded as work in progress and that complete accounts most likely will follow, expanding over the short descriptions given here; a few of the problems, those at the level of small research papers are indicated with an asterisk (*). In several passages I have pointed out problems that need work. Ideally, new questions will occur to the reader.

Preface to the Second Printing

In preparing the second printing of this book, several typographical and formatting errors have been corrected. The objectives of the book expressed in the original preface remain unchanged. Owing to space constraint limitations the amount of material covered remains approximately the same. The reader interested in the connection of these ideas with turbulence will find some leads in the article 'Mixing, chaotic advection, and turbulence', *Annual Reviews of Fluid Mechanics*, **22**, 207–53 (1990); a succinct summary of extensions of many of the ideas outlined in this book is presented in 'Chaos, Symmetry, and Self-Similarity: Exploiting Order and Disorder in Mixing Processes', *Science*, **257**, 754–60 (1992). I should appreciate comments from readers pertaining to related articles in the area of fluid mixing as well as possible extensions or shortcomings of the ideas presented in this work.

Notes

1 Even the terminology is complicated. For example, in chemical engineering the terms mixing, agitation, and blending are common (Hyman, 1963; McCabe and Smith, 1956, Chap. 2, Section 9; Ulbrecht and Patterson, 1985). The terms mixing, advection, and stirring appear in geophysics; e.g., Eckart, 1948; Holloway and Kirstmannson, 1984. Inevitably, different disciplines have created their own terminology (e.g., classical reaction engineering, combustion, polymer processing, etc.).

2 Kinematics appears as an integral part of books in continuum mechanics but much less so in modern fluid mechanics. There are exceptions of course: Chapters V and VI of the work of Tietjens based on the lecture notes by Prandtl contain and unusually long description of deformation and motion around a point (Prandtl and Tietjens, 1934).

Acknowledgments

This book grew out from a probably unintelligible course given in Santa Fé, Argentina, in July 1985, followed by a short course given in Amherst, Massachusetts, also in 1985. Most of the material was condensed in eight lectures given at the California Institute of Technology in June 1986, where the bulk of the material presented here was written.

The connection between stretching and folding, and mixing and chaos, became transparent after a conversation with H. Aref, then at Brown University, during a visit to Providence, in September 1982. I would particularly like to thank him for communicating his results regarding the 'blinking vortex' prior to publication (see Secton 7.3), and also for many research discussions and his friendship during these years. I would like to thank also the many comments of P. Holmes of Cornell University, on a rather imperfect draft of the manuscript, the comments of J. M. Greene of G. A. Technologies, who provided valuable ideas regarding symmetries as well as to the many comments and discussions with S. Wiggins and A. Leonard, both at the California Institute of Technology, during my stay at Pasadena. I am also grateful to H. Brenner of the Massachusetts Institute of Technology, S. Whitaker of the University of California at Davis, W. R. Schowalter of Princeton University, C. A. Truesdell of Johns Hopkins University, W. E. Stewart of the University of Wisconsin, R. E. Rosensweig and Exxon Research and Engineering, and J. E. Marsden, of the University of California at Berkeley, for various comments and support. I am also particularly thankful to those who supplied photographs or who permitted reproductions from previous publications (G. M. Corcos of the University of California at Berkeley, R. Chevray of Columbia University, P. E. Dimotakis and L. G. Leal, both at the California Institute of Technology, R. W. Metcalfe of the University of Houston, D. P. McKenzie, of the University of Texas at Austin, A. E. Perry of the University of Melbourne, I. Sobey of Oxford University, and P. Wellander of the University of Washington).

I am particularly indebted to all my former and present students, but particularly to R. Chella and D. V. Khakhar, for work prior to 1986, and to J. G. Franjione, P. D. Swanson, C. W. Leong, T. J. Danielson, and F. J. Muzzio, who supplied many of the figures and material used in Chapters 7–9. I am also indebted to H. A. Kusch for help with the proofs and to H. Rising for many discussions during the early stages of this work. Finally, I would like to express my gratitude to D. Tillwick who helped me with the endless task of typing and proofing, to D. Tranah, from Cambridge University Press, for making this project an enjoyable one, and to my wife, Alicia, for help and support in many other ways.

Amherst, Massachusetts

Introduction

1.1. Physical picture

In spite of its universality, mixing does not enjoy the reputation of being a very scientific subject and, generally speaking, mixing problems in nature and technology are attacked on a case-by-case basis. From a theoretical viewpoint the entire problem appears to be complex and unwieldy and there is no idealized starting picture for analysis; from an applied viewpoint it is easy to get lost in the complexities of particular cases without ever seeing the structure of the entire subject.

Figure 1.1.1, which we will use repeatedly throughout this work, describes the most important physics occurring during mixing. In the simplest case, during mechanical mixing, an initially designated material region of fluid stretches and folds throughout the space. This is indeed the goal of visualization experiments where a region of fluid marked by a suitable tracer moves with the mean velocity of the fluid. This case is also closely approximated by the mixing of two fluids with similar properties and no interfacial tension (in this case the interfaces are termed passive, see Aref and Tryggvason (1984)).

Obviously, an exact description of the mixing is given by the location of the interfaces as a function of space and time. However, this level of description is rare because the velocity fields usually found in mixing processes are complex. Moreover, relatively simple velocity fields can produce very efficient area generation in such a way that the combined action of stretching and folding produces exponential area growth. Whereas this is a desirable goal in achieving efficient mixing it also implies that initial errors in the location of the interface are amplified exponentially fast and numerical tracing becomes hopeless. More significantly, this is also a signature of *chaotic* flows and it is important to study the conditions under which they are produced (more rigorous definitions of *chaos* are given in Chapters 5 and 6). However, without the action of molecular diffusion, an instantaneous cut of the fluids reveals a *lamellar structure* (Figure 1.1.1). A measure of the state of mechanical mixing is given by

Figure 1.1.1. Basic processes occurring during mixing of fluids: (*a*) corresponds
to the case of two similar fluids with negligible interfacial tension and negligible
interdiffusion; an initially designated material region stretches and folds by
the action of a flow; (*b*) corresponds to a blob diffusing in the fluid; in this
case the boundaries become diffuse and the extent of the mixing is given by
level curves of concentration (a profile normal to the striations is shown at
the right); in (*c*) the blob breaks due to interfacial tension forces, producing
smaller fragments which might in turn stretch and break producing smaller
fragments. Case (*b*) is an excellent approximation to (*a*) if diffusion is small
during the time of the stretching and folding. In (*a*) the blob is *passive*, in (*c*)
the blob is *active*.

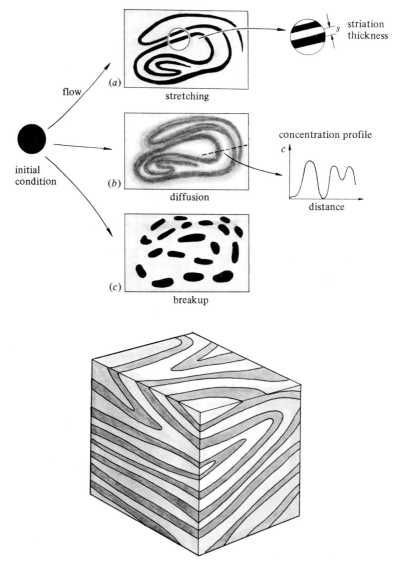

the thicknesses of the layers, say s_A and s_B and $\frac{1}{2}(s_A + s_B)$ is called the *striation thickness* (see Ottino, Ranz, and Macosko, 1979). The amount of interfacial area per unit volume, interpreted as a structured continuum property, is called the *intermaterial area density*, a_v. Thus, if S designates the area within a volume V enclosing the point $\mathbf{x}$ at time t,

$$a_v(\mathbf{x}, t) = \lim_{V \to 0} \frac{S}{V}.$$

Some of the above concepts require modification if the fluids are miscible or immiscible. If the fluids are immiscible, at some point in the mixing process the striations or blobs do not remain connected and break into smaller fragments (Figure 1.1.1(c)). At these length scales the interfaces are not *passive* and instead of being convected (passively) by the flow, they modify the surrounding flow, making the analysis considerably more complicated (in this case the interfaces are termed *active*, see Aref and Tryggvason, 1984). If the fluids are miscible we can still track material volumes in terms of a (hypothetical) non-diffusive tracer which moves with the mean mass velocity of the fluid or any other suitable reference velocity. Designated surfaces of the tracer remain connected and diffusing species traverse them in both directions.[1] However, during the mixing process, connected iso-concentration surfaces might break and cuts might reveal islands rather than striations (Figure 1.1.1(b)). In this case the specification of the concentration fields of $n-1$ species constitutes a complete description of mixing but it is also clear that this is an elusive goal.

Thus, it is apparent that the goal of mixing is reduction of length scales (thinning of material volumes and dispersion throughout space, possibly involving breakup), and in the case of miscible fluids, uniformity of concentration.[2] With this as a basis, we discuss a few of the ideas used to describe mixing and then move to examples before returning to the problem formulation in Section 1.4.

1.2. Scope and early works

A cursory examination of an eclectic and fairly arbitrary listing of some of the earliest references in the literature gives an idea of the scope of mixing processes and the ways in which mixing problems have been attacked in the past.[3]

G. I. Taylor (1934) The formation of emulsions in definable fields of flow, *Proc. Roy. Soc.*, **A146**, 501–23.

A. Brothman, G. N. Wollan, and S. M. Feldman (1945) New analysis provides formula to solve mixing problems, *Chem. Metal. Eng.*, **52**, 102–6.

W. R. Hawthorne, D. S. Wendell, and H. C. Hottel (1948) Mixing and combustion in turbulent gas jets, p. 266–88 in *Third Symp. on Combustion and Flame and Explosion Phenomena*, Baltimore: Williams & Wikens.

C. Eckart (1948) An analysis of the stirring and mixing processes in incompressible fluids, *J. Marine Res.*, VII, 265–75.

R. S. Spencer and R. M. Wiley (1951) The mixing of very viscous liquids, *J. Coll. Sci.*, **6**, 133–45.

P. V. Danckwerts (1952) The definition and measurement of some characteristics of mixtures, *Appl. Sci. Res.*, A3, 279–96.

P. V. Danckwerts (1953) Continuous flow systems-distribution of residence times, *Chem. Eng. Sci.*, **2**, 1–13.

P. Welander (1955) Studies on the general development of motion in a two-dimensional, ideal fluid, *Tellus*, **7**, 141–56.

S. Corrsin (1957) Simple theory of an idealized turbulent mixer, *A.I.Ch.E. J.*, **3**, 329–30.

W. D. Mohr, R. L. Saxton, and C. H. Jepson (1957) Mixing in laminar flow systems, *Ind. Eng. Chem.*, **49**, 1855–57.

Th. N. Zweitering (1959) The degree of mixing in continuous flow systems *Chem. Eng. Sci.*, **11**, 1–15.

It seems at first strange to start a discussion on mixing with Taylor's 1934 paper. However, there are several reasons. The first one is that the problem is of practical importance and was attacked with the best tools of the time, both theoretically and experimentally. The second one is that the natural extension of his ideas to mixing remain largely unfulfilled. Taylor's concerns are obvious from the title of the paper. He distilled the essence of the problem, in general a complex one, and reduced the question to a *local* analysis: the deformation, stretching, and breakup of a droplet in two prototypical flows – planar hyperbolic flow and simple shear flow.[4] Presumably, the long range goal was to mimic a complex velocity field in terms of populations of these two flows.[5] This is similar, in spirit, to the approach adopted in this work, in two respects: (i) analysis of simple building blocks which give useful powerful insights into the behavior of complex problems, and (ii) decomposition of a problem in *local* and *global* components (this idea is reconsidered in Chapter 9).

Brothman, Wollan, and Feldman (1945) had more practical and pressing needs in mind and tried to attack the problem of mixing in a general and abstract way. They spoke of fluid deformation and [fluid] rearrangement, and regarded mixing as a three-dimensional shuffling process. They worked out probabilistic arguments directed predominantly to mixing in closed systems, such as stirred tanks, and obtained kinetic expressions for the creation of interfacial area. Eckart (1948) had in mind substantially larger length scales and started his analysis with the continuum field equations and calculated the 'mixing times' of thermal and saline spots in oceans without resorting to any mechanistic description of the process. A conceptually similar problem in terms of scales, mixing in atmospheric flows, was addressed by Welander (1955). Remarkably, he did so by considering the possibility of applying Hamiltonian mechanics to ideal fluids and stressed the need for studying the stretching and folding of material elements in the flow and devised formulas to follow the process. The growth of a material line by fractal construction is also explained in his work as well as a treatment of motion of point vortices from a Hamiltonian viewpoint. He also performed experiments and one of his visualizations is reproduced in Figure 1.2.1.[6]

The interactions between turbulence and chemical reactions is of utmost importance in combustion. Although the approach to these problems in the 50's and 60's was largely statistical this was not always the case and it is comforting to know that in well thought out experimental papers such as the one by Hawthorne, Wendell, and Hottel (1948) one finds descriptions emphasizing the geometrical aspects of the problem. To quote

Figure 1.2.1. Reproduction of one of the early mixing experiments of Welander (1955); evolution of an initial condition in a rotating flow. He used butanol floated on water and the initial condition (square) was made of methyl-red; unfortunately few additional details regarding the experiment were given in the original paper.

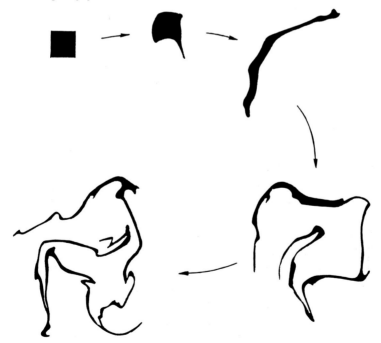

from their paper,

> According to the physical picture of the turbulent flame, eddies . . . are being drawn in from the surrounding atmosphere and being broken up into particles of various sizes . . . The total area of flame envelope is many times the area available in a diffusion flame but nevertheless the final intimate mixing of the gas and oxygen must occur between eddies as a result of molecular diffusion.[7]

Subseqently, statistical theory took over and the geometrical aspects of the problem were somewhat lost. Of the many possible offsprings of the statistical theory of turbulence to mixing (e.g., Batchelor, 1953) we might mention the short but influential paper by Corrsin (1959) which proposed a simple model to calculate the rate of decrease of concentration fluctuations in an ideal, yet subsequently widely used mixer.[8] Another important concept based on statistical reasoning is that of the mixing length theory (Prandtl, 1925; Schlichting, 1955, Chap. XIX), which has found application in an enormous range of problems ranging from chemical engineering (Bird, Stewart, and Lightfoot, 1960) to astrophysics (Chan and Sofia, 1987, Wallerstein, 1988). Even though the statistical treatment does not lend itself easily to visualization, notable exceptions exist and many of the early works focused on the stretching of material lines and surfaces, for example, Batchelor (1952; also Corrsin, 1972) analyzed the problem theoretically, whereas Corrsin and Karweit presented experimental results (Corrsin and Karweit, 1969). Other work focused on local deformation to capture the details of the turbulent motion at small scales. For example, Townsend (1951) performed an analysis of deformation and diffusion of small heat spots in order to interpret experimental Eulerian data in homogeneous decaying turbulence.

Concurrently, at the other end of the spectrum, the stretching of material lines and volumes was also a concern in the mixing of liquids in low Reynolds number flows. Spencer and Wiley (1951) focused on the mixing of very viscous liquids and stressed the idea of being able to describe the growth of interfacial area between two fluids and the need to relate the results to the fluid mechanics. However, even though the mathematical apparatus, largely developed in continuum mechanics, was already in place for such a program, it was not until much later that such developments took place and most of what followed from Spencer and Wiley's work was confined to deformation in shear flows. Two other points worth mentioning from their work, which have a clear relationship with dynamical systems, are the identification of stretching and cutting or folding as the primary mixing mechanism, the so-called 'baker's transformation' characteristic of chaotic systems (see Chapter 5), and the idea of representing mixing processes in terms of matrix transformations, which is obviously related to mappings (see Chapter 5) and transition matrices (see p. 164 in Reichl, 1980). A closely related study by Mohr, Saxton, and Jepson (1957) also focused on viscous liquids in the context of polymer processing. They considered the stretching of a filament of a fluid in the bulk of another one in a shear flow using simple arguments to account for the viscosity of the fluids but without taking into account the interfacial tension. The problem is similar in spirit to the one treated by Taylor (1934) and much work could follow along these lines. Nevertheless this simplified treatment forms the basis for most of the subsequent developments in the mixing of viscous fluids.[9]

It is probably fair to say that most of the previous works have the geometrical interpretation given in Figure 1.1.1. However, a point of departure from this picture of significant consequence in chemical engineering took place with the papers by Zweitering (1959) and Danckwerts (1958). In this case the approach became more 'lumped' or macroscopic and the emphasis shifted to continuous flow systems and the characterization of mixing by the temporal distribution of exit times.[10]

Whereas the objective of most of the above works was to relate the fluid mechanics to the mixing or some knowledge of the process to the output, Danckwerts (1953) focused primarily on the characterization of the mixed state, i.e., he devised numbers or indices to indicate to the user how well mixed a system is (e.g., how well mixed is the system of Figure 1.3.1? Figure 1.3.3? Figure 1.3.4?). Even though we are going to say little about 'the measurement of mixing', this is probably the place to stress our opinion on a few points: (i) the measure should be selected according to the specific application and it is futile to devise a single measure to cover all contingencies, and (ii) the measurement has to be relatable to the fluid mechanics.

Examples

(i) The striation thickness, s, Figure 1.1.1, is important in processes involving diffusing and reacting fluids and represents the distance that the molecules must diffuse in order to react with each other. In simple cases, s can be calculated exactly with a knowledge of the velocity field (see Chella and Ottino (1985a) and examples in Chapter 4).

(ii) Molten polymers are often mixed (an operation often referred to as blending) to produce materials with unique properties. For example, in the manufacture of barrier polymer films it might be desirable to produce structures with low effective permeability. This requires that the clusters of the more permeable materials are disconnected and do not form a percolating structure. However, the details of effective diffusion near the percolation point depend on the ramification of the clusters (Sevick, Monson, and Ottino, 1988). Even though such measurements can be extracted from electron micrographs (e.g., Figure 1.3.4, see color plates) via digital image analysis (Sax and Ottino, 1985), to date there are no models allowing the computation of such details from the fluid mechanics of the process.

(iii) A model for the structure of the Earth's upper mantle (Allègre and Turcotte, 1986) postulates that the oceanic crust becomes entrained in the convective mantle where it is subsequently stretched into filaments by buoyancy induced motions. It follows that, on the average, the 'oldest' layers are the thinnest and that the diffusion processes concurrent with the stretching become important over geological time scales when the striation thickness is of the order of 0.1–1 m (the typical diffusion coefficients are of the order 10^{-14}–10^{-16} cm^2/s, which implies diffusion on time scales of the order of 10^{16}–10^{20} s. By comparison the time scale based on the age of the Earth is 1.4×10^{17} s). In this case a model describing the entrainment of material in the convective mantle coupled to a model describing the evolution of the striation thickness as function of time is capable of describing the gross characteristics of the process.

(iv) Consider Eulerian concentration measurements in a turbulent mixing layer. In principle, the fluctuations can be taken as an indication of the mixing between the streams and an index such as Danckwerts's intensity of segregation[2] can be computed. In the ideal case of a non-invasive probe with an infinitely fast response and vanishingly small resolution volume we obtain an indication of the thickness of the striations passing by the point as a function of time. However, even if this were possible, the statistics of the fluctuations would be very complicated and hard to connect to the fluid mechanics of the process itself. What is worse, however, is the inability of the measurements to give a correct global picture of large scale structures, the so-called coherent structures (Roshko, 1976). In this case a flow visualization study based, for example, on shadowgraphs is infintely more revealing with regard to the structure of the flow (Brown and Roshko, 1974) (see also Figure 1.3.5; see color plates).

1.3. Applications and geometrical structure

It is clear that even restricting our attention to fluid–fluid systems,[11] miscible or immiscible, diffusive or non-diffusive, reacting or not, the scope of mixing problems is enormous and it is not possible to develop a complete and useful theory encompassing all the above situations. It is nevertheless evident that, in spite of the enormous range of length and time scales, the underlying geometrical structure associated with the process of reduction of length scales is that of Figure 1.1.1. In this section we highlight this aspect by means of a few examples.

Mixing is relevant in processes ranging from geological length scales (10^6 m) and exceedingly low Reynolds numbers (10^{-20}), such as in the mixing processes occurring in the Earth's mantle, to Reynolds number of order 10^{11} corresponding to mixing in oceans and the atmosphere.[12] An example of a simulation of mixing in the Earth's mantle is shown in Figure 1.3.1, in which the flow is modelled as a two-dimensional layer heated from below.[13] In these cases, actual mixing experiments are of course impossible. However, laboratory models of large scale circulation in oceans can be carried out with liquids in containers placed on a rotating turntable sometimes involving combinations of sources and sinks. Figure 1.2.1, from the early paper by Welander (1955), shows an example (it is worth noting that the output of similar experiments, were described as 'chaotic', e.g., Veronis, 1973).

Undoubtedly, *chaotic* is an apt description of stretching in truly turbulent flows. Figure 1.3.2 shows the stretching of material lines in a turbulent, nearly isotropic, flow (Corrsin and Karweit, 1969), where the expectation[14] is that of exponential growth (Batchelor, 1952). Note, however, the inherent limitation of experimental techniques in resolving the smallest scales.

The deformation of material lines has been studied also in the case of chaotic Stokes's flows. Experiments focusing on deformation of material lines were carried out by Chaiken *et al.* (1986) and Chien, Rising, and Ottino (1986). In the case of Chaiken *et al.* the flow consists in an eccentric journal bearing time-periodic two-dimensional flow which is described in detail in Chapter 7. Figure 1.3.3 shows the shape adopted by a material line of a tracer by the periodic discontinuous operation of the inner and outer cylinders in a counter-rotating sense. Note the absence of 'corners' and 'branches' in the folded structure even though the flow is essentially discontinuous (compare with Figure 1.2.1).

Figures 1.2.1 and 1.3.3 show complex stretched and folded structures, characteristic of mixing in two-dimensional flows. In both cases the

structure formed is lamellar and an indication of the state of mixing is
provided by the striation thickness. However, in other cases the structures
obtained are considerable more complex. If the fluids are immiscible and
sufficiently different, interfacial tension plays a dominant role at small

Figure 1.3.1. Deformation of a tracer in a numerical experiment of motion in
the Earth mantle. The sides of the rectangle are insulating but the bottom is
subjected to a constant heat flux while the temperature of the top surface is
kept constant. The motion is produced by buoyancy and internal heating
effects (the fluid is heated half from below and half from within). The Rayleigh
number is 1.4×10^6, the time scale of the numerical simulation corresponds
to 155 Myear, and the thickness of the layer is 700 km. An instantaneous
picture of the streamlines reveals five cells. (Reproduced with permission from
Hoffman and McKenzie (1985).)

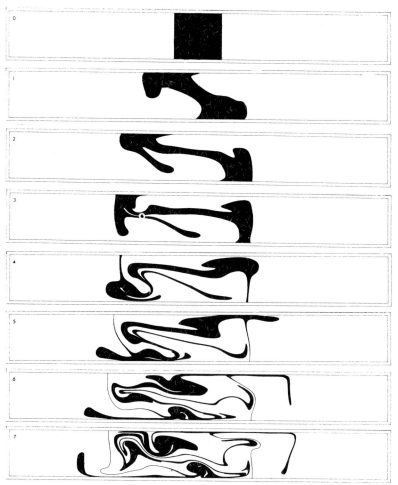

Figure 1.3.2. Growth of a 'material line' composed of small hydrogen bubbles produced by a platinum wire stretched across an decaying turbulent flow behind a grid placed at the extreme left. The Reynolds number based on the wire diameter is 1,360. (Reproduced with permission from Corrsin and Karweit (1969).)

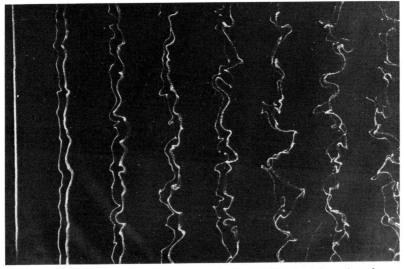

Figure 1.3.3. Mixing in a creeping flow. The figure shows the deformation of a material region in a journal bearing flow when it operates in a time-periodic fashion (experiment from Chaiken *et al.* (1986)), for a complete description see Section 7.4. (Reproduced with permission.)

scales. Figure 1.3.4 (see color plates) shows an image processed two-dimensional structure produced by mixing – and preserved by quenching – of two immiscible molten polymers. In this case there has been a complex process of breakup and coalescence. As indicated earlier, the character-

Figure 1.3.6. Concentration of a turbulent round jet fluid injected into water at Reynolds number 2,300, measured by laser induced fluorescence; the cut is along a plane including the axis of a symmetry of the jet. (Reproduced with permission from Dimotakis, Miake-Lye, and Papantoniou (1983).)

ization of this structure depends on the intended application of the blend and a large number of mixing measures are possible.[15]

The interplay between chemical reactions and mixing is nowhere more evident than in the case of fast reactions (Chapter 9). In many cases of interest the flows are turbulent and careful studies have been carried out to probe the interplay between the fluid mechanics and the transport processes and reaction by using prototypical flows such as turbulent shear flows and wakes, perturbed or not. The reaction itself can be used to map out the interface of reaction. For example, Koochesfahani and Dimotakis (1986; see also 1985) have used laser induced fluorescence and high speed real-time digital image acquisition techniques to visualize the interface of reaction between two reacting liquids undergoing a diffusion controlled reaction (see also Chapter 9). A similar technique can be used in the case of diffusing scalar. For example, Figure 1.3.5 (see color plates) was obtained by measuring the concentration of a fluorescent dye, initially located in one of the streams of the mixing layer, whereas Figure 1.3.6 shows the mixing of turbulent jet containing a fluorescent dye with a clear surrounding fluid. Lamellar structures (Chapter 9) are clearly seen, even at Kolmogorov length scales (Dimotakis, Miake-Lye, and Papantoniou, 1983).

Another instance of interplay between mixing with diffusion and reaction, but at smaller length scales and lower Reynolds numbers (approximately 200–500), well known in polymer engineering, is provided by the impingement mixing of polymers (Lee *et al.*, 1980) where the objective is to mix two viscous liquids (reactive monomers) in short time scales (order 10^{-2}–10^{-1} s) with reaction time scales of the order of 10^1–10^2 s. The geometrical picture is similar to the previous cases; in this case the mixing requirement is to produce striations of the order of 20–50 μm so that the reaction can take place under kinetically controlled conditions.[16] Mixing at even smaller scales might take place due to spontaneous emulsification (Fields, Thomas, and Ottino, 1987; Wickert, Macosko, and Ranz, 1987).

1.4. Approach

In spite of its overwhelming diversity, fluid mixing is basically a process involving a reduction of length scales accomplished by stretching and folding of material lines or surfaces. In some cases the material surface or line in question is placed in the flow and then subseqently stretched (e.g., Figures 1.2.1, 1.3.1–3), in others, the surface is continuously fed into the flow (e.g., Figures 1.3.5–6). Thus, at the most elementary level (i.e., without averaging but at the continuum level) *mixing consists of stretching and folding of fluid filaments, and distribution throughout space,* accompanied by breakup if the fluids are sufficiently different, and simultaneous diffusion of species and energy (Figure 1.1.1). In the most general case, various chemical species might be reacting. We seek an understanding of this process in terms of simple problems which can serve as a 'window' for more complicated situations. Our approach is to combine the kinematical foundations of fluid mechanics with dynamical systems concepts, especially chaotic dynamics. The objective throughout is to gain insight into the working of mixing flows. The goal is not to construct detailed models of specific problems but rather to provide prototypes for a broad class of problems. Nevertheless, we expect that the insight gained by the analysis will be important in practical applications such as the design of mixing devices and understanding of mixing experiments.

Mixing is also inherently related to flow visualization. However, contrary to popular perception the 'unprocessed' Eulerian velocity field gives very little information about mixing and the typical ways of visualizing a flow (streamlines, pathlines, and to a lesser degree, streaklines) are insufficient to completely understand the process. As we shall see, *our problem begins rather than ends with the specification of* $\mathbf{v}(\mathbf{x}, t)$. The solution of

$$d\mathbf{x}/dt = \mathbf{v}(\mathbf{x}, t)$$

with $\mathbf{x} = \mathbf{X}$ at time $t = 0$, $\mathbf{x} = \mathbf{\Phi}_t(\mathbf{X})$, which is called the *flow* or *motion*,[17] provides the starting point for our analysis. In even the simplest cases this 'solution' might be extremely hard to obtain. Actually, the impossibility of integrating the velocity field in the conventional sense is the subject of much of Chapters 5 and 6, where the modern notion of *integrability* is introduced. The kinematical foundations lie in an understanding of the point transformation $\mathbf{x} = \mathbf{\Phi}_t(\mathbf{X})$.

We consider the following sub-problems:

(1) Within the framework of $\mathbf{x} = \mathbf{\Phi}_t(\mathbf{X})$; mixing of a single fluid or similar fluids.

The basic objective here is to compute the length (or area) corresponding to a set of initial conditions. As we shall see only in a few cases can this be done exactly and in most of these the length stretch is mild. The best achievable mixing corresponds to exponential stretching nearly everywhere and occurs in some regions of chaotic flows. However, under these conditions the (exact) calculation of the length and location of lines and areas is hopelessly complicated. As we shall see in Chapter 5 even extremely simplified flows might be inherently chaotic and a *complete* characterization is not possible. For example, from a dynamical systems viewpoint we shall see that if the system possesses horseshoes we have infinitely many periodic points and with it the implication that we cannot possibly calculate precisely all of them. Fortunately, as far as mixing is concerned we are interested in low period events, since we want to achieve mixing quickly. Nevertheless there is always the intrinsic limitation of being unable to calculate precise information (most practical problems involve stretchings of order 10^4 or higher) such as length stretch and location of material surfaces. Note that this is true even though none of the flows discussed in Chapters 7 and 8 is turbulent in an Eulerian sense. Rather, the previous findings should be used to establish the limits of what might constitute reasonable answers in more complicated flows (real turbulent flows come immediately to mind). *The problem here is how to best characterize the mixing, knowing beforehand that a complete characterization is impossible.*

(2) Within the framework of a family of flows $\mathbf{x}_s = \mathbf{\Phi}_{s_t}(\mathbf{X}_s)$, $s = 1, \ldots, N$; each of the motions is assumed to be topological (see Section 2.3); mixing of similar diffusing and reacting fluids.

This case corresponds to the case of mixing of two streams, composed of possibly several species that are rheologically identical, i.e., they have the same density, viscosity, etc., and have no interfacial tension. Concurrently with the mechanical mixing there is mass diffusion, and possibly, chemical reaction. However, for simplicity, we will assume that neither the diffusion nor the reaction affects the fluid motion.[18] This case is discussed in Chapter 9 and corresponds to the case of *lamellar structures*.

(3) Mixing of different fluids; case in which the motions are non-topological, i.e., there is breakup and/or fusion of material elements.

In this case, the mixing of two or more fluids leads to breakup and coalescence of material regions. This problem is complicated and only a few special cases belonging to this category are discussed in Chapter 9, by decomposing the problem into *local* and *global* components.

An outline of the organization of the rest of the chapters is the following: Chapter 2 describes the kinematical foundations and Chapter 3 presents a brief overview of fluid mechanics. Whereas the material of Chapter 2 is indispensable, large parts of Chapter 3 were added to provide balance. Chapter 4 focuses on a few examples which can be solved in detail and ends in a rather defeatist note to provide a bridge for the study of chaos. Chapter 5 presents a general discussion of dynamical systems and Chapter

6 focuses on Hamiltonian systems. Similar comments apply in this case. We use more heavily the material of Chapter 6 but omission of Chapter 5 would result in serious imbalance and a misleading representation of facts. Chapters 7 and 8, by far the longest in this work, give examples of chaotic mixing systems in an increasing order of complexity. Chapter 7 discusses two-dimensional flows, Chapter 8 focuses on three-dimensional flows. Chapter 9 discusses briefly the case of diffusing and reacting fluids and active microstructures.

Notes

1 Obviously, one of the diffusing species can be temperature, as is the case of mixing of fluids with different initial temperatures or processes involving exothermic chemical reactions.

2 A gross, but popular, measure of the concentration variation is given by the *intensity of segregation*, I. If $c(\mathbf{x})$ denotes the concentration at point $\mathbf{x}$ and $\langle \cdot \rangle$ denotes a volume average, I is defined as $[\langle (c - \langle c \rangle)^2 \rangle]^{1/2}$ (Danckwerts, 1952).

3 Many of these references inspired additional work. Some, however, were largely ignored.

4 Subsequently, this problem took a life of its own and much research followed. See for example Rallison (1984).

5 This idea was not widely followed and most of the mixing work in the area of drop breakup and coalescence in complex flow fields resorts to population balances where breakup and coalescence are taken into account in a probabilistic sense.

6 This paper contains many good ideas, however, it has remained largely ignored by the mixing community.

7 In the past few years there has been a revival of this idea (e.g., Spalding, 1976, 1978b; Ottino, 1982).

8 The theory also found use in two-phase mixing. An early reference is Shinnar (1961).

9 The analysis of mixing of viscous fluids has been largely confined to polymers (Middleman, 1977; Tadmor and Gogos, 1979). A follow up paper, written in the context of the mixing of glasses, is Cooper (1966). Even though Cooper's treatment of the kinematics is at the same level as Mohr, Saxton, and Jepson (1957), there is substantially more, since it deals explicitly with mass diffusion. This paper, however, has remained largely ignored.

10 This approach works well for pre-mixed reactors with slow reactions, but it is not suited for diffusion controlled reactions, such as in combustion. Curiously enough, the participation of chemical engineers in subjects dealing with non-pre-mixed reactors has been relatively minor and in spite of complex chemistry, the area has become largely the domain of mechanical and aerospace engineering researchers. See the discussion (pp. 100–102) following the paper by Danckwerts (1958). Much work followed along these lines. For a summary, see Nauman and Buffham (1983).

11 With the possible exception of the very last example of the last chapter, we do not consider fluid–solid systems.

12 See for example Veronis, 1973; Rhines, 1979; 1983, and the articles by Holland ('Ocean circulation models', pp. 3–45), Veronis ('The use of tracer in circulation studies', pp. 169–188), and Rhines ('The dynamics of unsteady currents', pp. 189–318), in Goldberg *et al.*, 1977.

13 The fluid dynamical aspects are discussed by McKenzie, Roberts, and Weiss (1974). See also McKenzie (1983).

14 This expectation was not confirmed in their study. It is apparent that there is a need for more experiments adopting primarily a 'Lagrangian viewpoint'.

15 For example, average cluster size and cluster size distribution, interfacial perimeter per unit area, etc. (see Sax and Ottino, 1985).

16 Impingement mixing is similar to 'the stopped flow' method devised to study the kinetics of very fast reactions. A review of the technique by the inventors of the method is given by Roughton and Chance (1963), Chap. XIV.

17 *Flow* is the term preferred in dynamical systems, *motion* the term of choice in continuum mechanics. The reader should be warned about possible confusion of terms; the word *flow* is used often in the conventional fluid mechanical sense, the term *mixing* has a precise definition in dynamical systems and ergodic theory (see Waters, 1982, p. 40). To avoid confusion we will often use the term *fluid mixing*.

18 That is, the fluid mechanics governs the transport processes but not the other way around. There is no clean way of incorporating these couplings at the present time.

2

Flow, trajectories, and deformation

In the first part of this chapter we record the basic kinematical foundations of fluid mechanics, starting with the primitive concept of particle and motion, and the classical ways of visualizing a flow. In the second part we give the basic equations for the deformation of infinitesimal material lines, planes, and volumes, both with respect to spatial, $\mathbf{x}$, and material, $\mathbf{X}$, variables, and present equations for deformation of lines and surfaces of finite extent.

2.1. Flow

The physical idea of *flow* is represented by the map or point transformation (Arnold, 1985, Chap. 1)

$$\mathbf{x} = \mathbf{\Phi}_t(\mathbf{X}) \qquad \text{with } \mathbf{X} = \mathbf{\Phi}_{t=0}(\mathbf{X}), \qquad (2.1.1)$$

i.e., the initial condition of particle $\mathbf{X}$ (a means of identifying a point in a continuum, in this case labelled by its initial position vector) occupies the position $\mathbf{x}$ at time t (see Figure 2.1.1). We say that $\mathbf{X}$ is mapped to $\mathbf{x}$ after a time t.[1]

In continuum mechanics (2.1.1) is called the *motion* and is usually assumed to be invertible and differentiable. In the language of dynamical systems a mapping

$$\mathbf{\Phi}_t(\mathbf{X}) \rightarrow \mathbf{x} \qquad (2.1.2)$$

is called a C^k diffeomorphism if it is 1–1 and onto, and both $\mathbf{\Phi}_t(\cdot)$ and its inverse are k-times differentiable. If $k = 0$ the transformation is called a homeomorphism. In fluid mechanics, k is usually taken equal to three (see Truesdell, 1954; Serrin, 1959). Also, the transformation (2.1.1) is required to satisfy

$$0 < J < \infty$$
$$J = \det(\partial x_i / \partial X_j) \qquad (2.1.3)$$

or alternatively,

$$J = \det(\mathbf{D}\mathbf{\Phi}_t(\mathbf{X}))$$

where D denotes the operation $\partial(\ )_i/\partial X_j$, i.e., derivatives with respect to the reference configuration, in this case **X**. If the Jacobian J is equal to one the flow is called *isochoric*.

The requirement (2.1.3) precludes two particles, $\mathbf{X}_1$ and $\mathbf{X}_2$, from occupying the same position **x** at a given time, or one particle splitting into two; i.e., *non-topological* motions such as breakup or coalescence are not allowed (Truesdell and Toupin, 1960, p. 510).

In the language of dynamical systems, the set of diffeomorphisms (2.1.1) for all particles **X** belonging to the body V_0 is called the flow (i.e., a one-parameter set of diffeomorphisms) and is represented by

$$\{\mathbf{x}\} = \{\boldsymbol{\Phi}_t(\mathbf{X})\} = \boldsymbol{\Phi}_t\{\mathbf{X}\}$$

where $\{\mathbf{X}\}$ is the set of particles belonging to V_0 ($V_0 = \{\mathbf{X}\}$ and $V_t = \{\mathbf{x}\}$). Thus, we say that V_0, is mapped into V_t at time t,

$$\{V_t\} = \boldsymbol{\Phi}_t\{V_0\}$$

or that the material line L_0 is mapped into L_t at time t,

$$\{L_t\} = \boldsymbol{\Phi}_t\{L_0\}.$$

Figure 2.1.1. Deformation of lines and volumes by a flow $\mathbf{x} = \boldsymbol{\Phi}_t(\mathbf{X})$.

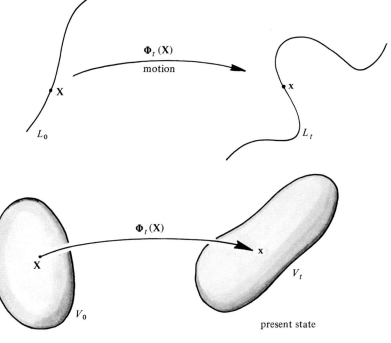

Note that flows can be composed according to

$$\Phi_{t+r}(\mathbf{X}) = \Phi_t(\Phi_r(\mathbf{X})),$$

i.e., $\mathbf{X}$ is taken to position $\Phi_t(\mathbf{X})$ and then to $\Phi_{t+r}(\mathbf{X})$. The flow can also be reversed,

$$\Phi_{t-t}(\mathbf{X}) = \Phi_t(\Phi_{-t}(\mathbf{X})) = \Phi_t(\Phi_t^{-1}(\mathbf{X})) = \mathbf{X}$$

i.e., $\mathbf{X}$ is taken to $\mathbf{x}$ and then back to $\mathbf{X}$.[2]

2.2. Velocity, acceleration, Lagrangian and Eulerian viewpoints

The velocity is defined as

$$\mathbf{v} \equiv (\partial\Phi_t(\mathbf{X})/\partial t)\big|_{\mathbf{X}} = \mathbf{v}(\mathbf{X}, t)$$

and it is the velocity of the particle $\mathbf{X}$. The acceleration, $\mathbf{a}$, is defined as

$$\mathbf{a} \equiv (\partial^2\Phi_t(\mathbf{X})/\partial t^2)\big|_{\mathbf{X}} = \mathbf{a}(\mathbf{X}, t).$$

Any function G (scalar, vector, tensor) can therefore be viewed in two different ways:

$$G(\mathbf{X}, t) \equiv \text{Lagrangian or material}$$

i.e., follows the motion of a particular fluid particle, or

$$G(\mathbf{x}, t) \equiv \text{Eulerian}^3 \text{ or spatial}$$

i.e., the property of the particle $\mathbf{X}$ that happens to be at the spatial location $\mathbf{x}$ at time t.

Thus, $\mathbf{v}(\mathbf{X}, t)$ is the Lagrangian velocity and $\mathbf{v}(\mathbf{x}, t)$[4] is the Eulerian velocity. In most classical problems in fluid mechanics it is enough to obtain the spatial description. The material (or Lagrangian) derivative is defined as

$$DG/Dt \equiv (\partial G/\partial t)\big|_{\mathbf{X}}$$

representing the change of property G with time while following the motion of particle $\mathbf{X}$, whereas the standard time derivative is

$$\partial G/\partial t \equiv (\partial G/\partial t)\big|_{\mathbf{x}},$$

and represents the change at a fixed position $\mathbf{x}$. The relationship between the two is easily obtained from the chain rule

$$DG/Dt = \partial G/\partial t + \mathbf{v} \cdot \nabla G,$$

where ∇ is defined as $\nabla = (\partial/\partial x_i)\mathbf{e}_i$ (see Appendix). The expression allows the computation of the acceleration at $(\mathbf{x}, t)$ without computing the motion first.

The expression for the material time derivative of the Jacobian of the flow is known as Euler's formula (Serrin, 1959, p. 131; Chadwick, 1976, p.65).

$$DJ/Dt = J(\nabla \cdot \mathbf{v}) = J \, \text{tr}(\nabla\mathbf{v})$$

and is the basis of a kinematical result known as the *transport theorem*.

Consider the integral

$$\int_{V_t} G(\mathbf{x}, t)\, dv$$

where V_t represents a material volume, that is, a volume composed always of the same particles $\mathbf{X}$ belonging to the body V_t. Making reference to Figure 2.1.1 $\{V_t\} = \mathbf{\Phi}_t\{V_0\}$, i.e., the flow $\mathbf{\Phi}_t$ transforms $\{V_0\}$ into $\{V_t\}$ at time t. The integral can be written, using the definition of the Jacobian, as

$$\int_{V_t} G(\mathbf{x}, t)\, dv = \int_{V_0} G(\mathbf{X}, t)J\, dV$$

where dV represents a volume in the reference configuration V_0. The time derivative,

$$\frac{d}{dt} \int_{V_t} G(\mathbf{x}, t)\, dv$$

(which is a material derivative since all the particles in V_t remain there) can be written as

$$\frac{d}{dt} \int_{V_0} G(\mathbf{X}, t)J\, dV = \int_{V_0} \frac{D(G(\mathbf{x}, t)J)}{Dt}\, dV,$$

and since the domain V_0 is not a function of time, expanding the material derivative we obtain

$$\frac{d}{dt} \int_{V_t} G(\mathbf{x}, t)\, dv = \int_{V_t} \left[\frac{DG}{Dt} + G(\nabla \cdot \mathbf{v})\right] dv.$$

By means of the divergence theorem,

$$\frac{d}{dt} \int_{V_t} G(\mathbf{x}, t)\, dv = \int_{V_t} \frac{\partial G}{\partial t}\, dv + \oint_{\partial V_t} G\mathbf{v} \cdot \mathbf{n}\, ds$$

where $\mathbf{n}$ is the outward normal to the boundary of V_t, denoted ∂V_t. In general this result holds for any arbitrary control volumes V'_t moving with velocity $\mathbf{v}_v$.

$$\frac{d}{dt} \int_{V'_t} G(\mathbf{x}, t)\, dv = \int_{V'_t} \frac{\partial G}{\partial t}\, dv + \oint_{\partial V'_t} G\mathbf{v}_v \cdot \mathbf{n}\, ds.$$

Problem 2.2.1
Show that if $\mathbf{A}(t)$ is invertible, $d(\det \mathbf{A})/dt = (\det \mathbf{A})\, \mathrm{tr}[(d\mathbf{A}/dt) \cdot \mathbf{A}^{-1}]$.

Problem 2.2.2
Show that if $\{V_t\} = \mathbf{\Phi}_t\{V_0\}$ then $\{\partial V_t\} = \mathbf{\Phi}_t\{\partial V_0\}$, i.e., the boundary is mapped into the boundary. Usually, this is taken to be that the surface of a material body consists of the same particles (von Mises and

Friedrichs, 1971). This result is called 'Lagrange's theorem' by Prandtl and Tietjens (1934, p. 97).

2.3. Extension to multicomponent media

In the case of multicomponent media we envision material surfaces moving with the mean mass velocity (see Chapter 9). If the system has several components, $s = 1, \ldots, N$, we assume the existence of a set of motions $\Phi_t^{(s)}$, and Equation (2.1.1) is generalized as

$$\mathbf{x}_s = \Phi_t^{(s)}(\mathbf{X}_s) \text{ (no sum)}$$

where $\mathbf{X}_s$ represents a particle of species-s and $\mathbf{x}_s$ its position at time t.[5] Each species is assigned a density $\rho_s = \rho_s(\mathbf{X}_s, t)$ such that $\rho = \sum \rho_s$, where the sum runs from 1 to N. Individual velocities are defined as

$$\mathbf{v}_s \equiv (\partial \Phi_t^{(s)}(\mathbf{X}_s)/\partial t)|_{\mathbf{X}_s} = \mathbf{v}_s(\mathbf{X}_s, t)$$

and the average mass velocity is defined as

$$\mathbf{v} \equiv \sum (\rho_s/\rho)\mathbf{v}_s.$$

The time derivative of any function G following the motion of the species denoted s, is given by

$$DG^{(s)}/Dt = \partial G/\partial t + \mathbf{v}_s \cdot \nabla G,$$

and the relative velocities are defined by

$$\mathbf{u}_s = \mathbf{v}_s - \mathbf{v}.$$

The simplest constitutive equation for $\mathbf{u}_s$ is

$$\mathbf{u}_s = -\omega_s^{-1} D \nabla \omega_s$$

(i.e., dilute solution or equimolecular counter-diffusion, Bird, Stewart, and Lightfoot, 1960, p. 502) where ω_s is the mass fraction ($= \rho/\rho_s$) and D is the diffusion coefficient.

Other quantities, such as individual deformation tensors for species-s, etc., can be defined analogously (Bowen, 1976) but they are not used in this work.

2.4. Classical means for visualization of flows

There are several ways of visualizing a flow. In this section we record the three classical ones.

2.4.1. Particle path, orbit, or trajectory

Given the Eulerian velocity field $\mathbf{v} = \mathbf{v}(\mathbf{x}, t)$, the particle path of $\mathbf{X}$ is given by the solution of $d\mathbf{x}/dt = \mathbf{v}(\mathbf{x}, t)$ with $\mathbf{x} = \mathbf{X}$ at $t = 0$. Physically it corresponds to a long time exposure photograph of an illuminated fluid particle.

As seen in texts of differential equations, the solution to the above problem, $\mathbf{x} = \mathbf{\Phi}_t(\mathbf{X})$, is unique and continuous with respect to the initial data if $\mathbf{v}(\mathbf{x})$ has a Lipschitz constant, $K > 0$.[6] Under these conditions, if we denote $\mathbf{x}_1 = \mathbf{\Phi}_t(\mathbf{X}_1)$ and $\mathbf{x}_2 = \mathbf{\Phi}_t(\mathbf{X}_2)$, we have the trajectories evolve according to

$$|\mathbf{x}_1 - \mathbf{x}_2| \leqslant |\mathbf{X}_1 - \mathbf{X}_2| \exp(Kt), \qquad K > 0.$$

As we shall see there are many systems (Chapter 5) that diverge from the initial conditions at an exponential rate, i.e., the non-strict inequality becomes an equality.[7]

2.4.2. Streamlines

The streamlines correspond to the solution of the system of equations

$$d\mathbf{x}/ds = \mathbf{v}(\mathbf{x}, t)$$

where the time t is treated as constant and s is a parameter (that is, we take a 'picture' of the vector field $\mathbf{v}$ at time t). Physically, we can mimic the streamlines by labelling a collection of fluid particles and taking two successive photographs at times t and $t + \Delta t$. Joining the displacements gives $\mathbf{v}$ in the neighborhood of the point $\mathbf{x}$. The streamlines are tangential to the instantaneous velocity at every point, except at points where $\mathbf{v} = 0$.

2.4.3. Streaklines

The picture at time t of the streakline passing through the point $\mathbf{x}'$ is the curved formed by all the particles $\mathbf{X}$ which happened to pass by $\mathbf{x}'$ during the time $0 < t' < t$. Physically, it corresponds to the curve traced out by a non-diffusive tracer (i.e., the particles $\mathbf{X}$ of the tracer move according to $\mathbf{x} = \mathbf{\Phi}_t(\mathbf{X})$) injected at the position $\mathbf{x}'$.

Example 2.4.1

Compute the pathlines, streamlines, and streaklines corresponding to the unsteady Eulerian velocity field

$$v_1 = x_1/(1 + t), \qquad v_2 = 1. \quad [8]$$

To compute the pathlines we solve

$$dx_1/dt = x_1/(1 + t), \qquad dx_2/dt = 1$$

with the condition $x_1 = X_1$, $x_2 = X_2$, at $t = 0$.

$$x_1 = X_1(1 + t), \qquad x_2 = X_2 + t$$

and, eliminating t, we obtain

$$x_1 - X_1 x_2 = X_1(1 - X_2),$$

i.e., the particles move in straight lines. The streamlines are given by the solution of

$$dx_1/ds = x_1/(1 + t), \qquad dx_2/ds = 1$$

with the condition $x_1 = x_1^0$, $x_2 = x_2^0$, at $s = 0$, while holding t constant. Thus, the streamline passing by $x_1 = x_1^0$, $x_2 = x_2^0$, is given by

$$x_1 = x_1^0 \exp[s/(1 + t)], \qquad x_2 = x_2^0 + s,$$

and eliminating the parameter s,

$$(1 + t) \ln(x_1/x_1^0) = x_2 - x_2^0,$$

which shows that the streamlines are *time dependent*. To get the streakline passing through x_1', x_2' we first invert the particle paths at time t'

$$X_1 = x_1'/(1 + t'), \qquad X_2 = x_2' - t',$$

which indicates that the particle X_1, X_2 will be found at the position x_1', x_2' at time t'. The place occupied by this particle at any time t is found again from the particle path as

$$x_1 = x_1'(1 + t)/(1 + t'), \qquad x_2 = x_2' - t' + t,$$

and is interpreted as: the particle which occupied position x_1', x_2' at time t' will be found in position x_1, x_2 at time t. Eliminating t' we get the locus of the streakline passing by x_1', x_2':

$$x_1 x_2 - x_1(1 + x_2' + t) + x_1'(1 + t) = 0$$

which shows that the streaklines *are also functions of time* (plots corresponding to this example are given by Truesdell and Toupin, 1960, p. 333).

2.5. Steady and periodic flows

A flow is steady if it lacks explicit time dependence, i.e., $\mathbf{v} = \mathbf{v}(\mathbf{x})$. Note that the concept of steadiness depends on the frame of reference. An unsteady flow in one frame can be steady in another frame (moving frames are studied in Chapter 3). When the flow is steady in a given frame F, the streamlines and pathlines coincide when viewed in the frame F. Furthermore, the streaklines coincide with both streamlines and pathlines provided that the position of the dye-injection apparatus is fixed with respect to the frame F. It is obvious that these statements are a property of dynamical systems in general and they are not confined to fluid mechanical systems.

A point $\mathbf{x}$ such that $\mathbf{v}(\mathbf{x}) = \mathbf{0}$ for all t is called a *fixed or singular point* (or alternatively, equilibrium point in dynamical systems or stagnation point in fluid mechanics). A point $\mathbf{P}$ is periodic, of period T, if

$$\mathbf{P} = \mathbf{\Phi}_T(\mathbf{P})$$

for $t = T$ but not for any $t < T$. That is, the material particle which happened to be at the position $\mathbf{P}$ at a time $t = 0$, without loss of generality, will be located in exactly the same spatial position after a time T,[9] i.e., if we view the flow at different times we have the sequence $\mathbf{P}_0$, $\mathbf{P}_1, \mathbf{P}_2, \ldots$, $\mathbf{P}_n \equiv \mathbf{P}$. Note that the concept of periodicity depends also on the frame of reference.[10]

Example 2.5.1
Consider the streamlines, pathlines, and streaklines in the shear flow

$$V_1 = 1 + \tanh x_1, \qquad V_2 = 0$$

subjected to a time dependent perturbation

$$v_1' = 2a \operatorname{sech} 2\pi x_2 \tanh 2\pi x_2 \sin[2\pi(x_1 - t)],$$
$$v_2' = 2a \operatorname{sech} 2\pi x_2 \cos[2\pi(x_1 - t)],$$

where a is the amplitude of the fluctuations. This problem was analyzed by Hama (1962) to serve as a warning in the interpretation of flow visualization studies since it is a case where the streaklines are considerably more complicated than the streamlines and pathlines. The motion of the fluid particles is governed by the system of equations

$$dx_1/dt = v_1(x_1, x_2, t), \qquad dx_2/dt = v_2(x_1, x_2, t)$$

where $v_1 = V_1 + v_1'$ and $v_2 = V_2 + v_2'$. Examples of computed streaklines and pathlines are shown in Figure E2.5.1 (computation by Franjione, 1987).

Example 2.5.2
Calculate the possible streamlines corresponding to the class of flows $\mathbf{v} = (\nabla\mathbf{v})^{\mathrm{T}} \cdot \mathbf{x}$ with $\nabla \cdot \mathbf{v} = 0$ (this material is used repeatedly throughout the book).

For simplicity consider first two-dimensional flows. By continuity ($\nabla \cdot \mathbf{v} = 0$, the sum of the eigenvalues being equal to zero), we can infer some of the character of the three-dimensional flow. We consider a cut of $\mathbf{v} = (\nabla\mathbf{v})^{\mathrm{T}} \cdot \mathbf{x}$ by setting $x_3 \equiv 0$. In general, the two-dimensional version of $(\nabla\mathbf{v})^{\mathrm{T}}$, $(\nabla\mathbf{v})^{\mathrm{T}}_{2\mathrm{d}} \equiv \mathbf{L}$, has all non-zero components, but by a suitable transformation $\mathbf{R}, \mathbf{R} \cdot \mathbf{L} \cdot \mathbf{R}^{-1}$, $\mathbf{L}$ can be written in one of the three possible ways (we follow Hirsch and Smale, 1974, Chap. 5, Section 4):

$$\begin{bmatrix} \lambda & 0 \\ 0 & \mu \end{bmatrix}, \qquad \begin{bmatrix} a & -b \\ b & a \end{bmatrix}, \qquad \begin{bmatrix} \lambda & 0 \\ 1 & \lambda \end{bmatrix}.$$

The character of the flow is given by the eigenvalues of $\mathbf{L}$,

$$\lambda^2 - \mathrm{tr}(\mathbf{L})\lambda + \det(\mathbf{L}) = 0.$$

The discriminant is:

$$\Delta = [\mathrm{tr}(\mathbf{L})]^2 - 4\det(\mathbf{L})$$

and the eigenvalues are

$$\tfrac{1}{2}(\mathrm{tr}(\mathbf{L}) \pm \Delta^{1/2}).$$

Thus, $\Delta > 0$ corresponds to real eigenvalues, and $\mathrm{tr}(\mathbf{L}) < 0$ correspond to eigenvalues with negative real part. If none of the eigenvalues lies on the imaginary axis, the flow is called *hyperbolic*.

The possibilities are the following (see Figure E2.5.2(a)):

I: All real eigenvalues of different signs

$$\begin{bmatrix} \lambda & 0 \\ 0 & \mu \end{bmatrix}$$
with $\lambda < 0 < \mu$ corresponds to a saddle.

Figure E2.5.1. Streaklines and pathlines in Hama flow corresponding to an amplitude $a = 0.05$. The vertical scale corresponds to $x_2 = -0.2$ to $x_2 = 0.2$. (a) Streaklines injected at -0.15, -0.10, -0.05, 0, 0.05, 0.10, 0.15, the time goes for 5 units; (b) pathlines of the particles injected at time $t = 0$ (tip of the streaklines), injected at -0.15, -0.10, -0.05, 0, 0.05, 0.10, 0.15, the total time goes for 5 units.

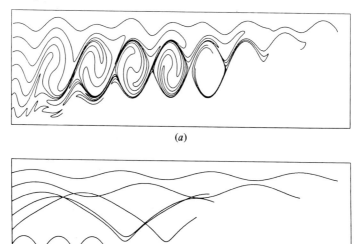

(a)

(b)

Figure E2.5.2. (*a*) Summary of the portraits of the linear two-dimensional velocity field $\mathbf{v} = \mathbf{L} \cdot \mathbf{x}$. The parabola corresponds to $\det(\mathbf{L}) = (1/4)[\text{tr}(\mathbf{L})]^2$. The discriminant is $\Delta = [\text{tr}(\mathbf{L})]^2 - 4 \det \mathbf{L}$. (*b*) Examples of three-dimensional velocity fields with $\text{tr}(\mathbf{L}) = 0$ in the neighborhood of $\mathbf{v}(\mathbf{P}) = \mathbf{0}$ (see also Figure 5.6.1).

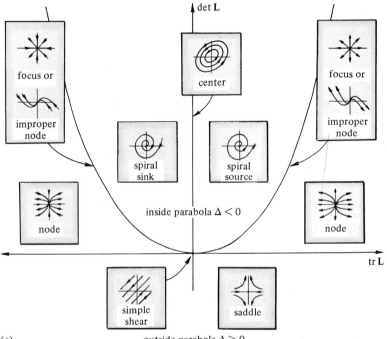

(*a*)

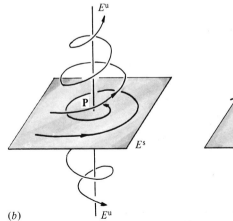
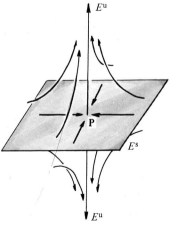

(*b*)

II: All eigenvalues have negative real parts

II(a), **L** *diagonal*

$$\begin{bmatrix} \lambda & 0 \\ 0 & \lambda \end{bmatrix} \qquad \text{with } \lambda < 0 \text{ corresponds to a focus.}$$

II(b), **L** *diagonalizable*

$$\begin{bmatrix} \lambda & 0 \\ 0 & \mu \end{bmatrix} \qquad \text{with } \lambda < \mu < 0 \text{ corresponds to a stable node.}$$

II(c), **L** *non-diagonalizable*

$$\begin{bmatrix} \lambda & 0 \\ 1 & \lambda \end{bmatrix} \qquad \text{with } \lambda < 0 \text{ corresponds to an improper node.}$$

II(d), eigenvalues of **L** *are complex conjugate*

$$\begin{bmatrix} a & -b \\ b & a \end{bmatrix} \qquad \text{with } a < 0 \text{ corresponds to a spiral sink.}$$

III: All eigenvalues have positive real parts

Same as **II** but with all arrows in Figure E2.5.2(a) reversed.

IV: All eigenvalues are pure imaginary

$$\begin{bmatrix} 0 & -b \\ b & 0 \end{bmatrix} \qquad \text{corresponds to a center.}$$

$$\begin{bmatrix} 0 & b \\ 0 & 0 \end{bmatrix} \qquad \text{corresponds to a simple shear.}$$

Two three-dimensional cases are shown in Figure E2.5.2(b). Note that if $\mathbf{v} \in \mathbb{R}^2$ and $\text{tr}((\nabla\mathbf{v})_{2\text{d}}^{\text{T}}) = 0$, only centers, saddles, and simple shear are allowed (see Figure P2.5.3).

Problem 2.5.1

In this book the acceleration is given by $\mathbf{a} = \partial\mathbf{v}/\partial t + \mathbf{v}\cdot\nabla\mathbf{v}$. Why not $\mathbf{a} = \partial\mathbf{v}/\partial t + \nabla\mathbf{v}\cdot\mathbf{v}$?

Problem 2.5.2

Consider a motion $\mathbf{x} = \mathbf{\Phi}_t(\mathbf{X})$ given by:

$$x_1 = X_1(1 + t), \qquad x_2 = X_2(1 + t)^2, \qquad x_3 = X_3(1 + t^2).$$

Compute the Lagrangian velocity, $\mathbf{v}(\mathbf{X}, t)$, and acceleration, $\mathbf{a}(\mathbf{X}, t)$. Compute the Eulerian velocity, $\mathbf{v}(\mathbf{x}, t)$, and acceleration $\mathbf{a}(\mathbf{x}, t)$. Verify that $\mathbf{a} = \partial\mathbf{v}/\partial t + \mathbf{v}\cdot\nabla\mathbf{v}$. Similarly, consider a motion given by

$$x_1 = X_1 \exp(-t), \qquad x_2 = X_2 \exp(-t), \qquad x_3 = X_3$$

and verify that $DJ/Dt = J(\nabla \cdot \mathbf{v})$. Is the flow steady?

Problem 2.5.3

Consider the flow $v_1 = Gx_2, v_2 = KGx_1$, where $-1 < K < 1$. Show that the streamlines are given by $x_2^2 - Kx_1^2 = $ constant, which corresponds to ellipses with axes ratio $(1/|K|)^{1/2}$, if $K < 0$, and to hyperbolas forming an angle $\beta = \arctan(1/K)^{1/2}$ between the axis of extension and x_2, if $K > 0$ (see Figure P2.5.3). Prove that this flow is the most general representation of a linear isochoric two-dimensional flow.[11]

Problem 2.5.4

Show that the pathlines corresponding to the flow $v_1 = ax_1, v_2 = -a(x_2 - bt)$ represent a circular motion about a center moving with a velocity $\mathbf{v} = (bt, b/a)$. Find the streamlines and particle paths corresponding to $\mathbf{v} = (x_1 t, -x_2)$. Show that the streamlines are given by $x_1 x_2^t = $ const. (for this and other examples, see Patterson, 1983).

Problem 2.5.5

Consider the one-dimensional time-periodic Eulerian velocity field, $v_x = U \cos[k(x - ct)]$. Find the velocity and time averaged velocity experienced by a material particle.

Problem 2.5.6

Given the Eulerian velocity field

$$\mathbf{v} = \mathbf{x} \cdot \mathbf{L}, \qquad \mathbf{L} = \text{const.},$$

obtain the Langrangian velocity field. Generalize for $\mathbf{L} = \mathbf{L}(t)$.

Figure P2.5.3. Portraits of two-dimensional isochoric linear velocity fields; $v_1 = Gx_2, v_2 = KGx_1$, as a function of K. (a) $K = -1$, pure rotation; (b) $K = 0$, unidirectional shear; and (c) $K = 1$, orthogonal stagnation flow.

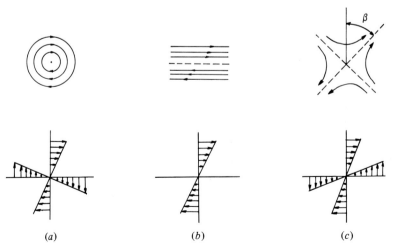

(a) (b) (c)

Similarly, consider the Lagrangian velocity field

$$\mathbf{v} = \mathbf{X} \cdot \mathbf{K}, \qquad \mathbf{K} = \text{const.}$$

Obtain the Eulerian velocity field. Is the velocity field steady? In both cases verify that

$$\mathbf{a} = \partial \mathbf{v}/\partial t + \mathbf{v} \cdot \nabla \mathbf{v}.$$

2.6. Deformation gradient and velocity gradient

The basic measure of deformation with respect to $\mathbf{X}$ (reference configuration) is the deformation gradient, $\mathbf{F}$:

$$\mathbf{F} = (\nabla_{\mathbf{X}} \boldsymbol{\Phi}_t(\mathbf{X}))^{\mathrm{T}}$$

i.e.,

$$F_{ij} = (\partial x_i / \partial X_j),$$

or

$$\mathbf{F} = \mathbf{D} \boldsymbol{\Phi}_t(\mathbf{X})$$

where $\nabla_{\mathbf{X}}$ and $\mathbf{D}$ denote differentiation with respect to $\mathbf{X}$. According to (2.1.3) $\mathbf{F}$ is non-singular. The basic measure of deformation with respect to $\mathbf{x}$ (present configuration) is the velocity gradient $\nabla \mathbf{v}$ (∇ denotes differentiation with respect to $\mathbf{x}$).

2.7. Kinematics of deformation-strain

By differentiation of $\mathbf{x}$ with respect to $\mathbf{X}$ we obtain

$$dx_i = (\partial x_i / \partial X_j) dX_j$$

or

$$d\mathbf{x} = \mathbf{F} \cdot d\mathbf{X} \qquad (2.7.1)$$

which gives the deformation of an infinitesimal filament of length $|d\mathbf{X}|$ and orientation $\mathbf{M}$ ($= d\mathbf{X}/|d\mathbf{X}|$) from its reference state to the present state, $d\mathbf{x}$, with length $|d\mathbf{x}|$ and orientation $\mathbf{m}$ ($= d\mathbf{x}/|d\mathbf{x}|$). That is,

$$\mathbf{F} \cdot d\mathbf{X} \to d\mathbf{x}$$

(see Figure 2.7.1).[12] This relation forms the basis of deformation of a material filament. The corresponding relation for the areal vector of an infinitesimal material plane is given by

$$d\mathbf{a} = (\det \mathbf{F})(\mathbf{F}^{-1})^{\mathrm{T}} \cdot d\mathbf{A} \qquad (2.7.2)$$

and can be obtained similarly. In this case the area in the present configuration is $d\mathbf{a} = |d\mathbf{a}|$ and the orientation $\mathbf{n}$ ($= d\mathbf{a}/|d\mathbf{a}|$), the area in the reference configuration is $dA = |d\mathbf{A}|$ and the initial orientation $\mathbf{N}$ $(= d\mathbf{A}/|d\mathbf{A}|)$ (see Figure 2.7.1). The volumetric change from dV to dv is given by

$$dv = (\det \mathbf{F})\, dV. \quad {}^{13} \qquad (2.7.3)$$

Problem 2.7.1

Obtain (2.7.2) by defining $d\mathbf{a} = d\mathbf{x}_1 \times d\mathbf{x}_2$, $d\mathbf{A} = d\mathbf{X}_1 \times d\mathbf{X}_2$ and using $d\mathbf{x} = \mathbf{F} \cdot d\mathbf{X}$.

Problem 2.7.2

Obtain (2.7.3).

The measures of strain here are the length stretch, λ, and the area stretch, η. They are defined as

$$\lambda \equiv \lim_{|d\mathbf{X}| \to 0} \frac{|d\mathbf{x}|}{|d\mathbf{X}|}, \qquad \eta \equiv \lim_{|d\mathbf{A}| \to 0} \frac{|d\mathbf{a}|}{|d\mathbf{A}|},$$

Figure 2.7.1. Deformation of infinitesimal elements, lines surfaces, and volumes.

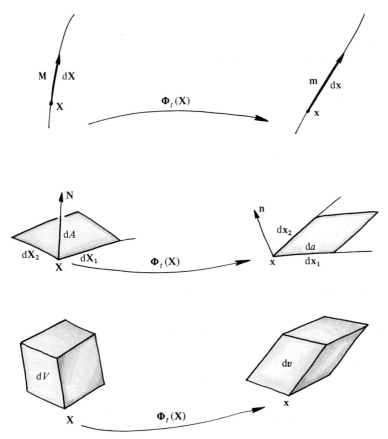

and can be obtained from

$$\lambda = (\mathbf{C}:\mathbf{MM})^{1/2} \tag{2.7.4}$$

$$\eta = (\det \mathbf{F})(\mathbf{C}^{-1}:\mathbf{NN})^{1/2} \tag{2.7.5}$$

where $\mathbf{C}$ ($\equiv \mathbf{F}^{\mathrm{T}} \cdot \mathbf{F}$) is called the right Cauchy–Green strain tensor.[14] The vectors $\mathbf{M}$ and $\mathbf{N}$ are defined by

$$\mathbf{M} \equiv d\mathbf{X}/|d\mathbf{X}|$$
$$\mathbf{N} \equiv d\mathbf{A}/|d\mathbf{A}|.$$

For comparison, the volumetric change $J \equiv (dv/dV)$ is given by

$$J = \det \mathbf{F}. \tag{2.7.6}$$

The orientations of the vectors $d\mathbf{x}$ and $d\mathbf{a}$ are given by

$$\mathbf{m} \equiv d\mathbf{x}/|d\mathbf{x}|$$
$$\mathbf{n} \equiv d\mathbf{a}/|d\mathbf{a}|.$$

The relationship of $\mathbf{m}$ and $\mathbf{M}$, and $\mathbf{n}$ and $\mathbf{N}$, to the motion is given by:

$$\mathbf{m} = \mathbf{F} \cdot \mathbf{M}/\lambda \tag{2.7.7}$$

$$\mathbf{n} = (\det \mathbf{F})(\mathbf{F}^{-1})^{\mathrm{T}} \cdot \mathbf{N}/\eta. \tag{2.7.8}$$

Since $\mathbf{F}$ is non-singular, the polar decomposition theorem states that it can be written as

$$\mathbf{F} = \mathbf{R} \cdot \mathbf{U} = \mathbf{V} \cdot \mathbf{R}$$

where $\mathbf{U}$ and $\mathbf{V}$ are positive definite and $\mathbf{R}$ is proper orthogonal. A tensor $\mathbf{T}$ is positive definite if $\mathbf{T}:\mathbf{uu} > 0$ for all $\mathbf{u} \neq 0$. $\mathbf{R}$ is proper orthogonal if $\mathbf{R} \cdot \mathbf{R}^{\mathrm{T}} = 1$ and $\det \mathbf{R} = +1$. Thus, locally the motion is a composition of rotation and stretching.

Example 2.7.1

Consider a flow $\mathbf{\Phi}_t(\mathbf{X})$ that transforms the body V_0, into V_t; i.e.,

$$\mathbf{\Phi}_t\{V_0\} = \{V_t\}$$

and a set of motions transforming $\{V_0\}$ into $\{V_1\}$, $\{V_1\}$ into $\{V_2\}, \ldots$

$$\mathbf{\Phi}_t^{(1)}\{V_0\} = \{V_1\}$$
$$\mathbf{\Phi}_t^{(2)}\{V_1\} = \{V_2\}$$
$$\mathbf{\Phi}_t^{(3)}\{V_2\} = \{V_3\}$$
$$\vdots$$
$$\mathbf{\Phi}_t^{(n)}\{V_{n-1}\} = \{V_n\} = \{V_t\}$$

or equivalently, in a different notation,

$$\Phi_t^{(1)}\{x_0\} = \{x_1\}$$
$$\Phi_t^{(2)}\{x_1\} = \{x_2\}$$
$$\Phi_t^{(3)}\{x_2\} = \{x_3\}$$
$$\vdots$$
$$\Phi_t^{(n)}\{x_{n-1}\} = \{x_n\} = \{x_t\}$$

which shows that $\{x_{n-1}\}$ is the reference configuration for the flow $\Phi_t^{(n)}$.

The motions are composed of

$$\Phi_t^{(n)} \cdot \Phi_t^{(n-1)} \cdot \ldots \cdot \Phi_t^{(2)} \cdot \Phi_t^{(1)}(\ ) = \Phi_t(\ )$$

and the deformation gradients are composed as

$$F^{(n)} \cdot F^{(n-1)} \cdot \ldots \cdot F^{(2)} \cdot F^{(1)}(\ ) = F(\ ).$$

It is easy to see that the lineal stretch is given by

$$\lambda^2 = (F^{(1)})^T \cdot (F^{(2)})^T \cdot \ldots \cdot (F^{(n-1)})^T \cdot (F^{(n)})^T \cdot F^{(n)} \cdot F^{(n-1)} \cdot \ldots \cdot F^{(2)} \cdot F^{(1)} : MM.$$

Problem 2.7.3
Obtain (2.7.4) through (2.7.8).

Problem 2.7.4
Prove that $DF/Dt = (\nabla v)^T \cdot F$.

Problem 2.7.5
Calculate the optimum orientation M for maximum stretching in a given time t, in a simple shear flow $v_1 = \dot\gamma x_2, v_2 = 0, v_3 = 0$.

2.8. Motion around a point

The Taylor series expansion of the relative velocity field around a point P (which represents either a fixed position x_P or a particle X_P) is

$$v = v_P + dx \cdot (\nabla v)_P + \text{higher order terms}$$

where dx is a vector centered on P. The velocity gradient ∇v can be decomposed uniquely into its symmetric and antisymmetric parts, $\nabla v = D + \Omega$,

$$D \equiv \tfrac{1}{2}(\nabla v + (\nabla v)^T), \text{ the stretching tensor (symmetric)}$$

$$\Omega \equiv \tfrac{1}{2}(\nabla v - (\nabla v)^T), \text{ the vorticity or spin tensor (antisymmetric)}.$$

Let us now consider the physical meaning of D and Ω.

The relative velocity, $v - v_P$, at dx is

$$v_{rel} = dx \cdot (D + \Omega).$$

The component in the direction dx is

$$v_{rel} \cdot (dx/|dx|) = (D + \Omega) : dxdx/|dx|$$

and
$$\mathbf{v}_{rel} \cdot \mathbf{n} = (\mathbf{D}:\mathbf{nn})|d\mathbf{x}|,$$
where $\mathbf{n} = d\mathbf{x}/|d\mathbf{x}|$. Note that since $\boldsymbol{\Omega}:\mathbf{nn} = 0$, $\boldsymbol{\Omega}$ does not contribute to the velocity in the direction normal to the sphere $|d\mathbf{x}| = $ constant and the contribution of $\boldsymbol{\Omega}$ is wholly tangential to the sphere (see Figure 2.8.1).

Problem 2.8.1
Where is $\mathbf{v}_{rel} \cdot \mathbf{n}$ maximum?

Problem 2.8.2
Prove that $\boldsymbol{\Omega}:\mathbf{nn} = 0$.

Problem 2.8.3
Show that the vector $\boldsymbol{\Omega} \cdot \mathbf{n}$ can be written as $\frac{1}{2}(\boldsymbol{\omega} \times \mathbf{n})$ where $\boldsymbol{\omega} = 2(\Omega_{23}, \Omega_{31}, \Omega_{12})$ and $\boldsymbol{\omega} = \nabla \times \mathbf{v}$ ($\boldsymbol{\omega}$ is called the vorticity vector, see Section 3.8). Present an argument to show that $\boldsymbol{\omega}$ represents angular rotation with speed $|\boldsymbol{\omega}/2|$ in a plane perpendicular to $\boldsymbol{\omega}$.

Problem 2.8.4
Show that $\mathbf{v}_{rel}$ can be written as
$$\mathbf{v}_{rel} = \tfrac{1}{2}\nabla(d\mathbf{x} \cdot \mathbf{D} \cdot d\mathbf{x}) + \tfrac{1}{2}\boldsymbol{\omega}_P \times d\mathbf{x}.$$

Problem 2.8.5
Show that near a surface with normal $\mathbf{n}$ such that $\mathbf{v} = 0$ (and $\boldsymbol{\omega} \cdot \mathbf{n} = 0$) $\mathbf{D}$ can be written as
$$\mathbf{D} = (\boldsymbol{\omega} \times \mathbf{n})\mathbf{n} + \mathbf{n}(\boldsymbol{\omega} \times \mathbf{n})$$
if $\nabla \cdot \mathbf{v} = 0$ (Caswell, 1967; Huilgol, 1975).

Figure 2.8.1. Velocity field around a point.

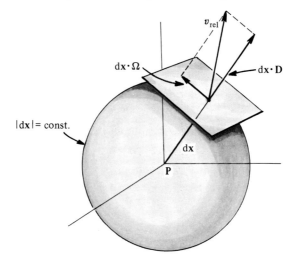

Problem 2.8.6
Study the possibility of expanding the relative velocity field near a surface as
$$\mathbf{v} = \phi(\mathbf{x})\mathbf{\Gamma} \cdot \mathbf{x}$$
where $\phi(\mathbf{x})$ is a scalar function and $\mathbf{\Gamma}$ is a matrix (Perry and Fairlie, 1974).

Problem 2.8.7
Obtain the general form of the velocity near a solid surface.

2.9. Kinematics of deformation: rate of strain

The companion equations to (2.7.4)–(2.7.8) are

$$D(\mathbf{dx})/Dt = \mathbf{dx} \cdot \nabla\mathbf{v} \tag{2.9.1}$$

$$D(\mathbf{da})/Dt = \mathbf{da}\, D(\det \mathbf{F})/Dt - \mathbf{da} \cdot (\nabla\mathbf{v})^{\mathsf{T}} \tag{2.9.2}$$

$$D(dv)/Dt = (\nabla \cdot \mathbf{v})dv. \tag{2.9.3}$$

The specific rate of stretching of λ and η are given by:

$$D(\ln \lambda)/Dt = \mathbf{D} : \mathbf{mm} \tag{2.9.4}$$

$$D(\ln \eta)/Dt = \nabla \cdot \mathbf{v} - \mathbf{D} : \mathbf{nn} \tag{2.9.5}$$

and the volumetric expansion by

$$D(\ln J)/Dt = \nabla \cdot \mathbf{v} \tag{2.9.6}$$

The Lagrangian histories $D(\ln \lambda)/Dt$ and $D(\ln \eta)/Dt$ will appear repeatedly throughout this work and are called *stretching functions*. They are typically denoted by α.

The companion equations to (2.7.7) and (2.7.8) are:

$$D\mathbf{m}/Dt = \mathbf{m} \cdot \nabla\mathbf{v} - (\mathbf{D} : \mathbf{mm})\mathbf{m} \tag{2.9.7}$$

for the rate of change of the orientation of a material filament, and

$$D\mathbf{n}/Dt = (\mathbf{D} : \mathbf{nn})\mathbf{n} - \mathbf{n} \cdot (\nabla\mathbf{v})^{\mathsf{T}} \tag{2.9.8}$$

for the rate of change of the orientation of the areal plane.

Note
Equations (2.9.4)–(2.9.6) can be written as

$D(\ln \lambda)/Dt = \nabla\mathbf{v} : \mathbf{mm}$ *projection of $\nabla\mathbf{v}$ onto line with orientation* $\mathbf{m}$

$D(\ln \eta)/Dt = \nabla\mathbf{v} : (\mathbf{1} - \mathbf{nn})$ *projection of $\nabla\mathbf{v}$ onto plane with orientation* $\mathbf{n}$

$D(\ln J)/Dt = \nabla\mathbf{v} : \mathbf{1}$ *projection of $\nabla\mathbf{v}$ onto volume,*

offering a somewhat different interpretation of the equations.

Problem 2.9.1
Show that Equation (2.9.5) can be written as $D\ln(\eta/\rho)/Dt = -\mathbf{D} : \mathbf{nn}$ where ρ is the density.

Problem 2.9.2
Starting with

$$D(\ln \eta)/Dt = \nabla \cdot \mathbf{v} - \mathbf{D} : \mathbf{nn},$$

show that

$$D(\ln \eta)/Dt = (\nabla_S \cdot \mathbf{v}_S) - \kappa_A v_n,$$

where the velocity is written as $\mathbf{v} = \mathbf{v}_S + \mathbf{N} v_n$, such that v_n is the speed normal to the surface, ∇_S is the surface gradient operator, and κ_A is the mean curvature, $\kappa_A = -\nabla_S \cdot \mathbf{n}$.

2.10. Rates of change of material integrals

In Section 2.2 we computed the rate of change of the integral over a material volume. It is relatively easy to generate similar versions for material lines and surfaces (see Figure 2.10.1). For example, for a material line joining $\mathbf{x}_1 \; (= \mathbf{\Phi}_t(\mathbf{X}_1))$ and $\mathbf{x}_2 \; (= \mathbf{\Phi}_t(\mathbf{X}_2))$ with configuration L_t, the

Figure 2.10.1. Deformation of finite material lines and surfaces.

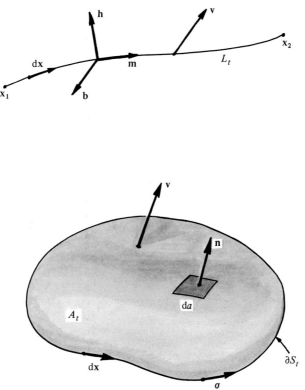

suitable formula is

$$\frac{d}{dt}\int_{L_t} \mathbf{G}\cdot d\mathbf{x} = \int_{L_t} \frac{D\mathbf{G}}{Dt}\cdot d\mathbf{x} + \int_{L_t} \mathbf{G}\cdot(\nabla\mathbf{v})^{\mathrm{T}}\cdot d\mathbf{x} \qquad (2.10.1)$$

where $\mathbf{G}$ denotes a scalar, vector or tensor property. Similarly, for a material surface, with configuration A_t,

$$\frac{d}{dt}\int_{A_t} \mathbf{G}\cdot d\mathbf{a} = \int_{A_t} \frac{D\mathbf{G}}{Dt}\cdot d\mathbf{a} + \int_{A_t}(\nabla\cdot\mathbf{v})\cdot\mathbf{G}\cdot d\mathbf{a} - \int_{A_t} \mathbf{G}\cdot(\nabla\mathbf{v})\cdot d\mathbf{a}$$
$$(2.10.2)$$

Special cases correspond to the evolution of length of material lines and area of material surfaces. Thus, the evolution of length can be computed as

$$\frac{d}{dt}[\mathrm{Length}(L_t)] = \frac{d}{dt}\int_{L_t}|d\mathbf{x}| = \int_{L_0}\frac{D\lambda}{Dt}|d\mathbf{X}|$$

and since

$$\int_{L_0}\lambda(\mathbf{D}:\mathbf{mm})|d\mathbf{X}| = \int_{L_t}(\mathbf{D}:\mathbf{mm})|d\mathbf{x}| = \int_{L_t}(\mathbf{m}\cdot(\nabla\mathbf{v}))\cdot\mathbf{m}\,|d\mathbf{x}|.$$

Noting that

$$d\mathbf{v}/|d\mathbf{x}| = (d\mathbf{x}/|d\mathbf{x}|)\cdot\nabla\mathbf{v} = \mathbf{m}\cdot\nabla\mathbf{v},$$

we obtain,

$$\frac{d}{dt}[\mathrm{Length}(L_t)] = \int_{L_t}\frac{d\mathbf{v}}{|d\mathbf{x}|}\cdot\mathbf{m}\,|d\mathbf{x}| = \mathbf{v}\cdot\mathbf{m}\Big|_{\mathbf{x}_1}^{\mathbf{x}_2} - \int_{L_t}\mathbf{v}\cdot\mathbf{h}\kappa_L|d\mathbf{x}| \quad (2.10.3)$$

where the $\mathbf{h}$ is the normal of the Frenet triad ($\mathbf{m}$, tangent; $\mathbf{h}$, normal; $\mathbf{b}$, binormal. See Aris, 1962, p. 40 on) and the curvature κ_L is defined as $d\mathbf{m}/|d\mathbf{x}| = \kappa_L\mathbf{h}$.

Special case
If $\mathbf{x}_1 = \mathbf{x}_2$ or if

$$\mathbf{v}\cdot\mathbf{m}\Big|_{\mathbf{x}_1}^{\mathbf{x}_2} = 0$$

(the first case corresponds to a loop, the second to a line attached to non-moving walls for example),

$$\frac{d}{dt}[\mathrm{Length}(L_t)] = -\int_{L_t}\mathbf{v}\cdot\mathbf{h}\kappa_L|d\mathbf{x}|.$$

The companion equation for the rate of area increase is computed to be

$$\frac{d}{dt}[\mathrm{Area}(A_t)] = \int_{\partial A_t}(\boldsymbol{\sigma}\times\mathbf{n})\cdot\mathbf{v}\,|d\mathbf{x}| - \int_{A_t}(\mathbf{v}\cdot\mathbf{n})\kappa_A|d\mathbf{a}| \qquad (2.10.4)$$

where κ_A is the mean curvature,

$$\kappa_A = -\nabla_S\cdot\mathbf{n}$$

with ∇_S being the surface gradient operator ($\nabla_S(\ ) = \nabla(\ ) - \partial(\ )//\partial n$, where n is a coordinate normal to the surface). The vector $\boldsymbol{\sigma}$ is a tangent vector to the boundary of the surface, ∂A_t, such that it is right-hand oriented with respect to $\mathbf{n}$. Thus, $\boldsymbol{\sigma} \times \mathbf{n}$ is tangent to A_t.

Special case
Closed surface, i.e., $\partial A_t = \phi$.

$$\frac{d}{dt}[\text{Area}(A_t)] = -\int_{A_t} (\mathbf{v} \cdot \mathbf{n})\kappa_A |da|.$$

Problem 2.10.1
Denote by $G(\mathbf{x}, t)$ a scalar function. Show that

$$\Phi = \int_{L_t} G |d\mathbf{x}|$$

evolves as

$$\frac{d\Phi}{dt} = \int_{L_t} \left[\frac{DG}{Dt} + G(\mathbf{D}:\mathbf{mm}) \right] |d\mathbf{x}|.$$

Problem 2.10.2
Obtain the evolution of a finite area (Equation (2.10.4)) starting with

$$D(\ln \eta)/Dt = \nabla \cdot \mathbf{v} - \mathbf{D}:\mathbf{nn}.$$

Hint: A rough derivation, without using the theory of calculus on surfaces, can be obtained by using the identity

$$\nabla \times [\mathbf{f} \times \mathbf{v}] = (\nabla \mathbf{f}) \cdot \mathbf{v} - (\nabla \mathbf{v}) \cdot \mathbf{f} + \mathbf{f}(\nabla \cdot \mathbf{v}) - \mathbf{v}(\nabla \cdot \mathbf{f}).$$

Another possibility is to integrate

$$D(\ln \eta)/Dt = (\nabla_S \cdot \mathbf{v}_S) - \kappa_A v_n.$$

Problem 2.10.3
Using Equation (2.9.4), prove (2.10.3).

2.11. Physical meaning of $\nabla\mathbf{v}$, $(\nabla\mathbf{v})^T$, and $\mathbf{D}$

Consider a case such that

$$\lim_{t \to \infty} \frac{D\mathbf{m}}{Dt} = 0, \qquad \mathbf{m}_{ss} = \lim_{t \to \infty} \mathbf{m}$$

where the subscript ss denotes steady-state orientation. Then, by (2.9.7) we have

$$\mathbf{m}_{ss} \cdot (\nabla\mathbf{v}) = (\mathbf{D}:\mathbf{m}_{ss}\mathbf{m}_{ss})\mathbf{m}_{ss}$$

or

$$\mathbf{m}_{ss} \cdot (\nabla\mathbf{v}) = (\nabla\mathbf{v}:\mathbf{m}_{ss}\mathbf{m}_{ss})\mathbf{m}_{ss}$$

which shows that $(\nabla\mathbf{v}:\mathbf{m}_{ss}\mathbf{m}_{ss}) = \gamma_{ss}$, is an eigenvalue of $\nabla\mathbf{v}$

$$\mathbf{m}_{ss,i}\cdot(\nabla\mathbf{v}) = \gamma_{ss,i}\mathbf{m}_{ss,i} \qquad (2.11.1)$$

$\gamma_{ss,i}$, and $\mathbf{m}_{ss,i}$ the eigenvectors ($i = 1$ to 3), where $\mathbf{m}_{ss}$ represents the steady-state orientation of the material filament.[15]

Similarly, for areal vectors we have

$$\lim_{t\to\infty} \frac{D\mathbf{n}}{Dt} = 0, \qquad \mathbf{n}_{ss} = \lim_{t\to\infty} \mathbf{n}$$

and the steady-state orientation corresponds to the eigenvalues problem

$$\mathbf{n}_{ss,i}\cdot(\nabla\mathbf{v})^T = \gamma_{ss,i}\mathbf{n}_{ss,i}.$$

On the other hand, the solution of the eigenvalue problem

$$\mathbf{D}\cdot\mathbf{d}_i = \gamma_i\mathbf{d}_i, \quad |\mathbf{d}_i| = 1$$

gives eigenvectors $\mathbf{d}_i$ which are the *maximum directions of stretching*. The physical meaning of $\mathbf{d}_i$ can be appreciated in the following way. If $\mathbf{m}$ coincides with $\mathbf{d}_i$ we have,

$$D\mathbf{d}_i/Dt = \mathbf{d}_i\cdot(\mathbf{D}+\mathbf{\Omega}) - (\mathbf{D}:\mathbf{d}_i\mathbf{d}_i)\mathbf{d}_i$$

and using the result of Problem 2.8.2 we obtain

$$D\mathbf{d}_i/Dt = \mathbf{d}_i\cdot\mathbf{\Omega}$$

i.e., the rate of change of the $\mathbf{d}_i$s is due only to $\mathbf{\Omega}$, giving also an alternative interpretation to the spin tensor.

Example 2.11.1
Apply the above ideas to the linear flow $v_1 = Gx_2$, $v_2 = KGx_1$, where $-1 < K < 1$, considered in Problem 2.5.3. Note that the maximum directions of stretching are independent of K.

Problem 2.11.1
Compute the rate of rotation of the maximum and minimum directions of stretching in the shear flow $v_1 = \dot{\gamma}x_2$, $v_2 = 0$, $v_3 = 0$. Do they actually rotate?

Problem 2.11.2
Find the error in the following reasoning (given in Ottino, Ranz, and Macosko, 1981): Since $D(\ln\lambda)/Dt = \mathbf{D}:\mathbf{mm}$ and $D(\ln\eta)/Dt = \nabla\cdot\mathbf{v} - \mathbf{D}:\mathbf{nn}$, for $\mathbf{n} = \mathbf{m}$, we have $D(\ln\lambda)/Dt + D(\ln\eta)/Dt = \nabla\cdot\mathbf{v}$. Hence, if $\nabla\cdot\mathbf{v} = 0$, we obtain $\lambda\eta = 1$.

Problem 2.11.3
Explain why it is possible to 'integrate' $D(d\mathbf{x})/Dt = d\mathbf{x}\cdot\nabla\mathbf{v}$ but not $D(\mathbf{d}_i)/Dt = \mathbf{d}_i\cdot\mathbf{\Omega}$.

Bibliography

Even though the kinematical foundation of fluid mechanics dates from the seventeenth century, little is given nowadays in standard works in fluid mechanics. For modern accounts the reader should consult works in continuum mechanics. The presentation given here is based on a much larger body of work. The most comprehensive account is given by Truesdell and Toupin (1960, pp. 226–793). A brief and lucid account is given by Chadwick (1976) in *Continuum mechanics*. A more advanced treatment is given in Chapter II of Truesdell (1977). Other accessible accounts, slanted towards solid mechanics and fluid mechanics are given by Malvern (1969) and Aris (1962), respectively, whereas for historical references the reader should consult Truesdell (1954). A classical work in fluid mechanics with an unusually long discussion on the kinematics of flow around a point is *Fundamentals of hydro- and aerodynamics* by Tietjens, based on the lectures by Prandtl (see Prandtl and Tietjens, 1934). One of the best visual demonstrations of Lagrangian and Eulerian descriptions is given in the movie *Eulerian and Lagrangian descriptions in fluid mechanics*. One clear visual treatment of deformation around a point is given in the movie *Deformation in continuous media*. Both films are by J. L. Lumley. The scripts are given in *Illustrated experiments in fluid mechanics*, produced by the National Committee for Fluid Mechanics Films, MIT Press, Cambridge, 1972. The 'kinematical reversibility' of creeping flows is illustrated in the film *Low Reynolds number creeping flows*, by Taylor, also in *Illustrated experiments in fluid mechanics*, and in the article 'An unmixing demonstration', by Heller (1960). Both works focused on Couette flows, which are the exception rather than the rule among all two-dimensional flows. In most cases reversibility is impossible in practice due to unbounded growth of initial errors.

Notes

1 Throughout this work $x \in \mathbb{R}^3$. However, most of the results of this chapter are valid even if $x \in \mathbb{R}^n$. The extension, however, does not carry over to vorticity since the relation between an antisymmetric tensor and an axial vector is valid only in $\mathbb{R}^3$ (Truesdell, 1954, pp. 58, 59).

2 See Arnold, 1985, p. 4. Distinguish carefully between the label of the particle, $\mathbf{X}$, and its position, $\mathbf{x}$.

3 It is well documented that the association of names is incorrect but it is apparently too late to set the record straight. Both descriptions are actually due to Euler. For the historical account of credits see Truesdell and Toupin, 1960, p. 327 and Truesdell, 1954, p. 30.

4 In the framework of Equation (2.1.1) $v(x, t)$ should be interpreted as the velocity of the particle X which happens to be at x at time t. The solution of most fluid mechanical problems yields the Eulerian velocity since, usually, the equations are solved after being formulated in this viewpoint. Historically, the very earliest formulations of fluid mechanics were formulated using this viewpoint, taking $v(x, t)$ as a primitive quantity (for an authoritative account, see Truesdell, 1954, p. 37). For an alternative definition of the Eulerian velocity, considering it as the primitive quantity, see Problem 3.3.1.

5 Particles of different species can coexist at the same position x (see Bowen, 1976, for various extensions of this concept).

6 Recall that the non-autonomous case can always be transformed into an autonomous one by defining $t = x_4$.

7 There have been claims that the velocity field of an invisicid fluid in turbulent motion might not be Lipschitz. An early indication is given by Onsager, 1949).

8 For reasons that will be painfully apparent in Chapters 5 and 6, the examples cannot be, in general, much more complicated than this one.

9 As in Equation (2.1.1), we are labelling particles by their initial positions, however, in this case one has to distinguish carefully between the *label* of the particle and its *placement* at an arbitrary time, since it can be a source of confusion.

10 It is common practice in continuum mechanics to use capital letters to denote reference state and lower case letters to denote present state (see Figure 2.1.1). However, as is also universal practice in dynamical systems in the case of periodic flows, we denote X as x_0, and subsequent states as x_1, x_2, etc.

11 This flow can be realized by means of a four-roller apparatus; the concept is due to Giesekus (1962). See also Section 9.3.1 and Bentley and Leal (1986a).

12 Equation (2.7.1) can be interpreted as a matrix multiplication where F operates on a column vector dX. We shall not make any distinction between row and column vectors by means of a transpose since the meaning is always clear from the context.

13 This relation gives the physical meaning to the Jacobian of the flow ($= \det F$).

14 The so-called left Cauchy–Green strain tensor is $F \cdot F^T$. The tensor C^{-1} is called the Piola tensor.

15 Note that a steady state orientation need not exist. For example the rate of rotation can be constant or periodic in time (see Chapter 4). Also, note that the steady state orientation might be unstable.

3

Conservation equations, change of frame, and vorticity

In this chapter we record for further use and completeness the equations of conservation of mass and linear momentum and laws of transformation for velocity, acceleration, velocity gradient, etc., for frame transformations involving translation and rotation. We conclude the chapter by studying the equations of motion in terms of vorticity and the streamfunction.

3.1. Principle of conservation of mass

Integral version

The mass contained in a material volume V_t is given by

$$M(V_t) = \int_{V_t} \rho(\mathbf{x}, t)\, dv,$$

where $\rho(\mathbf{x}, t)$ is the density (see Figure 2.1.1). The principle of conservation of mass states that:

$$d(M(V_t))/dt = 0 \qquad \text{or} \qquad M(V_t) = M(V_0),$$

where V_0 is the reference configuration.

Microscopic versions

Denoting the reference mass density, $\rho(\mathbf{x}, t = 0)$ as $\rho_0(\mathbf{X})$, we have

$$\int_{V_t} \rho(\mathbf{x}, t)\, dv = \int_{V_0} \rho(\mathbf{\Phi}_t(\mathbf{X}), t)J\, dV = \int_{V_0} \rho_0(\mathbf{X})\, dV,$$

and since V_0 is arbitrary,

$$\rho(\mathbf{x}, t) = \rho_0(\mathbf{X})/J \qquad (Lagrangian\ version). \qquad (3.1.1)$$

Taking the material derivative of $\rho_0(\mathbf{X})$ we obtain

$$D(\rho_0(\mathbf{X}))/Dt = 0 = (D\rho/Dt)J + \rho DJ/Dt.$$

Using Euler's formula and the condition $J \neq 0$, we obtain,

$$D\rho/Dt = -\rho(\nabla \cdot \mathbf{v}). \qquad (3.1.2)$$

Expanding $D\rho/Dt$ we obtain

$$\partial\rho/\partial t = -\nabla \cdot (\rho\mathbf{v}), \qquad (Eulerian\ version). \qquad (3.1.3)$$

Both Equations (3.1.2) and (3.1.3) are known as the continuity equation or mass balance.

Problem 3.1.1
Using the continuity equation prove that

$$\frac{d}{dt} \int_{V_t} \rho G(\mathbf{x}, t) \, dv = \int_{V_t} \rho \frac{DG}{Dt} \, dv$$

where G is any scalar, vector, or tensor function.

3.2. Principle of conservation of linear momentum

Integral version (or Euler's axiom)
This principle states that for a material volume V_t, linear momentum is conserved, i.e.,

$$\underset{\substack{\text{rate of change of} \\ \text{momentum}}}{\frac{d}{dt} \int_{V_t} \rho \mathbf{v} \, dv} = \underset{\substack{\text{forces acting on} \\ \text{body}}}{\int_{V_t} \rho \mathbf{f} \, dv} + \oint_{\partial V_t} \mathbf{t} \, ds \qquad (3.2.1)$$

$$\underset{\text{acceleration}}{} \quad \underset{\text{body forces}}{} \quad \underset{\text{contact forces}}{}$$

where $\mathbf{t}(\mathbf{n}, \mathbf{x}, t)$ is as yet an undefined vector called the *traction* which depends on the placement $\mathbf{x}$ on the boundary, ∂V_t, the instantaneous orientation, $\mathbf{n}$, and time, t (see Figure 3.2.1). The vector $\mathbf{f}$ is the body force, which in this work is assumed to be independent of the configuration of the body, V_t. Using the transport theorem (Section 2.2) we obtain

$$\int_{V_t} \rho \frac{D\mathbf{v}}{Dt} \, dv = \int_{V_t} \rho \mathbf{f} \, dv + \oint_{\partial V_t} \mathbf{t} \, ds \qquad (3.2.2)$$

for any region V_t, at any time (where V_t can be an arbitrary control volume).

Problem 3.2.1
Prove that the Principle of Conservation of Linear Momentum implies the continuum version of 'Newton's third law': $\mathbf{t}(\mathbf{x}, \mathbf{n}, t) = -\mathbf{t}(\mathbf{x}, -\mathbf{n}, t)$.

3.3 Traction $\mathbf{t}(\mathbf{n}, \mathbf{x}, t)$

It can be proved that $\mathbf{t}(\mathbf{n}, \mathbf{x}, t) = \mathbf{T}^{\mathrm{T}} \cdot \mathbf{n}$, where $\mathbf{T}$ is a tensor (note convention). This implies that the information about tractions at the point $\mathbf{x}$ and any surface with normal $\mathbf{n}$ is contained in the tensor $\mathbf{T}$. The proof consists of three steps which we will briefly repeat here.

(i) Cauchy's theorem
Consider a (small) region V_t surrounding a particle $\mathbf{X}$. Then, if L represents some length scale of V_t we have:

$$\text{Volume } (V_t) = \text{const.}_1 \, L^3$$
$$\text{Surface } (\partial V_t) = \text{const.}_2 \, L^2.$$

The mean value theorem states that for any continuous function $G(\mathbf{X}, t)$ defined over V_t and ∂V_t we can find $\mathbf{X}'$ and $\mathbf{X}''$ (at any time t) such that

$$\int_{V_t} G \, dv = \text{const.}_1 \, L^3 \, G(\mathbf{X}', t), \qquad \text{where } \mathbf{X}' \text{ belongs to } V_t$$

$$\oint_{\partial V_t} G \, ds = \text{const.}_2 \, L^2 \, G(\mathbf{X}'', t), \qquad \text{where } \mathbf{X}'' \text{ belongs to } \partial V_t.$$

Figure 3.2.1. (a) Material region in present configuration, V_t, indicating normal **n** and traction **t**; (b) construction for balance of angular momentum.

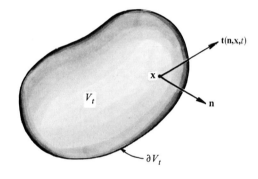

(a)

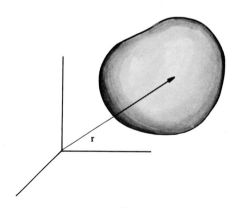

(b)

Applying the theorem to Equation (3.2.2) we obtain

$$\rho \frac{D}{Dt} \{\mathbf{v}(\mathbf{X}', t) \text{ const.}_1 L^3\} = \rho \mathbf{f} \text{ const.}_1 L^3 + \oint_{\partial V_t} \mathbf{t} \, ds.$$

Then, dividing by L^2 and letting L go to zero (preserving geometrical similarity),

$$\lim_{L \to \infty} \left\{ \frac{1}{L^2} \oint_{\partial V_t} \mathbf{t} \, ds \right\} = 0.$$

That is, the tractions are locally in equilibrium (see Serrin, 1959, p. 134).

(ii) *Traction on an arbitrary plane (Cauchy's tetrahedron construction)*
By means of this construction, which we will not repeat here (see Serrin, 1959, p. 134, for details), Cauchy was able to prove that the components of the traction $\mathbf{t}(\mathbf{n})$, $[\mathbf{t}([\mathbf{n}])] = (t^{n_1}, t^{n_2}, t^{n_3})$, are related to the normal $\mathbf{n}$, $[\mathbf{n}] = (n_1, n_2, n_3)$, by the matrix multiplication:[1]

$$t^{n_i} = T_{ji}n_j \qquad \text{or equivalently} \qquad [\mathbf{t}([\mathbf{n}])] = [\mathbf{n}][\mathbf{T}], \ [\mathbf{t}([\mathbf{n}])] = [\mathbf{T}^{\mathsf{T}}][\mathbf{n}]$$

where the brackets [ ] represent the display of the components of the vector $\mathbf{n}$ and the matrix representation of the tensor $\mathbf{T}$.

(iii) *The third step is to prove that* $[\mathbf{T}]$ *is indeed the matrix representation of a tensor* $\mathbf{T}$ (see Appendix)
In order to prove that the T_{ji}s are the components of a tensor $\mathbf{T}$, we need to prove that the components T_{ji} transform as a tensor. Since $\mathbf{n}$ is just a free vector, which is objective,

$$\mathbf{n}' = \mathbf{Q} \cdot \mathbf{n},$$

and since the traction $\mathbf{t}$ transforms as

$$\mathbf{t}' = \mathbf{Q} \cdot \mathbf{t},$$

then, $\mathbf{T}$ transforms as[2]

$$\mathbf{T}' = \mathbf{Q} \cdot \mathbf{T} \cdot \mathbf{Q}^{\mathsf{T}}.$$

Problem 3.3.1
Use Cauchy's construction in conjunction with the 'mass balance' to show that the mass flux in the direction normal to a plane $\mathbf{n}$ is given by $\mathbf{j} \cdot \mathbf{n}$, where $\mathbf{j}$ is the mass flux vector. This allows the definition of the Eulerian velocity as $\mathbf{v} = \mathbf{j}/\rho$ which can now be regarded as the primitive quantity, rather than the velocity obtained by means of time differentiation of the motion (Section 2.2).

3.4. Cauchy's equation of motion

Using Gauss's theorem, Equation (3.2.2) can be written as

$$\int_{V_t} \rho \frac{D\mathbf{v}}{Dt} \, dv = \int_{V_t} \rho \mathbf{f} \, dv + \int_{V_t} \nabla \cdot \mathbf{T} \, dv. \qquad (3.4.1)$$

Since V_t is arbitrary, invoking the usual conditions,

$$\rho \frac{D\mathbf{v}}{Dt} = \rho \mathbf{f} + \nabla \cdot \mathbf{T}, \qquad (3.4.2)$$

we obtain Cauchy's Equation of Motion or Cauchy's First Law of Motion. This equation is in terms of spatial co-ordinates $(\mathbf{x}, t)$. For a formulation in material co-ordinates $(\mathbf{X}, t)$ see Truesdell and Toupin, 1960, p. 553.

Problem 3.4.1

Show that (3.4.2) can be written as

$$\frac{\partial(\rho v_i)}{\partial t} = \rho f_i + \frac{\partial(T_{ji} - \rho v_j v_i)}{\partial x_j}$$

and interpret the result physically.

3.5. Principle of conservation of angular momentum

In the simplest case (without body couples, or equivalently, for non-polar materials, see Truesdell and Toupin, 1960, p. 538; Serrin, 1959, p. 136) this principle states that (see Figure 3.2.1)

$$\frac{d}{dt} \int_{V_t} \rho \mathbf{r} \times \mathbf{v} \, dv = \int_{V_t} \rho \mathbf{r} \times \mathbf{f} \, dv + \oint_{\partial V_t} \mathbf{r} \times \mathbf{t} \, ds. \qquad (3.5.1)$$

The interrelation between linear momentum, angular momentum (*without body couples*, as above), and the symmetry of the stress tensor is the following:

---

(1) Principle of conservation + Principle of conservation $\rightarrow \mathbf{T} = \mathbf{T}^\mathsf{T}$
     of linear momentum   of angular momentum

(2) Principle of conservation + $\mathbf{T} = \mathbf{T}^\mathsf{T} \rightarrow$ Principle of conservation
     of linear momentum   of angular momentum

(3) Principle of conservation + $\mathbf{T} = \mathbf{T}^\mathsf{T} \rightarrow$ Principle of conservation
     of angular momentum   of linear momentum

For example, (2) is proved in Serrin (1959, p. 136). We will assume that

$\mathbf{T} = \mathbf{T}^T$ but we note that non-symmetric tensors are indeed possible and have highly non-trivial consequences as well as practical importance (Brenner, 1984; Rosensweig, 1985).

The tensor $\mathbf{T}$ is normally written as

$$\mathbf{T} = -p\mathbf{1} + \boldsymbol{\tau}$$

where $p = -\operatorname{tr}(\mathbf{T})/3$, so that $\operatorname{tr}(\tau) \equiv 0$. For a fluid at rest $\tau \equiv 0$, and $\mathbf{T} = -p\mathbf{1}$. These conditions identify also an inviscid fluid. The simplest constitutive equation involving viscosity is the *incompressible* Newtonian fluid which is defined as $\tau = 2\mu\mathbf{D}^3$ and where μ is the shear viscosity. Replacing $\mathbf{T} = -p\mathbf{1} + 2\mu\mathbf{D}$ into (3.4.2) and assuming $\nabla\mu = \nabla \cdot \mathbf{v} = 0$, we obtain the Navier–Stokes equation

$$\rho \frac{D\mathbf{v}}{Dt} = \rho\mathbf{f} - \nabla p + \mu\nabla^2\mathbf{v}. \tag{3.5.2}$$

If $\mu = 0$, the equation is traditionally called Euler's equation.

Problem 3.5.1
The *normal stress* on a plane $\mathbf{n}$ is $\mathbf{T}:\mathbf{nn}$ (i.e., components T_{jj}). The *shear stress* on a plane $\mathbf{n}$ is $\mathbf{T}:\mathbf{nt}$ where $\mathbf{t}$ is orthogonal to $\mathbf{n}$ (i.e., components $T_{ij}, i \neq j$). Prove that if $\mathbf{T} = -p\mathbf{1}$ there are no shear stresses on any plane and that the normal stresses are non-zero and independent of $\mathbf{n}$.

3.6. Mechanical energy equation and the energy equation

The *mechanical energy equation* is not an independent principle but a consequence of the Principle of Linear Momentum. Taking the scalar product of Cauchy's Equation of Motion with $\mathbf{v}$ and integrating over V_t we obtain

$$\int_{V_t} \rho\mathbf{v} \cdot \frac{D\mathbf{v}}{Dt}\, dv = \int_{V_t} \rho\mathbf{v} \cdot \mathbf{f}\, dv + \int_{V_t} (\nabla \cdot \mathbf{T}) \cdot \mathbf{v}\, dv. \tag{3.6.1}$$

For a symmetric $\mathbf{T}$,

$$\nabla \cdot (\mathbf{T} \cdot \mathbf{v}) = (\nabla \cdot \mathbf{T}) \cdot \mathbf{v} + \mathbf{T}:\nabla\mathbf{v},$$

and since $\boldsymbol{\Omega}$ is antisymmetric,

$$\mathbf{T}:\nabla\mathbf{v} = \mathbf{T}:(\mathbf{D} + \boldsymbol{\Omega}) = \mathbf{T}:\mathbf{D}.$$

Replacing into (3.6.1) and using Gauss's theorem we obtain

$$\frac{d}{dt}\int_{V_t} \tfrac{1}{2}\rho\mathbf{v}^2\, dv = \int_{V_t} \rho\mathbf{f} \cdot \mathbf{v}\, dv + \oint_{\partial V_t} (\mathbf{T} \cdot \mathbf{v}) \cdot \mathbf{n}\, ds - \int_{V_t} (\mathbf{T}:\mathbf{D})\, dv \tag{3.6.2}$$

$$\frac{d}{dt}\int_{V_t} \tfrac{1}{2}\rho\mathbf{v}^2\, dv = \int_{V_t} \rho\mathbf{f} \cdot \mathbf{v}\, dv + \oint_{\partial V_t} \mathbf{t} \cdot \mathbf{v}\, ds - \int_{V_t} (\mathbf{T}:\mathbf{D})\, dv.$$

| rate of change of kinetic energy | rate of work of body forces | rate of work of contact forces | expansion of work and viscous dissipation |

An alternative way of expressing the mechanical energy equation is

$$\frac{dK}{dt} = \frac{dW}{dt} - \int_{V_t} (\mathbf{T}:\mathbf{D})\, dv$$

where

$$K = \int_{V_t} \tfrac{1}{2}\rho \mathbf{v}^2\, dv$$

is the kinetic energy in V_t, and

$$\frac{dW}{dt} = \int_{V_t} \rho\mathbf{f}\cdot\mathbf{v}\, dv + \oint_{\partial V_t} \mathbf{t}\cdot\mathbf{v}\, ds$$

is the rate of work on ∂V_t and within V_t. The *energy equation* is the first law of thermodynamics for a continuum. If we define:

$$E = \int_{V_t} \rho\varepsilon\, dv,$$

where ε is internal energy per unit mass and $\mathbf{q}$ the heat flux vector, then the Principle of Conservation of Energy states:

$$\frac{d}{dt}(K + E) = \int_{V_t} \rho\mathbf{f}\cdot\mathbf{v}\, dv + \oint_{\partial V_t} (\mathbf{T}\cdot\mathbf{v})\cdot\mathbf{n}\, ds - \oint_{\partial V_t} \mathbf{q}\cdot\mathbf{n}\, ds. \quad (3.6.3)$$

Combining (3.6.2) and (3.6.3) we obtain,

$$\frac{dE}{dt} = \int_{V_t} (\mathbf{T}:\mathbf{D})\, dv - \oint_{\partial V_t} \mathbf{q}\cdot\mathbf{n}\, ds \quad (3.6.4)$$

i.e., the rate of generation of internal energy within the volume is due to viscous dissipation–expansion within the volume and energy input–output through the boundaries.

Problem 3.6.1

With the aid of the mechanical energy equation, show that any fluid of a Newtonian incompressible fluid completely enclosed within a rigid non-moving boundary and acted upon by a conservative body force must approach zero velocity everywhere for long times.

Problem 3.6.2

Show that $\mathbf{T}:\mathbf{D}$ gives rise to $p(\nabla\cdot\mathbf{v})$ (expansion work) and $\tau:\mathbf{D}$ (viscous dissipation). Using the results of Section 3.7, prove that the viscous dissipation is frame indifferent.

Problem 3.6.3

Verify that the viscous dissipation produced by a Newtonian fluid is positive.

Problem 3.6.4

By manipulation of $\mathbf{T}:\mathbf{D}$ show that for the case of an inviscid fluid ($\mathbf{T} = -p\mathbf{1}$) obeying the ideal gas law, the time integral of the rate of work, $\int (d\mathrm{W}/dt)dt$, is given by the popular equation $\int p\,dv$. Indicate any other necessary assumptions.

3.7. Change of frame

Consider a point $\mathbf{P}$ as seen in two frames, F and F', related by

$$\mathbf{x}' = \mathbf{x_0}(t) + \mathbf{Q}(t)\cdot\mathbf{x} \tag{3.7.1}$$

where $\mathbf{x}$ indicates the position in F and $\mathbf{x}'$ in F', $\mathbf{x_0}(t)$ is an arbitrary vector, and $\mathbf{Q}(t)$ represents a (time dependent) proper orthogonal transformation, i.e.,

$$\mathbf{Q}(t)^\mathrm{T}\cdot\mathbf{Q}(t) = 1 \qquad \text{with } \det[\mathbf{Q}(t)] = 1.$$

If $\mathbf{O}$ represents the center of co-ordinates of the frame F, then $\mathbf{x_0}(t)$ represents its position as seen from F' (see Figure 3.7.1).

3.7.1. Objectivity

Denote generic scalars, vectors, and tensors, as f, $\mathbf{w}$, and $\mathbf{S}$, respectively. f, $\mathbf{w}$, and $\mathbf{S}$ are *objective* if they transform according to:

(i) $f' = f$

(ii) $\mathbf{w}' = \mathbf{Q}\cdot\mathbf{w}$

(iii) $\mathbf{S}' = \mathbf{Q}\cdot\mathbf{S}\cdot\mathbf{Q}^\mathrm{T}$.

Figure 3.7.1. Change of frame.

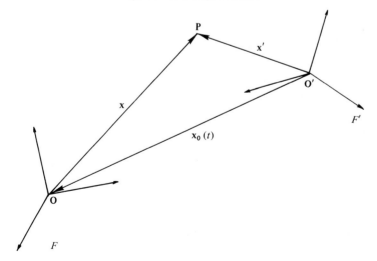

For example (ii) indicates that **w** transforms as a 'true' vector, and (iii) can be regarded as the definition of a second order tensor (see Appendix). However, as we shall see not every $[1 \times 3]$ matrix transforms as a 'true' vector, i.e., according to (ii) above, and not every $[3 \times 3]$ matrix qualifies as a tensor, i.e., they do not transform according to the transformation rule (iii). However, due to usage, the words 'vector' and 'tensor' are used for quantities that do not transform as above under the change of frame (3.7.1).

If the quantities f, **w**, and **S**, transform according to (i)–(iii) under the frame transformation (3.7.1), then they are called *frame indifferent*.

3.7.2. Velocity

The velocity of a particle **X** is the time derivative of the position vector **x**. Thus, the velocity measured in F' is given by dx'/dt:

$$dx'/dt = dx_0(t)/dt + d(\mathbf{Q}(t) \cdot \mathbf{x})/dt.$$

Since $d(\mathbf{Q}^T \cdot \mathbf{Q})/dt = d\mathbf{1}/dt = 0$,

$$[d\mathbf{Q}/dt] \cdot \mathbf{Q}^T + \mathbf{Q} \cdot [d\mathbf{Q}^T/dt] = 0,$$

and since $d\mathbf{Q}^T/dt = (d\mathbf{Q}/dt)^T$, then

$$[d\mathbf{Q}/dt] \cdot \mathbf{Q}^T + \mathbf{Q} \cdot (d\mathbf{Q}/dt)^T = [d\mathbf{Q}/dt] \cdot \mathbf{Q}^T + ([d\mathbf{Q}/dt] \cdot \mathbf{Q}^T)^T$$

and we conclude that

$$[d\mathbf{Q}/dt] \cdot \mathbf{Q}^T = -([d\mathbf{Q}/dt] \cdot \mathbf{Q}^T)^T,$$

i.e., $[d\mathbf{Q}/dt] \cdot \mathbf{Q}^T$ is antisymmetric.

Since $dx/dt = \mathbf{v}$, replacing **x** as

$$\mathbf{x} = \mathbf{Q}^T \cdot \mathbf{x}' - \mathbf{Q}^T \cdot \mathbf{x}_0$$

we obtain

$$\mathbf{v}' = \mathbf{Q} \cdot \mathbf{v} + \mathbf{c}(t) + \mathbf{A}(t) \cdot \mathbf{x}' \tag{3.7.2}$$

where

$$\mathbf{c}(t) = dx_0(t)/dt - \mathbf{A}(t) \cdot \mathbf{x}_0(t)$$

and

$$\mathbf{A}(t) = [d\mathbf{Q}/dt] \cdot \mathbf{Q}^T.$$

Thus, the important result is that the velocity vector is, in general, not frame indifferent. In principle, this result can be used to establish the conditions under which a velocity field is steady in a moving frame ($\partial \mathbf{v}'/\partial t = 0$).

Special case

If $\mathbf{x}_0 = \text{constant}$ and $\mathbf{Q} = \text{constant}$, then $\mathbf{v}' = \mathbf{Q} \cdot \mathbf{v}$. This implies that the two frames are fixed without relative motion.

Problem 3.7.1
Show that if $\mathbf{A}$ is an antisymmetric tensor $\mathbf{A} \cdot \mathbf{b}$ can always be written as $\boldsymbol{\alpha} \times \mathbf{b}$ where $\boldsymbol{\alpha}$ is a vector. Interpret $\boldsymbol{\alpha}$ in physical terms.

3.7.3. Acceleration

Using (3.7.2) and after similar manipulations as in the previous section, we get

$$\mathbf{a}' = \mathbf{Q} \cdot \mathbf{a} + 2\mathbf{A} \cdot \mathbf{v}' + [(d\mathbf{A}/dt) - \mathbf{A} \cdot \mathbf{A})] \cdot \mathbf{x}' + (d\mathbf{c}/dt - \mathbf{A} \cdot \mathbf{c}) \quad (3.7.3)$$

i.e., the acceleration is not frame indifferent.

Special case
$\mathbf{a}' = \mathbf{Q} \cdot \mathbf{a}$ if and only if $\mathbf{A} = 0$ and $d\mathbf{c}/dt = 0$ which implies $\mathbf{Q} = $ constant (no rotation) and $d\mathbf{x}_0(t)/dt = $ constant (linear speed of separation between F and F'). Such a transformation is called a *Galilean transformation*.

Problem 3.7.2
Prove that the Navier–Stokes equation is Galilean invariant.

Problem 3.7.3
Show that the deformation tensor transforms as $\mathbf{F}' = \mathbf{Q} \cdot \mathbf{F}$.

Problem 3.7.4
Show the transformation rules for the gradient operator ∇.

Problem 3.7.5
Verify that the length stretch is frame indifferent using the equations of Section 2.7.

Problem 3.7.6
Check that the velocity gradient transforms as

$$\nabla' \mathbf{v}' = \mathbf{Q} \cdot \nabla \mathbf{v} \cdot \mathbf{Q}^{\mathrm{T}} - \mathbf{A}$$

where $\mathbf{A}$ is antisymmetric. Show also that the stretching tensor and the spin tensor transform components as

$$\mathbf{D}' = \mathbf{Q} \cdot \mathbf{D} \cdot \mathbf{Q}^{\mathrm{T}}, \qquad \text{indifferent}$$
$$\boldsymbol{\Omega}' = \mathbf{Q} \cdot \boldsymbol{\Omega} \cdot \mathbf{Q}^{\mathrm{T}} - \mathbf{A}, \qquad \text{non-indifferent.}$$

Problem 3.7.7
Compute the acceleration in frame F' using $\partial \mathbf{v}'/\partial t + \mathbf{v}' \cdot \nabla' \mathbf{v}'$. Compare the result with that obtained by computing $D\mathbf{v}'/Dt$.

Problem 3.7.8
Show that the isovorticity map of a flow is not changed by Galilean transformations.

Problem 3.7.9

Assuming that if in a frame F we have

$$\rho\mathbf{a} = \rho\mathbf{f} + \nabla\cdot\mathbf{T}$$

and in F' we have

$$\rho'\mathbf{a}' = \rho'\mathbf{f}' + \nabla'\cdot\mathbf{T}',$$

show that

$$(2\mathbf{v}'\cdot\mathbf{A}^{\mathrm{T}} + \mathbf{x}'\cdot[d\mathbf{A}/dt - \mathbf{A}\cdot\mathbf{A}]^{\mathrm{T}} + d\mathbf{c}/dt - \mathbf{c}\cdot\mathbf{A}^{\mathrm{T}})$$
$$= (\mathbf{f}' - \mathbf{f}\cdot\mathbf{Q}^{\mathrm{T}}) + (1/\rho)\nabla'\cdot(\mathbf{T}' - \mathbf{Q}\cdot\mathbf{T}\cdot\mathbf{Q}^{\mathrm{T}}),$$

Note that the first term is independent of the material while the second is dependent on the nature of the material. Present an argument to show that $\mathbf{T}' = \mathbf{Q}\cdot\mathbf{T}\cdot\mathbf{Q}^{\mathrm{T}}$. Note that this reasoning assumes only that the linear momentum equation has the same form in any frame and does not invoke frame indifference for $\mathbf{T}$ (suggested by Serrin, 1977).

Problem 3.7.10

Show that the equation in the moving frame F' is identical to that in frame F if the 'new' body force, $\mathbf{f}'$, is:

$$\mathbf{f}' = \quad \mathbf{Q}\cdot\mathbf{f} \quad + \rho[2\mathbf{A}\cdot\mathbf{v}' + (d\mathbf{A}/dt - \mathbf{A}\cdot\mathbf{A})\cdot\mathbf{x}' + (d\mathbf{c}/dt - \mathbf{A}\cdot\mathbf{c})].$$

    'old force'   'extra body forces that arise due to motion of frame'

Interpret the additional terms from a physical viewpoint. It is customary to designate the additional terms as (Batchelor, 1967, p. 140)

$$2\mathbf{A}\cdot\mathbf{v}' = 2\mathbf{w}\times\mathbf{v}', \qquad\qquad \text{Coriolis 'force'}$$
$$-\mathbf{A}\cdot\mathbf{A}\cdot\mathbf{x}' = -\mathbf{w}\times(\mathbf{w}\times\mathbf{x}'), \qquad \text{centrifugal 'force'}$$
$$(d\mathbf{A}/dt)\cdot\mathbf{x}' = (d\mathbf{w}/dt)\times\mathbf{x}', \qquad \text{Euler's acceleration.}$$

Interpret $\mathbf{w}$.

Problem 3.7.11

Consider the linear velocity field $v_1 = Gx_2, v_2 = KGx_1$. Work out, explicitly, $\mathbf{v}'$, for a change of frame such that

$$[\mathbf{Q}] = \begin{bmatrix} \cos\omega t & -\sin\omega t \\ \sin\omega t & \cos\omega t \end{bmatrix}$$

Obtain $\nabla'\mathbf{v}'$ and verify that $\nabla'\mathbf{v}' = \mathbf{Q}\cdot\nabla\mathbf{v}\cdot\mathbf{Q}^{\mathrm{T}} - \mathbf{A}$. Obtain $\mathbf{D}':\mathbf{D}'$ and verify that it is independent of ω. Compute $\mathbf{\Omega}':\mathbf{\Omega}'$ and obtain its dependence with ω. Investigate the possibility of selecting $\omega(t)$ in such a way that $\mathbf{\Omega}':\mathbf{\Omega}' = 0$.

Problem 3.7.12

Consider $\theta(\mathbf{x}, t) = \theta_0\cos[k(x_1 - ct)]$. Sample θ in a trajectory $x_1 = R\cos\omega t$, $x_2 = R\sin\omega t$. Compute the time history of θ observed following this trajectory.

3.8. Vorticity distribution

It is well known that new insight is obtained when the behavior of fluid motion is examined in terms of vorticity, rather than velocity. Several results, hard to visualize or understand in terms of velocity and momentum, become clear in this type of formulation.

Origin of vorticity

We have seen that the velocity gradient can be written as the sum of a symmetric part ($\mathbf{D}$) and an antisymmetric part ($\boldsymbol{\Omega}$). Since in three-dimensions an antisymmetric tensor can be expressed in terms of an axial vector we obtain

$$\nabla \mathbf{v} = \mathbf{D} - \mathbf{1} \times \omega/2,$$

where ω is the vorticity, $\omega = \nabla \times \mathbf{v}$ (see Truesdell, 1954, p. 3 and p. 58). The vorticity is readily interpreted as twice the local angular velocity in the fluid (see Section 2.8). Note also that

$$\nabla \cdot \omega = \nabla \cdot \nabla \times \mathbf{v} = 0.$$

Vortex line

A vortex line is a line everywhere tangential to the local vorticity (i.e., same relationship as streamlines and velocity field). Obviously, vortex lines cannot cross for that would imply that a given fluid element has two different rates of rotation. Similarly, physical arguments indicate that a vortex line cannot end somewhere in the fluid. However, they can end at boundaries, form closed loops, and can also, as will be shown in Chapter 7, continuously wander throughout space without ever intersecting themselves (see Section 8.7).

Circulation

Making reference to Figure 3.8.1(*a*), the circulation on a closed curve $C = C(t)$ is defined as

$$\oint_C \mathbf{v} \cdot \mathbf{dx}.$$

By using Stokes's theorem we have

$$\oint_C \mathbf{v} \cdot \mathbf{dx} = \int_A \omega \cdot \mathbf{n} \, da$$

which gives further insight into the meaning of ω, i.e.,

$$|\omega| = \lim_{A \to 0} \left\{ \frac{1}{A} \int_A \omega \cdot \mathbf{n} \, da \right\}.$$

An important kinematical result is

$$\frac{d}{dt} \oint_C \mathbf{v} \cdot d\mathbf{x} = \oint_C \mathbf{a} \cdot d\mathbf{x}$$

(i.e., replace $\mathbf{G}$ by $\mathbf{v}$ in Equation (2.10.1) and consider L_t as a loop with perimeter C). Two special cases of this result are of interest. For incompressible Newtonian fluids the acceleration is given by the Navier–Stokes equation and therefore,

$$\frac{d}{dt} \oint_C \mathbf{v} \cdot d\mathbf{x} = v \oint_C \nabla^2 \mathbf{v} \cdot d\mathbf{x} = -v \oint_C \nabla \times \boldsymbol{\omega} \cdot d\mathbf{x}.$$

Thus, the circulation is constant if $v = 0$ or if the vorticity is uniform. For barotropic fluids, $p = f(\rho)$ (Serrin, 1959, p. 150), $\mathbf{a}$ is given in terms of a potential, and we obtain also

$$\frac{d}{dt} \oint_C \mathbf{v} \cdot d\mathbf{x} = 0.$$

A *vortex tube* is the surface formed by all the vortex lines passing through a given closed curve C (which is assumed to be reducible). Consider the construction of Figure 3.8.1(*b*) (at an arbitrary instant of time). Applying

Figure 3.8.1. (*a*) Vortex line, vortex tube, and circulation around a closed curve C; (*b*) construction to prove constancy of circulation in a vortex tube.

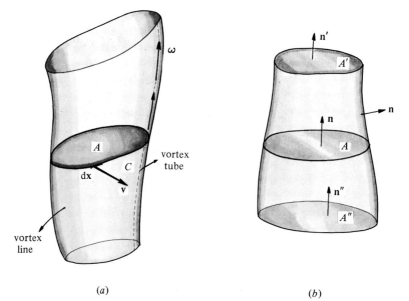

(*a*)   (*b*)

the divergence theorem we obtain:

$$\int_{\text{'top', }A'} \boldsymbol{\omega}\cdot\mathbf{n}'\,da - \int_{\text{'bottom', }A''} \boldsymbol{\omega}\cdot\mathbf{n}''\,da + \int_{\text{'lateral area'}} \boldsymbol{\omega}\cdot\mathbf{n}\,da = \int \nabla\cdot\boldsymbol{\omega}\,dv = 0$$

hence

$$\int_A \boldsymbol{\omega}\cdot\mathbf{n}\,da = \text{constant},$$

over a vortex tube, which implies that the circulation is constant.

3.9. Vorticity dynamics

Consider the Navier–Stokes equation, written as

$$\frac{\partial\mathbf{v}}{\partial t} + \mathbf{v}\cdot\nabla\mathbf{v} = \upsilon\nabla^2\mathbf{v} - \frac{\nabla p}{\rho} + \mathbf{f}$$

with $\mathbf{f}$ such that $\mathbf{f} = -\nabla\varphi$ (i.e., $\mathbf{f}$ is conservative given by a potential φ). Using the identity

$$\mathbf{v}\times(\nabla\times\mathbf{v}) = \tfrac{1}{2}\nabla(\mathbf{v}\cdot\mathbf{v}) - \mathbf{v}\cdot\nabla\mathbf{v}$$

and defining the speed $q = (\mathbf{v}\cdot\mathbf{v})^{1/2}$ we obtain

$$\frac{\partial\mathbf{v}}{\partial t} - \mathbf{v}\times\boldsymbol{\omega} = -\nabla\left[\varphi + \frac{p}{\rho} + \frac{q^2}{2}\right] + \upsilon\nabla^2\mathbf{v}. \tag{3.9.1}$$

This equation forms the basis of many important results. For example, we have:

(i) If $\upsilon = 0$ and the flow is steady and irrotational ($\boldsymbol{\omega} = 0$) the Equations (3.9.1) reduce to:

$$\nabla(\varphi + p/\rho + \tfrac{1}{2}q^2) = 0,$$

$$\varphi + p/\rho + \tfrac{1}{2}q^2 = \text{constant (everywhere in the flow).}$$

This is the simplest version of the 'Bernoulli equation'.

(ii) If the flow is steady and $\upsilon = 0$. In this case, using the definition of the material derivative we obtain,

$$D(\varphi + p/\rho + \tfrac{1}{2}q^2)/Dt = 0,$$

so $\varphi + p/\rho + \tfrac{1}{2}q^2 = \text{constant}$ over the pathline of a fluid particle[4] (and also, since the flow is steady, over streamlines).

(iii) If the flow is steady, multiplying Equation (3.9.1) by $\mathbf{v}$

$$\mathbf{v}\times\boldsymbol{\omega}\cdot\mathbf{v} = 0 = -\nabla(\varphi + p/\rho + \tfrac{1}{2}q^2)\cdot\mathbf{v} + \upsilon(\nabla^2\mathbf{v})\cdot\mathbf{v}$$

and using the definition of the material derivative,

$$\frac{D}{Dt}\left[\varphi + \frac{p}{\rho} + \frac{q^2}{2}\right] = \upsilon(\nabla^2\mathbf{v})\cdot\mathbf{v}$$

which implies that $\varphi + p/\rho + \tfrac{1}{2}q^2$ decreases following a fluid particle if $(\nabla^2\mathbf{v})\cdot\mathbf{v} < 0$, i.e., if the viscous forces decelerate the fluid particle.

(iv) If the flow is steady and two-dimensional and $v = 0$ (Crocco's theorem). Taking the dot product with $\mathbf{v}$ we obtain

$$(\mathbf{v} \times \boldsymbol{\omega}) \cdot \mathbf{v} = 0 = -\nabla(\varphi + p/\rho + \tfrac{1}{2}q^2) \cdot \mathbf{v}$$

which implies that $\varphi + p/\rho + \tfrac{1}{2}q^2$ is constant over streamlines (and pathlines). Further manipulation produces a very important result. Since $\nabla \times \nabla(\;\;) \equiv 0$ by taking the curl of Equation (3.9.1) we can obtain an equation in terms of vorticty and velocity gradients.

$$\nabla \times \left\{ \frac{\partial \mathbf{v}}{\partial t} - \mathbf{v} \times \boldsymbol{\omega} \right\} = \nabla \times \left\{ -\nabla\left[\varphi + \frac{p}{\rho} + \frac{q^2}{2}\right] + v\nabla^2 \mathbf{v} \right\}.$$

Using the identity

$$\nabla \times (\mathbf{v} \times \boldsymbol{\omega}) = \boldsymbol{\omega} \cdot \nabla \mathbf{v} - \boldsymbol{\omega}(\underline{\nabla \cdot \mathbf{v}}) - \mathbf{v} \cdot \nabla \boldsymbol{\omega} + \mathbf{v}(\underline{\nabla \cdot \boldsymbol{\omega}}),$$

and recognizing that the underlined terms are zero, we obtain

$$\frac{\partial \boldsymbol{\omega}}{\partial t} - \boldsymbol{\omega} \cdot \nabla \mathbf{v} + \mathbf{v} \cdot \nabla \boldsymbol{\omega} = v\nabla^2 \boldsymbol{\omega}$$

or alternatively,

$$\underset{\substack{\text{rate of change} \\ \text{of vorticity} \\ \text{following} \\ \text{a particle}}}{\frac{D\boldsymbol{\omega}}{Dt}} = \underset{\substack{\text{interaction} \\ \text{between velocity} \\ \text{gradients and} \\ \text{vorticity}}}{\boldsymbol{\omega} \cdot \nabla \mathbf{v}} + \underset{\substack{\text{diffusion} \\ \text{of} \\ \text{vorticity}}}{v\nabla^2 \boldsymbol{\omega}}. \qquad (3.9.2)$$

Remarks

Note that the vorticity equation does not involve pressure. Note also that $\boldsymbol{\omega} \cdot \nabla \mathbf{v} = 0$ in two cases: two-dimensional motion, where $\boldsymbol{\omega}$ is perpendicular to $\mathbf{v}$, and unidirectional motion, i.e., $\mathbf{v} = (v_1, 0, 0)$. Under these conditions the vorticity diffuses as a passive scalar. Note that the transport coefficient is the kinematical viscosity.

Problem 3.9.1
Verify that $\boldsymbol{\Omega} = -\mathbf{1} \times \boldsymbol{\omega}/2$.

Problem 3.9.2
Show that if $\nabla \cdot \mathbf{v} = 0$ and $\nabla \times \mathbf{v} = 0$, then the viscous forces in a Newtonian fluid, $\nabla^2 \mathbf{v}$, are zero.

Problem 3.9.3
There are cases where the intensification of vorticity due to velocity gradients and the vorticity diffusion balance in such a way that there is a steady state distribution of vorticity. Consider

$$v_z = \alpha z, \qquad v_r = -\tfrac{1}{2}\alpha r.$$

Find v_θ consistent with the Navier–Stokes's Equation for the cases:
(i) $\omega = \omega(r)$ (Burgers, 1948; Batchelor, 1979, p. 271).
(ii) $\omega_z = \omega_z(r, \theta, t)$ (Lundgren, 1982).

3.10. Macroscopic balance of vorticity

In order to obtain a macroscopic balance of vorticity, form the scalar product of Equation (3.9.2) with ω and integrate over material volume V_t. After some manipulations,

$$\frac{d}{dt} \int_{V_t} \tfrac{1}{2}\boldsymbol{\omega}^2 \, dv = \int_{V_t} (\nabla \mathbf{v} : \boldsymbol{\omega}\boldsymbol{\omega}) \, dv + v \int_{V_t} (\nabla \boldsymbol{\omega} : (\nabla \boldsymbol{\omega})^{\mathrm{T}}) \, dv - \tfrac{1}{2} v \oint_{\partial V_t} \nabla(\boldsymbol{\omega}\cdot\boldsymbol{\omega})\cdot \mathbf{n} \, ds.$$

| rate of accumulation | generation by stretching | dissipation (always positive) | generation at the boundaries |

$$(3.10.1)$$

Problem 3.10.1
Show that in the case of a fluid enclosed entirely by stationary rigid boundaries (or $\mathbf{v} \to 0$ as $|\mathbf{x}| \to \infty$), the energy dissipation decays as $\int \omega^2 \, dv$. *Hint*: Prove the kinematical identity

$$\nabla \cdot \mathbf{a} = D(\nabla \cdot \mathbf{v})/Dt + \mathbf{D} : \mathbf{D} - \tfrac{1}{2}\omega^2.$$

Problem 3.10.2
Expand the vorticity near a solid wall with normal $\mathbf{n}$. Assume that very near the wall ω is parallel to the wall. Show that to the first order approximation

$$\mathbf{v} = (\boldsymbol{\omega} \times \mathbf{n})z$$

where z measures distances normal to the wall. Assuming $\nabla \cdot \mathbf{v} = 0$ show that the normal component of the velocity field is proportional to z^2 (Lighthill, 1963, p. 64). Obtain the same result by direct expansion of the velocity field in a Taylor series expansion, forcing the series to satisfy impenetrability and no-slip at the wall.

3.11. Vortex line stretching in inviscid fluid

For $v = 0$ Equation (3.9.2) gives

$$D\boldsymbol{\omega}/Dt = \boldsymbol{\omega} \cdot \nabla \mathbf{v}. \qquad (3.11.1)$$

The solution of (3.11.1) can be computed explicitly (Serrin, 1959, p. 152). Define a vector $\mathbf{c}$ by $\boldsymbol{\omega} = \mathbf{c} \cdot \mathbf{F}^{\mathrm{T}}$, where $\mathbf{F}$ is the deformation tensor. Then,

$$D\boldsymbol{\omega}/Dt = \boldsymbol{\omega} \cdot \nabla \mathbf{v} = (D\mathbf{c}/Dt) \cdot \mathbf{F}^{\mathrm{T}} + \mathbf{c} \cdot (D\mathbf{F}^{\mathrm{T}}/Dt).$$

Since $D\mathbf{F}^T/Dt = \mathbf{F}^T \cdot \nabla\mathbf{v}$, and $\boldsymbol{\omega} = \mathbf{c} \cdot \mathbf{F}^T$, we obtain

$$\boldsymbol{\omega} \cdot \nabla\mathbf{v} = (D\mathbf{c}/Dt) \cdot \mathbf{F}^T + \boldsymbol{\omega} \cdot \nabla\mathbf{v},$$

which implies that $(D\mathbf{c}/Dt) \cdot \mathbf{F}^T = 0$ and $\mathbf{c} = \mathbf{c}(\mathbf{X})$. Since at $t = 0$, $\boldsymbol{\omega} = \boldsymbol{\omega}_0$, we obtain $\boldsymbol{\omega} = \boldsymbol{\omega}_0 \cdot \mathbf{F}^T$.

Note also that (3.11.1) is identical in form to the equation giving the material rate of change of $d\mathbf{x}$ (Section 2.9). Thus,

$$D(\boldsymbol{\omega} - \kappa \, d\mathbf{x})/Dt = (\boldsymbol{\omega} - \kappa \, d\mathbf{x}) \cdot \nabla\mathbf{v}$$

where κ is a constant. If at $t = 0$, $(\boldsymbol{\omega} - \kappa \, d\mathbf{x})$ is equal to $(\boldsymbol{\omega}_0 - \kappa \, d\mathbf{X})$ then

$$(\boldsymbol{\omega} - \kappa \, d\mathbf{x}) = (\boldsymbol{\omega}_0 - \kappa \, d\mathbf{X}) \exp\left\{ \int_0^t \nabla\mathbf{v}(\mathbf{X}, t') \, dt' \right\}.$$

Hence if $(\boldsymbol{\omega}_0 - \kappa \, d\mathbf{X}) = \mathbf{0}$ initially, then $(\boldsymbol{\omega} - \kappa \, d\mathbf{x}) = \mathbf{0}$, i.e., vortex lines move as material lines (Figure 3.11.1). For a vortex line initially coincident with material filament, we obtain

$$|\boldsymbol{\omega}|/|\boldsymbol{\omega}_0| = \lambda/\lambda_0.$$

Thus, the intensification of vorticity is proportional to the length strength.[5] Many of these results can be easily generalized to the case of barotropic fluids (see Batchelor, 1967). For example, (3.11.1) becomes

$$D(\boldsymbol{\omega}/\rho)Dt = (\boldsymbol{\omega}/\rho) \cdot \nabla\mathbf{v}.$$

Figure 3.11.1. Stretching of material lines in inviscid flow; the material lines and the vortex lines coincide.

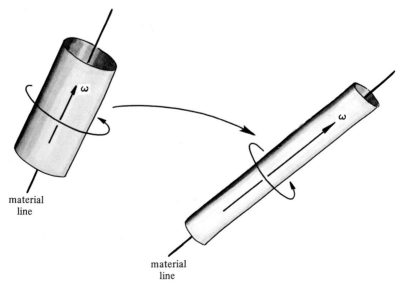

material
line

material
line

Other results follow within this framework. For example, it can be shown that the helicity (Moffat, 1969)

$$I = \int_V \mathbf{v} \cdot \boldsymbol{\omega} \, dv,$$

a measure of the degree of knottedness of the vortex lines, satisfies

$$DI/Dt = 0$$

provided that $\mathbf{n} \cdot \boldsymbol{\omega} = 0$ at the boundaries of V (or $\omega = \mathbf{0}(|\mathbf{x}|^{-4})$ as $|\mathbf{x}| \to \infty$). Furthermore I is non-zero if the vortex lines are knotted.

Problem 3.11.1
Using Equation (3.11.1) show that a body of fluid initially in irrotational motion continues to move irrotationally.

3.12. Streamfunction and potential function

The streamfunction allows the formulation of problems in such a way that the continuity equation is satisfied identically. Such a formulation is possible in two-dimensional flows and axisymmetric flows and for very special classes of three-dimensional flows (Truesdell and Toupin, 1960, p. 479). For example, for two-dimensional flow in rectangular coordinates we can write $\mathbf{v} = \nabla \times (\psi \mathbf{e}_z)$, or[6]

$$v_x = \partial\psi/\partial y, \qquad v_y = -\partial\psi/\partial x,$$

in such a way that $\nabla \cdot \mathbf{v} = 0$. The curve $\psi(x, y, t = \text{given}) = \text{constant}$, gives the instantaneous picture of the streamlines. Since $\mathbf{v} \cdot \nabla\psi = 0$, then if $\partial\psi/\partial t = 0$, ψ is constant following a fluid particle.[7]

The streamfunction and the vorticity are related by

$$\nabla^2\psi = -\omega.$$

In irrotational motions there are several simplifications. To start with, ψ satisfies Laplace's equation

$$\nabla^2\psi = 0.$$

Also, since the flow is irrotational the velocity can be obtained from a potential, $\phi(\mathbf{x}, t)$, as

$$\mathbf{v} = \nabla\phi,$$

($\nabla \times \nabla\phi = \omega = 0$). Also since $\nabla \cdot \mathbf{v} = 0$, the potential function satisfies

$$\nabla^2\phi = 0$$

and the streamfunction and the potential function are related by

$$\partial\phi/\partial x = \partial\psi/\partial y,$$
$$\partial\phi/\partial y = -(\partial\psi/\partial x),$$

which are precisely the Cauchy–Riemann conditions for a complex function, $w(z) = \phi(x, y) + i\psi\,(x, y)$, to be analytic. Thus, any complex analytic function is the solution of some fluid mechanical problem. However, finding the complex potential for a given fluid mechanical situation is a substantially more difficult problem.

Example 3.12.1

Consider the complex potential $w(z) = Uz + Ua^2/z$ which corresponds to potential flow past a cylinder of radius a (Figure E3.12.1(a)).[8] If the cylinder moves at speed U, we substract the contribution of the mean flow (Uz) and the potential function is therefore Ua^2/z. Thus, to an observer mounted on the cylinder the portrait of the streamlines is given by

$$-(Ua^2 \sin \theta/r) = \psi,$$

which is shown in Figure 3.12.1(b). The instantaneous speed of the particle at (r, θ) is given by

$$v_r = \partial\phi/\partial r \qquad \text{and} \qquad v_\theta = (1/r)\partial\phi/\partial\theta$$

where $\phi = Ua^2 \cos \theta/r$ and therefore $|\mathbf{v}|^2 = U^2a^4/r^4$ which indicates that all the particles located at $r = \text{const.}$ have the same speed. In particular for $r = a$ the fluid particles slip with speed U. Consider the cylinder at $x = -\infty$ and label a fluid particle in the as yet undisturbed flow by its initial coordinates X_1, Y_1. As the cylinder passes by, the particle moves as indicated in Figure E3.12.1(c), undergoing a trajectory which is the solution of a differential equation commonly known as the *elastica*. The particles suffer a *permanent* displacement in the x direction (d_x, given by a rather complicated expression) and a *temporary* displacement in the y direction, given by $d_y = \frac{1}{2}[(4a^2 + Y_1^2)^{1/2} - |Y_1^2|]$, which is maximum when the particle is abreast the cylinder. Note that to an observer moving with the cylinder the flow is steady.[9]

Example 3.12.2

Assume that we have an *unsteady* description of a flow field

$$v_1 = v_1(x_1, x_2, t) \qquad \text{and} \qquad v_2 = v_2(x_1, x_2, t).$$

How do we calculate the instantaneous picture of the streamlines? In this case

$$d\psi = (\partial\psi/\partial x_1)\,dx_1 + (\partial\psi/\partial x_2)\,dx_2 + (\partial\psi/\partial t)\,dt$$

and

$$d\psi = v_1\,dx_2 - v_2\,dx_1 + (\partial\psi/\partial t)\,dt.$$

Figure E3.12.1. Flow around a cylinder; (*a*) streamlines with respect to a fixed frame, the cylinder is at rest and the fluid moves at speed *U*; (*b*) instantaneous streamlines as they appear to an observer mounted on the cylinder; and (*c*) typical trajectories of fluid particles as the cylinder moves from left to right.

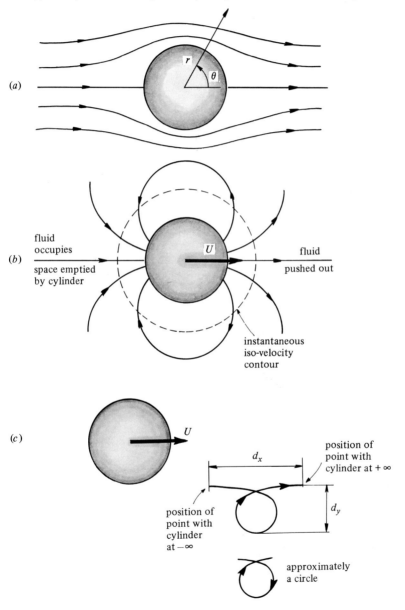

Thus, the portrait at a given time t' passing by the point x'_1, x'_2 is

$$\psi(x_1, x_2, t') - \psi(x'_1, x'_2, t') = \int_{x'_1, x'_2, t'}^{x_1, x_2, t'} (v_1 \, dx_2 - v_2 \, dx_1).$$

Such a method was used to analyze experimental data by Cantwell, Coles, and Dimotakis (1978).

Example 3.12.3
As will become more apparent in Chapter 7, the visualization of unsteady two-dimensional flows is extremely complicated even in cases where the velocity field is known explicitly (see, for example, Hama, 1962, and Example 2.5.1). An interesting approach to visualize flows was proposed by Cantwell (1978). The basic idea is to reduce, not exactly, but possibly *for limited scales of space and time*, the system

$$dx_1/dt = v_1(x_1, x_2, t), \qquad dx_2/dt = v_2(x_1, x_2, t)$$

to an autonomous system by studying the flow in a moving frame and with new variables τ, ξ_1, ξ_2, such that the velocity field is transformed into

$$d\xi_1/d\tau = U_1(\xi_1, \xi_2), \qquad d\xi_2/d\tau = U_2(\xi_1, \xi_2).$$

A list of problems (such as jets, boundary layer problems, etc.) for which this approach works is given by Cantwell (1981).[10]

Problem 3.12.1
Compute the location at time t of a material region of radius kR, $k > 1$, initially surrounding a sphere of radius R outside which there is a fluid moving in potential flow. Compare the result with that obtained for a sphere of radius R in a low Reynolds number flow.

Problem 3.12.2
Consider a flow in the frame F with the stream function ψ. Compute the stream function ψ' for a Galilean transformation between F and F'.

Bibliography

This chapter is a very succinct account of fluid mechanics suited only to our rather specific needs. The interested reader should consult the original works on which this exposition is based. The sections on conservation equation are inspired in the classical article, 'The mathematical principles of classical fluid mechanics', by Serrin (1959). The treatment of vorticity follows Batchelor (1979). For a treatment of change of frame and objectivity the reader should consult the book by Chadwick, *Continuum mechanics* (1976). A few sections, such as Example 3.12.1, are based on the article by Truesdell and Toupin (1960).

Notes

1 We have dropped the position **x** and time *t*. It should be understood that all quantities are evaluated at **x**, *t*.

2 Equivalently, the transformations of **t** as above can be considered an alternative statement of the Principle of Material Frame Indifference (see Section 3.7).

3 Note that in accordance with the results of Chapter 6, $\tau' = \mathbf{Q} \cdot \tau \cdot \mathbf{Q}^{\mathrm{T}}$. All constitutive equations must generate a frame indifference stress tensor (for details see Schowalter, 1979, Chapter 6; Chadwick, 1976).

4 In the language of Chapter 5 this quantity might be regarded as a 'constant' of the motion.

5 According to the classical picture of a turbulent flow, initially designated vortex lines stretch and fold as material lines until the folding reaches a scale where the diffusion of vorticity becomes dominant. At this point a statistical steady-state is reached.

6 In fluid mechanics books one can find suitable expressions in other co-ordinate systems.

7 As shown in Problem 4.7.1, if $\partial\psi/\partial t = 0$ the mixing is poor since fluid particles are trapped between level curves of ψ.

8 This example is based on Milne-Thomson, 1955, Section 9.21; and Truesdell and Toupin, 1960, p. 332, where additional references can be found.

9 Streamlines referred to a rotating cylinder are analyzed by Milne-Thompson (1955, Section 9.71) and Lamb (1932 edition, Section 72). A good discussion of streamlines and streaklines, especially from an experimental viewpoint, is given by Tritton, 1977, Chapter 6.

10 Obviously, if this is possible the flow cannot be chaotic (see Chapters 5 and 6).

4

Computation of stretching and efficiency

As we have seen in Chapter 2, the deformation tensor $\mathbf{F}$ and its associated tensors $\mathbf{C}$, $\mathbf{C}^{-1}$, etc., form the fundamental quantities for the analysis of deformation of infinitesimal elements. In most cases the flow $\mathbf{x} = \mathbf{\Phi}_t(\mathbf{X})$ is unknown and has to be obtained by integration from the Eulerian velocity field. If this can be done analytically, $\mathbf{F}$ can then be obtained by differentiation of the flow with respect to the material coordinates $\mathbf{X}$. The chapter starts with examples belonging to this class. However, this procedure is not always convenient, or even possible, for reasons that will be apparent in Chapters 5 through 8, and we explore briefly other approaches which are valid for relatively simple but, nevertheless, 'integrable' flows of widespread applicability. The flows of interest belong to two classes: (i) flows with a special form of $\nabla \mathbf{v}$, and (ii) flows with a special form of $\mathbf{F}$. In class (i) we have the subcases $D(\nabla \mathbf{v})/Dt = 0$ and $D(\nabla \mathbf{v})/Dt$ small. In the second class we have Constant Stretch History Motions (CSHM) and an important subset of viscometric flows, Steady Curvilineal Flows (SCF).

4.1. Efficiency of mixing

According to the physical picture described in Chapter 1, mixing involves stretching and folding of material elements. Thus, if we place a material filament $d\mathbf{X}$ and an element of area $d\mathbf{A}$, in an arbitrary initial location $\mathbf{P}$, the specific rates of generation of length, $D \ln \lambda/Dt$, and area, $D \ln \eta/Dt$, are given by the equations of Chapter 2. We say that the flow $\mathbf{x} = \mathbf{\Phi}_t(\mathbf{X})$, or a region of the flow, mixes well if the time averaged values of $D \ln \lambda/Dt$ and $D \ln \eta/Dt$, do not decay to zero, no matter what the initial placement $\mathbf{P}$ and the initial orientations $\mathbf{M}$ and $\mathbf{N}$. However, it is not reasonable to compare flows on the basis of the values of the stretchings $D \ln \lambda/Dt$ and $D \ln \eta/Dt$ since they are obviously dependent upon the units of time. It is therefore of importance to seek some rational way of quantifying the efficiency of the stretching process in order to have some basis of comparison for the mixing ability of different flows.

We note that the specific rates of stretching are bounded. Thus, for $D \ln \lambda/Dt$, using the Cauchy–Schwarz inequality we have:

$$D \ln \lambda/Dt = \mathbf{D}:\mathbf{mm} \leqslant |\mathbf{D}||\mathbf{mm}| = (\mathbf{D}:\mathbf{D})^{1/2},$$

since $|\mathbf{mm}| = 1$. For $D \ln \eta/Dt$ we obtain

$$D \ln \eta/Dt = \nabla \cdot \mathbf{v} - \mathbf{D}:\mathbf{nn} = [\mathbf{1}(\nabla \cdot \mathbf{v}) - \nabla\mathbf{v}]:\mathbf{nn} \leqslant |\mathbf{1}(\nabla \cdot \mathbf{v}) - \nabla\mathbf{v}||\mathbf{nn}|$$
$$= [(\nabla \cdot \mathbf{v})^2 + \mathbf{D}:\mathbf{D}]^{1/2},$$

or, if $D \ln \eta/Dt$ is written in a slightly different way,

$$D \ln \eta/Dt = \mathbf{D}:(\mathbf{1} - \mathbf{nn}) \leqslant |\mathbf{D}||\mathbf{1} - \mathbf{nn}| = 2^{1/2}(\mathbf{D}:\mathbf{D})^{1/2}.$$

These upper bounds provide a natural way of quantifying the efficiency of the stretching. Thus, we define the stretching efficiency, $e_\lambda = e_\lambda(\mathbf{X}, \mathbf{M}, t)$ of the material element $d\mathbf{X}$ ($\mathbf{M} = d\mathbf{X}/|d\mathbf{X}|$), placed at $\mathbf{X}$, as

$$e_\lambda \equiv (D \ln \lambda/Dt)/(\mathbf{D}:\mathbf{D})^{1/2} \leqslant 1. \qquad (4.1.1)$$

The stretching efficiencies, $e_\eta = e_\eta(\mathbf{X}, \mathbf{N}, t)$ of the area element $d\mathbf{A}$ placed at $\mathbf{X}$ ($\mathbf{N} = d\mathbf{A}/|d\mathbf{A}|$), are defined in a similar way. If the motion is isochoric we define

$$e_\eta \equiv (D \ln \eta/Dt)/(\mathbf{D}:\mathbf{D})^{1/2} \leqslant 1, \qquad (4.1.2)$$

and if it is not

$$e_\eta \equiv (D \ln \lambda/Dt)/[2^{1/2}(\mathbf{D}:\mathbf{D}^{1/2}] \leqslant 1.$$

In this work we will use the first definition since we will assume throughout that the flows are isochoric. We define the *asymptotic* efficiencies as

$$e_{\lambda\infty} = \lim_{t \to \infty} e_\lambda(\mathbf{X}, \mathbf{M}, t)$$

$$e_{\eta\infty} = \lim_{t \to \infty} e_\eta(\mathbf{X}, \mathbf{N}, t).$$

The *time averaged* efficiencies are defined as

$$\langle e_\lambda \rangle \equiv \frac{1}{t} \int_0^t e_\lambda(\mathbf{X}, \mathbf{M}, t')\, dt'$$

$$\langle e_\eta \rangle \equiv \frac{1}{t} \int_0^t e_\eta(\mathbf{X}, \mathbf{N}, t')\, dt'$$

and their corresponding asymptotic values are denoted as $\langle e_\lambda \rangle_\infty$ and $\langle e_\eta \rangle_\infty$, respectively. Efficiencies can also be defined with respect to $(\nabla\mathbf{v}:(\nabla\mathbf{v})^{\mathrm{T}})^{1/2}$; thus we have

$$e_\lambda^* \equiv (D \ln \lambda/Dt)/(\nabla\mathbf{v}:(\nabla\mathbf{v})^{\mathrm{T}})^{1/2} = \mathbf{D}:\mathbf{mm}/(\nabla\mathbf{v}:(\nabla\mathbf{v})^{\mathrm{T}})^{1/2}$$

$$e_\eta^* \equiv (D \ln \eta/Dt)/(\nabla\mathbf{v}:(\nabla\mathbf{v})^{\mathrm{T}})^{1/2} = (\nabla \cdot \mathbf{v} - \mathbf{D}:\mathbf{nn})/(\nabla\mathbf{v}:(\nabla\mathbf{v})^{\mathrm{T}})^{1/2}.$$

Similarly, the corresponding time averaged asymptotic efficiencies are denoted $\langle e_\lambda^* \rangle_\infty$ and $\langle e_\eta^* \rangle_\infty$. Other efficiencies, for finite material lines and surfaces will be used when convenient.

4.1.1. Properties of e_λ and e_η

Frame indifference
The efficiencies e_λ and e_η are frame indifferent, i.e., they are objective, $e_\lambda = e_\lambda'$, $e_\eta = e_\eta'$, under the frame transformation $\mathbf{x}' = \mathbf{x}_0(t) + \mathbf{Q}(t) \cdot \mathbf{x}$. Note that $\mathbf{D}$, $\mathbf{m}$, and $\mathbf{n}$, are frame indifferent. Since $\mathbf{D}$, $\mathbf{m}$, and $\mathbf{n}$ transform as $\mathbf{Q} \cdot \mathbf{D} \cdot \mathbf{Q}^T$, $\mathbf{Q} \cdot \mathbf{m}$, and $\mathbf{Q} \cdot \mathbf{n}$, respectively, then $\mathbf{D}:\mathbf{mm}$, $\mathbf{D}:\mathbf{nn}$, and $\mathbf{D}:\mathbf{D}$ are also frame indifferent. Additionally, since $\nabla \cdot \mathbf{v}$ is also frame indifferent, it follows that e_λ and e_η are frame indifferent. Note, however, that e_λ^* and e_η^* are not frame indifferent since $\nabla \mathbf{v}$ is not objective.

Physical meaning of e_λ and e_η
For purely viscous fluids, $(\mathbf{D}:\mathbf{D})^{1/2}$ is related to the viscous dissipation. For example, for incompressible Newtonian fluids $\boldsymbol{\tau}:\mathbf{D} = 2\mu(\mathbf{D}:\mathbf{D})$. In this case the efficiency can be thought of as the fraction of the energy dissipated locally that is used to stretch fluid elements.

Relationships among e_λ, e_η and e_λ^* and e_η^*
Since

$$\nabla \mathbf{v}:(\nabla \mathbf{v})^T = \mathbf{D}:\mathbf{D} - \boldsymbol{\Omega}:\boldsymbol{\Omega}$$

it follows that

$$e_i^* = e_i/(1 + W_K^2)^{1/2}, \qquad i = \lambda, \eta$$

where

$$W_K^2 = -(\boldsymbol{\Omega}:\boldsymbol{\Omega})/(\mathbf{D}:\mathbf{D})$$

is the kinematical vorticity number.[1] Although neither the e_i^* $(i = \lambda, \eta)$ nor W_K^2 are objective, the product $e_i^*(1 + W_K^2)^{1/2}$ is objective.

4.1.2. Typical behavior of the efficiency

According to the time history of the stretching efficiency we divide bounded flows into: flows without reorientation, where the efficiency $e(\mathbf{X}, t)$ decays as t^{-1}; flows with partial reorientation, where there is some periodic restoration, but where on the average the efficiency also decays as t^{-1}; and flows with strong reorientation, where the time average of the efficiency tends to a constant value (see Figure 4.1.1). We shall encounter examples of all these flows; the first two in this chapter, the third one in Chapters 7 and 8.

4.1.3. Flow classification

The ability of flows to stretch material lines and areas provides a useful means of flow classification. The objective of flow classification is to define a 'measure' which reflects the predominant characteristics of an arbitrary flow, and provides a basis for comparison of flows. Translating this rather loose concept into concrete parameters has led to the use of different criteria which have been used in different contexts. The idea is popular in rheology where it is interpreted in a 'local' sense and has applications such as drop breakup and drag reduction.

Tanner and Huilgol (1975) and Tanner (1976) proposed a scheme based on the asymptotic rate of growth of 'test' fluid microstructures. Their framework was extended by Olbricht, Rallison, and Leal (1982) to account for the microstructure shape, interactions between it and the surrounding fluid, and elastic forces resisting deformation, by means of a linearized dynamic analysis (see Chapter 9). However, these criteria are applicable only to flows with $D(\nabla v)/Dt = 0$. Astarita (1976) proposed an objective measure, applicable to arbitrary flows, related to the ratio of the magnitudes of the vorticity rate to the strain tensor. However, this measure is in general difficult to calculate. Another measure is the vorticity number, W_K^2, defined many years ago by Truesdell (1954). $W_K^2 = 0$ corresponds to purely irrotational motion, and the value increases with increasing vorticity becoming $W_K^2 = \infty$ for a purely rotational flow. This quantity however, suffers from the disadvantage of being, in general, non-frame indifferent (see Problems 3.7.11 and 4.1.2).

Problem 4.1.1
Prove that if $\nabla \cdot v = 0, v \in \mathbb{R}^n, e_i \leqslant [(n-1)/n]^{1/2}, i = \lambda, \eta$ (Khakhar, 1986).

Problem 4.1.2
Compute the vorticity number W_K^2 for the linear flow $v_1 = Gx_2, v_2 = KGx_1$. How does W_K^2 change under a change of frame?

Figure 4.1.1. Typical behavior of mixing efficiency; (*a*) flow with decaying efficiency; (*b*) flow with partial restoration; and (*c*) flow with strong reorientation.

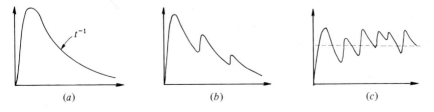

      (*a*)   (*b*)   (*c*)

Problem 4.1.3
Show that for a two-dimensional isochoric flow such that $D\mathbf{m}/Dt = 0$, $e_{\lambda\infty}$ can be written as $(1/2^{1/2})[1 - W_{\kappa}^2]^{1/2}$ (Franjione, 1986).

Problem 4.1.4
Consider a two-dimensional isochoric flow. Let θ denote the angle between a material filament and the maximum direction of stretching. Show that the instantaneous value of the stretching efficiency is given by $(2^{1/2}/2)\cos 2\theta$ (Franjione, 1987).

4.2. Examples of stretching and efficiency

Example 4.2.1 *Efficiency of axisymmetric extensional flows*
In this case
$$dx_1/dt = \dot{\varepsilon}x_1, \qquad dx_2/dt = (-\dot{\varepsilon}/2)x_2, \qquad dx_3/dt = (-\dot{\varepsilon}/2)x_3.$$
If at $t = 0$, $\mathbf{x} = \mathbf{X}$ we have $d\mathbf{x} = d\mathbf{X}$, $\mathbf{M} = d\mathbf{X}/|d\mathbf{X}|$,
$$x_1 = X_1 \exp(\dot{\varepsilon}t), \qquad x_2 = X_2 \exp[(-\dot{\varepsilon}/2)t], \qquad x_3 = X_3 \exp[(-\dot{\varepsilon}/2)t]$$
then $\mathbf{F}$ and $\mathbf{C} = \mathbf{F}^{\mathrm{T}} \cdot \mathbf{F}$ are computed to be
$$[\mathbf{F}] = \begin{bmatrix} \exp(\dot{\varepsilon}t) & 0 & 0 \\ 0 & \exp(-\dot{\varepsilon}t/2) & 0 \\ 0 & 0 & \exp(-\dot{\varepsilon}t/2) \end{bmatrix}$$
$$[\mathbf{C}] = \begin{bmatrix} \exp(2\dot{\varepsilon}t) & 0 & 0 \\ 0 & \exp(-\dot{\varepsilon}t) & 0 \\ 0 & 0 & \exp(-\dot{\varepsilon}t) \end{bmatrix}$$
The lineal stretch is given by
$$\lambda^2 = M_1^2 \exp(2\dot{\varepsilon}t) + M_2^2 \exp(-\dot{\varepsilon}t) + M_3^2 \exp(-\dot{\varepsilon}t)$$
and the specific rate of stretching by,
$$\frac{D\ln\lambda}{Dt} = \frac{\dot{\varepsilon}[2M_1^2 \exp(3\dot{\varepsilon}t) - M_2^2 - M_3^2]}{2[M_1^2 \exp(3\dot{\varepsilon}t) + M_2^2 + M_3^2]}.$$
The magnitude of $\mathbf{D}$, $(\mathbf{D}:\mathbf{D})^{1/2}$, is equal to $\dot{\varepsilon}(3/2)^{1/2}$. The long time value of the efficiency, $(D\ln\lambda/Dt)/(\mathbf{D}:\mathbf{D})^{1/2}$ is equal to $(2/3)^{1/2} \approx 0.816$ (which is the upper bound for three-dimensional flows), unless $\mathbf{M}$ is of the form $(0, M_2, M_3)$. Note that $\lambda = \lambda(\mathbf{X}, \mathbf{M}, t)$ in this case reduces to $\lambda = \lambda(\mathbf{M}, t)$. This is true for all linear flows.

Example 4.2.2 *Efficiency of simple shear flow*
In this case
$$dx_1/dt = \dot{\gamma}x_2, \qquad dx_2/dt = 0, \qquad dx_3/dt = 0.$$

With the same initial condition as before,

$$x_1 = \dot{\gamma}x_2 t + X_1, \qquad x_2 = X_2, \qquad x_3 = X_3.$$

Similarly, $\mathbf{C}$ is computed to be

$$[\mathbf{C}] = \begin{bmatrix} 1 & \dot{\gamma}t & 0 \\ \dot{\gamma}t & 1 + (\dot{\gamma}t)^2 & 0 \\ 0 & 0 & 1 \end{bmatrix}$$

and $D \ln \lambda / Dt$ is,

$$\frac{D \ln \lambda}{Dt} = \frac{\dot{\gamma}M_2(M_1 + M_2\dot{\gamma}t)}{[1 + M_2\dot{\gamma}t(2M_1 + M_2\dot{\gamma}t)]}$$

In this case $(\mathbf{D}:\mathbf{D})^{1/2}$ is equal to $\dot{\gamma}/(2^{1/2})$. The efficiency reaches a maximum value of $(2^{1/2})/2 \approx 0.707$ (which is the upper bound for two-dimensional flows). For long times, the efficiency decays as t^{-1}.

Example 4.2.3 Efficiency in linear two-dimensional flow
Consider again the two-dimensional flow of Problem 2.5.3. The motion is obtained by solving the second order system

$$dx_1/dt = Gx_2, \qquad dx_2/dt = KGx_1,$$

with the initial condition $x_1 = X_1$ and $x_2 = X_2$ (the algebraic manipulations can be reduced considerably using the formalism of Section 4.3). The specific rate of stretching of a material plane $\mathbf{N} = (N_1, N_2)$ is

$$\frac{D \ln \eta}{Dt} = \frac{\dot{\gamma}[2b_1b_2 \sin(\dot{\gamma}t) + (b_2^2 - b_1^2)\cos(\dot{\gamma}t)]}{4\left\{1 + (1 - |K|)\left[2b_1b_2 \sin^2\left(\frac{\dot{\gamma}t}{2}\right) + \frac{1}{2}(b_2^2 - b_1^2)\sin(\dot{\gamma}t)\right]\right\}}$$

for $K < 0$, and

$$\frac{D \ln \eta}{Dt} = \frac{\dot{\gamma}[b_2^2 \exp(\dot{\gamma}t) - b_1^2 \exp(-\dot{\gamma}t)]}{2\left[b_2^2 \exp(\dot{\gamma}t) + b_1^2 \exp(-\dot{\gamma}t) + 2b_1b_2\left(\dfrac{K-1}{K+1}\right)\right]}$$

for $K > 0$, where

$$\dot{\gamma} = 2G(|K|)^{1/2}$$

$$b_1 = \frac{1}{2^{1/2}}\left[\frac{N_1}{(|K|)^{1/2}} + N_2\right]$$

$$b_2 = \frac{1}{2^{1/2}}\left[\frac{N_1}{(|K|)^{1/2}} - N_2\right].$$

If $K < 0$, $D \ln \eta / Dt$ is periodic; if $K > 0$, $D \ln \eta / Dt$ attains a non-zero limit. However, of more interest than the instantaneous efficiencies are the time averaged values (the asymptotic values are related to the Liapunov exponents, see Chapter 5). In this case,

$$\langle e_\eta \rangle = \frac{(2K)^{1/2}}{(1+K)\dot{\gamma}t} \ln \left\{ \frac{1+K}{2} \left[b_2^2 \exp(\dot{\gamma}t) + b_1^2 \exp(-\dot{\gamma}t) + 2b_1 b_2 \left(\frac{K-1}{K+1} \right) \right] \right\}.$$

The time evolution of $\langle e_\eta \rangle$ for different values of $K > 0$ is given in Figure E4.2.1 for a material plane with initial orientation $\mathbf{N} = (1, 0)$. For $K > 0$ the average efficiency rises rapidly to a constant value $\langle e_\eta \rangle_\infty = (2K)^{1/2}/(1 + K)$ (for $\dot{\gamma}t = O(1)$, as expected) and it is relatively insensitive to the value of K ($\langle e_\eta \rangle_\infty$ is equal to 0.666 for $K = 0.5$ and 0.406 for $K = 0.1$). This observation is significant and indicates that *the efficiency can be significantly increased by introducing a small amount of irrotationality into the flow.* For $K = 0$, corresponding to simple shear flow, $\langle e_\eta \rangle$ decays as t^{-1}. For $K < 0$ the efficiency oscillates with a period proportional to $1/G|K|^{1/2}$. Thus, flows with elongational character are efficient but the efficiency of shear-like flows is poor unless special precautions are taken. One way to increase the efficiency of rotational flows is by periodic reorientation of filaments or areas to offset the tendency of the flow efficiency to decay as

Figure E4.2.1. Evolution of the time-averaged efficiency, $\langle e_\eta \rangle$, for various values of K in a linear flow (see Figure P2.5.3).

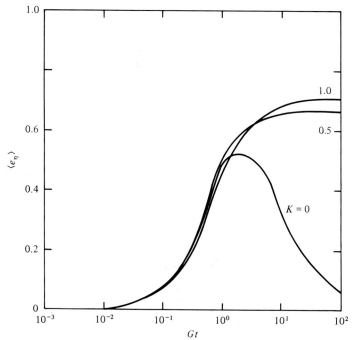

t^{-1}. Such a concept is explored in detail in Section 4.6 and forms the basis for the high efficiencies encountered in Chapters 7 and 8.

Example 4.2.4 Efficiency in the helical annular mixer

Here we consider the stretching in the mixer of the Figure E4.2.2. The flow is a combination of Couette and Poiseuille flow. For a Newtonian fluid the azimuthal (θ) and axial (z) components of the velocity field are uncoupled. The velocity field is of the form

$$v_r = 0, \qquad v_\theta = f(r), \qquad v_z = g(r),$$

and the motion is given by the solution of

$$dr/dt = 0, \qquad r\, d\theta/dt = f(r), \qquad dz/dt = g(r),$$

that is,

$$r = R, \qquad \theta = \Theta + [f(r)/r]t, \qquad z = Z + g(r)t,$$

where (R, Θ, Z) represents the initial position of the particle (r, θ, z). The deformation tensor is computed using

$$[\mathbf{F}] = \begin{bmatrix} \dfrac{\partial r}{\partial R} & \dfrac{1}{R}\dfrac{\partial r}{\partial \Theta} & \dfrac{\partial r}{\partial Z} \\[2ex] r\dfrac{\partial \theta}{\partial R} & \dfrac{r}{R}\dfrac{\partial \theta}{\partial \Theta} & r\dfrac{\partial \theta}{\partial Z} \\[2ex] \dfrac{\partial z}{\partial R} & \dfrac{1}{R}\dfrac{\partial z}{\partial \Theta} & \dfrac{\partial z}{\partial Z} \end{bmatrix}$$

Figure E4.2.2. Helical annular mixer. The flow is a combination of a Couette flow (transversal) and Poiseuille flow (axial). The length of the mixer is L.

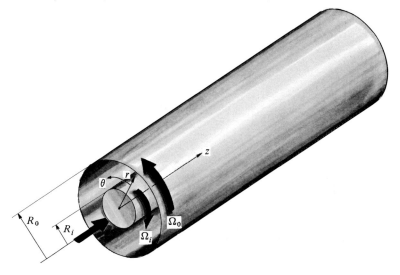

and

$$[\mathbf{F}] = \begin{bmatrix} 1 & 0 & 0 \\ tR[f(R)/R]' & 1 & 0 \\ tg'(R) & 0 & 1 \end{bmatrix}$$

where the primes represent differentiation with respect to R.

It is relatively straightforward to compute the area stretch, η, and to show that $D \ln \eta/Dt$ decays as t^{-1} and forms the basis for the understanding of more complex cases.[2,3] For example, the 'mixed cup average', i.e., the average weighted with respect to the axial flow, $\langle\langle \cdot \rangle\rangle = \int \cdot v_z \, da/(\int v_z \, da)$, of the stretching function $\alpha = D \ln \eta/Dt$, $\langle\langle \alpha \rangle\rangle$, decays as z^{-1}. For small gaps ($K \to 1$, $K = R_i/R_0$), the distribution of striation thicknesses, s, is nearly symmetric between the cylinders and the value of $s_{max}/\langle\langle s \rangle\rangle$ is 1.5. Also, as $K \to 1$, it can be shown that

$$(s/s_0)_{max} \approx 3(1 - K)/(2zAE)$$

i.e., the maximum striation thickness decreases linearly with axial distance. This result corresponds to the case of the inner cylinder rotating and the outer cylinder stationary; the striations are fed radially. The parameters A and E are defined as $A = L/R_0$, $E = V_\theta/V_{z,ave}$, where $V_\theta = \Omega_i R_0/(K^{-2} - 1)$ and $V_{z,ave}$ is the average axial velocity. For small gaps the mixing is dominated by the azimuthal flow rather than by the axial flow.

Example 4.2.5 Efficiency in the standard two-dimensional cavity flow

Here we consider stretching in the configuration shown in Figure E4.2.3, which is a rectangular cavity with width W and height H, and infinitely long in the x_3 direction (perpendicular to the plane x_1–x_2). A two-dimensional flow (v_1, v_2) is induced by the motion of the upper wall at $x_2 = H$ at a constant velocity U in the direction of negative x_1. We consider the case where the flow is described by the creeping flow equations

$$\mu\nabla^2\mathbf{v} = \nabla p, \qquad \nabla \cdot \mathbf{v} = 0$$

in a shallow cavity, i.e., $H/W \ll 1$. Defining dimensionless co-ordinates $x_1 \to x_1/W$, $x_2 \to x_2/H$ and dimensionless velocities v_1 and v_2 with respect to U, the boundary conditions are:

$$v_2 = 0 \qquad \text{(for } x_1 = 0, 1; \, x_2 \in [0, 1])$$
$$v_1 = 0 \qquad \text{(for } x_2 = 0; \, x_1 \in [0, 1])$$
$$v_1 = -1 \qquad \text{(for } x_2 = 1; \, x_1 \in [0, 1]).$$

In this case an exact solution is not available and we use the approximate solution of Chella and Ottino (1985a) obtained using the Kantorovich–Galerkin method (the solution compares very well with numerical simulations for large aspect ratios, $W/H > O(10)$). In order to compute the length stretch we use

$$L(t) = \int_{L(0)} (\mathbf{C} : \mathbf{MM})^{1/2} |d\mathbf{X}|,$$

where the integration (numerical, but based on an analytical expression for the velocity field) is carried out with respect to the *initial* configuration of the line, $L(0)$. For purposes of illustration, we consider two different orientations: $\mathbf{M} = (0, 1)$, a horizontal line spanning the cavity at $x_2 = H/2$, and $\mathbf{M} = (1, 0)$, a vertical line at $x_1 = W/2$, also spanning the entire cavity (see Figure 7.5.2 for experimental results).

A typical result is shown in Figure E4.2.3 where $[d \ln L(t)/dt]/(U/H)$, which can be interpreted as the efficiency of the stretching, is shown as

Figure E4.2.3. Specific rate of length generation for two different initial orientations, in a cavity of aspect ratio $W/H = 15$. Note that the effects of initial orientation disappear quickly.

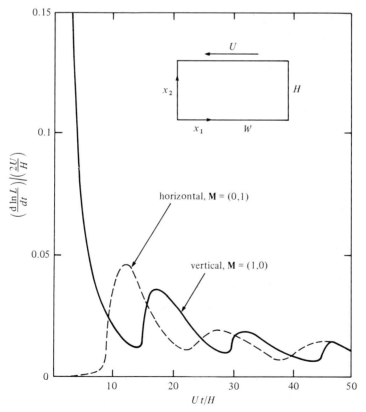

a function of time for the two initial placements of the line (note that $(\mathbf{D}:\mathbf{D})^{1/2} = O(2U/H)$). As expected, for short times the stretching of the vertical interface is more efficient since the line is placed perpendicular to the streamlines. However, rather quickly, the efficiency becomes independent of the initial orientation of the line and, on the average, the efficiency decays as t^{-1}. This is a case of mixing with *weak reorientation*. More complete computational results indicate the aspect ratio W/H, has a very small effect on the (dimensionless) stretch, which is mostly confined to changing the period of oscillation (of order W/U).

Example 4.2.6 Extensions of the cavity flow and the helical mixer flow
Several flows can be created by combining variations of cavity flows – square, trapezoidal, with one or two walls moving in *steady motion* – with Poiseuille or axial drag flows. One such case is shown in Figure E4.2.4 where an upper plate slides diagonally at an angle θ over a rectangular cavity.[4] Provided that the cross-sectional flow and axial flows can be decoupled (as in the case of a Newtonian fluid) the previous results of the cavity flow can be extended to this situation. The mixing characteristics of such a system are similar to those of the cavity flow; fluid particles are

Figure E4.2.4. Flow in a rectangular channel produced by a combination of pressure induced axial flow and a cross flow produced by sliding diagonally an upper plate. The cross section flow corresponds to the system of Figure 7.5.2(*d*).

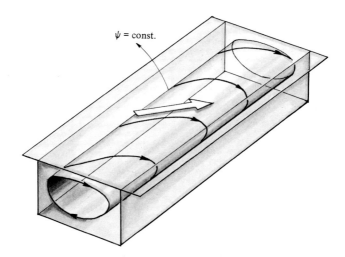

confined to cylindrical stream-surfaces characterized by a value of ψ. As in the helical annular mixer, the mixing is largely dominated by the azimuthal flow, since the axial flow is similar to the simple shear flow in terms of mixing and involves no reorientation.

Problem 4.2.1
Calculate the time necessary to achieve the maximum efficiency in a simple shear flow as a function of M_1, M_2, and M_3.

Problem 4.2.2
Prove that the relative flow around any material particle in a linear flow is identical to the flow itself.

Problem 4.2.3
Calculate the length stretch λ in a Couette flow by means of the formulas given in Chapter 2, by using both Cartesian and cylindrical co-ordinates.

Problem 4.2.4
Study the stretching efficiency in vortex decay (Batchelor, 1967, p. 204).
$$v_\theta = (C/2\pi r)[1 - \exp(-r^2/4vt)].$$
Show that the efficiency decays as $t^{-1/2}$.

Problem 4.2.5
Consider the velocity field
$$v_\theta = (Cr)/(2\pi r_1^2) \qquad \text{for } r \leqslant r_1$$
$$v_\theta = C/(2\pi r) \qquad \text{for } r > r_1.$$
Show that the striation thickness varies as
$$s \approx \pi r/[1 + (C^2 t^2/4\pi r^4)]^{1/2}.$$
Thus, $s \doteq r^3$ for $C^2 t^2/4\pi r^4 \gg 1$ and $d \ln s/dt \doteq t^{-1}$. Similarly, if r_s represents the radial position of the layer with striation thickness s, r_s moves outward as $r_s \doteq t^{1/3}$ (see Figure P4.2.1) (Ottino, 1982)). The striation thickness distribution is important in problems involving mixing with diffusion and reaction (see Chapter 9).

Problem 4.2.6
Study the efficiency in the linear three-dimensional flow
$$dx/dt = x \cdot \nabla v$$
with ∇v constant. We can formally solve for the motion as
$$x = X \cdot \exp(t\nabla v).$$
The deformation gradient is
$$F = [\exp(t\nabla v)]^T = \exp(t[\nabla v]^T)$$

and the length stretch,
$$\lambda^2 = \mathbf{F}^T \cdot \mathbf{F} : \mathbf{MM} = \exp(t\nabla v) \exp(t[\nabla v]^T) : \mathbf{MM}.$$
Examine the stretching behavior of this flow by studying the eigenvalues of ∇v (note that $\exp(\mathbf{Q} \cdot \nabla v \cdot \mathbf{Q}^T) = \mathbf{Q} \cdot \exp(\nabla v) \cdot \mathbf{Q}^T$). This flow is studied in more detail in Section 4.5. Compare also with the CSHM of Section 4.4.

Problem 4.2.7
Consider Poiseiulle flow between parallel lines of length L. Compute the stretching λ and efficiency e_λ as a function of the placement $\mathbf{M}$ and location r. What is the value of the efficiency at the center of the tube? Obtain the map of the efficiencies of the elements reaching $z = L$. Obtain the location r_{max} corresponding to maximum efficiency as function of L (Franjione and Ottino, 1987).

Problem 4.2.8
Compute the stretching for the velocity field
$$v_r = (K/\alpha r)(1 - \phi^2), \qquad v_\theta = 0, \qquad v_z = 0$$
where K is positive for diverging flow and negative for converging flow, and $\phi = \theta/\alpha$ (Hamel flow, see Figure P4.2.2). Obtain the motion by integrating the velocity field with the initial condition
$$(r, \theta, z)_{t=0} = (R, \Theta, Z).$$
Prove that $\mathbf{F}$ can be written as
$$[\mathbf{F}] = \begin{bmatrix} R/r & t*\phi(R/r) & 0 \\ 0 & r/R & 0 \\ 0 & 0 & 1 \end{bmatrix}$$

Figure P4.2.1. Stretching of a material line in a vortex after two and four turns of the vortex core.

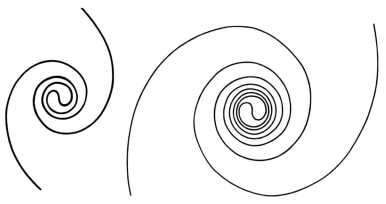

where $t^* = 2|K|t/(R\alpha)^2$. Obtain $\eta^2 = \mathbf{C}:\mathbf{NN}$. This problem is analyzed in some detail, by a different method, in Section 4.3.2.

4.3. Flows with a special form of ∇v

We consider the case of stretching of lines; the derivation for areas is similar. We know that the rate of change of d**x** is given by

$$D(d\mathbf{x})/Dt = d\mathbf{x}\cdot\nabla\mathbf{v} \qquad (4.3.1)$$

with d**x** = d**X** at $t = 0$. Formally, this equation can be solved as

$$d\mathbf{x} = d\mathbf{X}\cdot\exp\left[\int^t \nabla\mathbf{v}(\mathbf{X}, t)\, dt'\right]$$

$$d\mathbf{x} \equiv d\mathbf{X}\cdot\mathbf{F}^\mathsf{T},$$

but, if $D(\nabla\mathbf{v})/Dt \neq 0$ this route does not offer any advantage over the conventional approach, since in general, we cannot compute $\nabla\mathbf{v}(\mathbf{X}, t)$ without knowing the flow. However, for two special subcases this approach is useful, and even though their practical utility is limited, it is convenient to record them here.

(i) $D(\nabla\mathbf{v})/Dt = 0$. In this case the problem can be solved in terms of the eigenvectors of $\nabla\mathbf{v}$.

(ii) For slowly varying flows $\mathbf{F}(t)$ can be determined approximately by perturbation expansion.[5]

4.3.1. Flows with $D(\nabla\mathbf{v})/Dt = 0$

In this case consider the eigenvalue problems:

$$\mathbf{g}_i\cdot\nabla\mathbf{v} = \beta_i\mathbf{g}_i$$

and

$$\mathbf{g}^i\cdot(\nabla\mathbf{v})^\mathsf{T} = \beta_i\mathbf{g}^i$$

Figure P4.2.2. Hamel flow in a wedge.

which generate the dual basis $\{\mathbf{g}_i\}$ and $\{\mathbf{g}^i\}$, which we assume to be linearly independent. (Note that the eigenvalues for both problems are the same.) Taking the dot product of both sides of (4.3.1) with $\mathbf{g}'$,

$$(D(d\mathbf{x})/Dt)\cdot\mathbf{g}^i = d\mathbf{x}\cdot(\nabla\mathbf{v})\cdot\mathbf{g}^i$$

and since the $\mathbf{g}^i$ are independent of time,

$$D(d\mathbf{x}\cdot\mathbf{g}^i)/Dt = \mathbf{g}^i\cdot(\nabla\mathbf{v})^\mathrm{T}\cdot d\mathbf{x}$$

$$D(d\mathbf{x}\cdot\mathbf{g}^i)/Dt = \beta_i\mathbf{g}^i\cdot d\mathbf{x}$$

we can solve for $d\mathbf{x}\cdot\mathbf{g}^i$,

$$d\mathbf{x}\cdot\mathbf{g}^i = (d\mathbf{X}\cdot\mathbf{g}^i)\exp(\beta_i t).$$

Since a vector can be written in terms of the dual bases $\{\mathbf{g}^i\}$ and $\{\mathbf{g}_i\}$ as

$$d\mathbf{x} = \sum (d\mathbf{x}\cdot\mathbf{g}^i)\mathbf{g}_i$$

we have

$$d\mathbf{x} = \sum (d\mathbf{X}\cdot\mathbf{g}^i)\exp(\beta_i t)\mathbf{g}_i,$$

and in terms of the initial orientation, since $\mathbf{M} = d\mathbf{X}/|d\mathbf{X}|$

$$d\mathbf{x} = \sum a_i \exp(\beta_i t)\mathbf{g}_i|d\mathbf{X}|,$$

where $a_i = (d\mathbf{X}/|d\mathbf{X}|)\cdot\mathbf{g}^i$. Then, the length stretch is computed as

$$\lambda^2 = \sum\sum a_i a_j \exp[(\beta_i + \beta_j)t]g_{ij}$$

where $g_{ij} = \mathbf{g}_i\cdot\mathbf{g}_j$ (summations on i and j carried from 1 to 3). The results for area stretch are obtained in a similar manner except that the governing equation is

$$D(d\mathbf{a})/Dt = d\mathbf{a}\, D(\det\mathbf{F})/Dt - d\mathbf{a}\cdot(\nabla\mathbf{v})^\mathrm{T}.$$

Example 4.3.1 *Linear two-dimensional flow*

Consider again the flow of Problem 2.5.3. We can compute the stretching without solving any differential equations. In this case

$$[\nabla\mathbf{v}] = \begin{bmatrix} 0 & KG & 0 \\ G & 0 & 0 \\ 0 & 0 & 0 \end{bmatrix}$$

with eigenvalues $\beta_i = \pm G(K)^{1/2}$, 0, and eigenvectors

$$\mathbf{g}^i = [(K/(1+K))^{1/2}, \pm 1/(1+K)^{1/2}, 0], [0, 0, 1].$$

Assuming $K > 0$, the area stretch is computed to be (γ, b_1, b_2, defined earlier, Example 4.2.3)

$$\eta^2 = [(1+K)/2][b_2^2 \exp(\dot{\gamma}t) + b_1^2 \exp(-\dot{\gamma}t) + 2b_1 b_2(K-1)/(K+1)].$$

Remark

If $K = 0$, $\{\mathbf{g}_i\}$ and $\{\mathbf{g}^i\}$ are not independent and the method breaks down (e.g., the method does not work for shear flows).

Special case

When ∇v is symmetric (as for example in all irrotational flows), $\{g_i\} = \{g^i\}$ (i.e., orthogonal), and the equations simplify considerably.

Problem 4.3.1

Show that if $D(\nabla v)/Dt = 0$, then the length stretch is given by

$$\lambda^2 = \sum \sum b_i b_j \exp[(\beta_i + \beta_j)t] h_{ij}$$

where $h_{ij} = g^i \cdot g^j$, $b_i = (dX/|dX|) \cdot g_i$, and where the g^i are obtained from

$$g^i \cdot (\nabla v)^T = \beta_i g^i,$$

Assume $D(\det F)/Dt = 0$.

Problem 4.3.2

Show that

$$\lim_{t \to \infty} \frac{D \ln \lambda}{Dt} = \beta_{max}$$

where β_{max} is the largest positive real part of the eigenvalues of ∇v. Note that if the eigenvalues are of the form $\pm i\xi$, $D \ln \lambda/Dt$ oscillates. Since the eigenvalues of ∇v are the same as those of $(\nabla v)^T$, note also that the limit of $D(\ln \eta)/Dt$ is also given by β_{max}.

Problem 4.3.3

Study the general conditions for which $D(\nabla v)/Dt = 0$ for steady flows.

4.3.2 Flows with $D(\nabla v)/Dt$ small

In many cases of interest it is not possible to obtain $x = \Phi_t(X)$, and hence we do not have $\nabla v(X, t)$. However, if the flow is slow,[6] it might be possible to express ∇v in the form

$$\nabla v = \nabla v_0 + \varepsilon \nabla v_1 + \varepsilon^2 \nabla v_2 + \cdots$$

with the base flow, v_0, such that $D(\nabla v_0)/Dt = 0$. In such cases it is possible to obtain the stretching by means of a perturbation expansion. For example, to obtain the lineal stretch we write

$$dx = dx_0 + \varepsilon \, dx_1 + \varepsilon^2 \, dx_2 + \cdots$$

Using Equation (2.9.1) and grouping powers of ε we obtain

$$\varepsilon^0: \frac{D(dx_0)}{Dt} = dx_0 \cdot \nabla v_0$$

$$\varepsilon^1: \frac{D(dx_1)}{Dt} = dx_1 \cdot \nabla v_0 + dx_0 \cdot \nabla v_1$$

$$\vdots$$

$$\varepsilon^n: \frac{D(dx_n)}{Dt} = dx_n \cdot \nabla v_0 + \sum_{k=1}^{n} dx_{n-k} \cdot \nabla v_k.$$

As initial conditions for the dx_is we take:

$$dx_0 = dX \qquad \text{at } t = 0$$
$$dx_n = 0 \qquad \text{at } t = 0 \text{ for } n \geqslant 1.$$

Since the system of equations is linear they can be solved successively to obtain dx, up to any order in ε.

Example 4.3.2

Consider again the Hamel flow, solved earlier using the conventional approach to compare the exact and perturbation solutions. In this case we have

$$[\nabla v] = \begin{bmatrix} \mp \gamma(1 - \phi^2)\alpha & \pm 2\gamma\phi & 0 \\ 0 & \pm\gamma(1-\phi^2)\alpha & 0 \\ 0 & 0 & 0 \end{bmatrix} \left(\frac{R}{r}\right)^2$$

where

$$\gamma = |K|/(R\alpha)^2$$

and R is the initial radial co-ordinate of the particle X. Integrating v_r, i.e.,

$$d(r^2)/dt = 2K(1 - \phi^2)/\alpha$$

with the initial condition $r = R$, and expanding the series, we obtain

$$(R/r)^2 = 1/(1 - \varepsilon t^*) = 1 + \varepsilon t^* + (\varepsilon t^*)^2 + \cdots$$

where $t^* = 2\gamma t$ and the small parameter ε is

$$\varepsilon = \mp(1 - \phi^2)\alpha.$$

Thus, in this case ∇v_0 is taken to be

$$[\nabla v_0] = \begin{bmatrix} \varepsilon\gamma & \pm 2\gamma\phi & 0 \\ 0 & -\varepsilon\gamma & 0 \\ 0 & 0 & 0 \end{bmatrix}$$

and

$$\nabla v_n = (t^*)^n \nabla v_0.$$

Such an approach was used by Chella and Ottino (1985b). Figure P4.3.1 shows a comparison of the exact solution (obtained as indicated in Problem 4.2.8) and a five-term perturbation solution for two different sets of parameters.

Problem 4.3.4

Calculate the stretch of a filament with orientation $N = (N_r, 0)$ placed at $\phi = 0$ (this is the easiest case and requires no knowledge of the material of Chapter 2).

Problem 4.3.5
Show that near the wall, the stretch is similar to that of a shear flow *only* in the case of diverging flow ($K > 1$). Show also that away from the walls the stretch is *always* weaker than an elongational flow.

Problem 4.3.6
Calculate the vorticity number for Hamel flow. Compare your results with the limiting behavior obtained in the example above. Does the vorticity number provide a good measure of the balance between shear and elongation in the flow, in terms of stretching ability? Study the limit cases of $W_K^2 \to 0$ and $W_K^2 \to 1$. Does $W_K^2 = 1$ mean that the flow is elongational? Note also that W_K^2 does not distinguish between converging ($K < 0$) and diverging flows ($K > 0$).

Problem 4.3.7
Calculate the viscous dissipation following a fluid particle in Hamel flow.

Figure P4.3.1. Stretching in Hamel flow. Comparison between exact solution (full line) and perturbation solution (broken). The values of the parameters are $\mathbf{N} = (1, 0)$, $\phi = 0.5$, $a = 1/20$, $K < 0$. (Reproduced with permission from Chella and Ottino (1985b).)

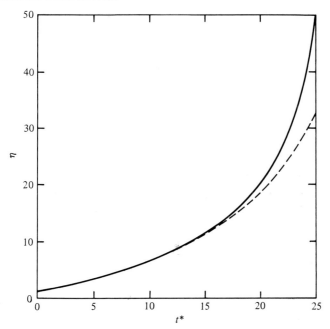

4.4. Flows with a special form of F; motions with constant stretch history

For our purposes a constant stretch history motion (CSHM) is defined such that the value of the deformation tensor following a material particle is given by[7]

$$\mathbf{F}(t) = \mathbf{Q}(t) \cdot \exp[t\mathbf{H}]$$

where $\mathbf{H}$ is a constant tensor and $\mathbf{Q}(t)$ an orthogonal tensor such that $\mathbf{Q}(0) = \mathbf{1}$ (Noll, 1962). $\mathbf{H}$ is related to $\nabla \mathbf{v}$ through the relation

$$\nabla \mathbf{v} = \mathbf{Q}(t) \cdot \mathbf{H} \cdot \mathbf{Q}^{\mathrm{T}}(t) - \mathbf{Z}$$

where $\mathbf{Z}$ is the antisymmetric tensor

$$\mathbf{Z} = [d\mathbf{Q}(t)/dt] \cdot \mathbf{Q}^{\mathrm{T}}(t).$$

Recalling the results of Section 3.7.2, $\mathbf{H}$ can be interpreted as the velocity gradient tensor with reference to a frame that moves with a fluid particle and rotates as $\mathbf{Q}(t)$. The tensor $\mathbf{H}$ is the fundamental quantity in the analysis of deformation in a CSHM; if $\mathbf{H}$ is known, then it is a trivial matter to calculate λ and η. However, in general, it is not an easy matter to calculate $\mathbf{H}$, starting with $\mathbf{v} = \mathbf{v}(\mathbf{x})$ without calculating the flow first. One important case where this can be done corresponds to an important subclass of viscometric flows (see Problem 4.4.2) termed *Steady Curvilineal Flows* (SCF).[8] Noll (1962) defined an SCF as a motion whose velocity field has the contravariant components

$$v^1 = 0, \qquad v^2 = v(x^1), \qquad v^3 = w(x^1)$$

in an orthogonal system $\{x^k\}$ for which the magnitudes of the natural basis of $\{x^k\}$, e_i (see Appendix), are constant along the trajectory $\xi = \xi(\mathbf{X}, t)$, with $\xi(\mathbf{X}, t = 0) = \mathbf{0}$, given by $\xi^1 = X^1, \xi^2 = X^2 + tv(X^1), \xi^3 = X^3 + tw(X^1)$. Thus, in an SCF the surfaces $x^1 = $ constant are material surfaces and they slide over one another without net stretching (Huilgol, 1975, Chap. 2). In the language of dynamical systems these surfaces are invariants of the motion. See Figure 4.4.1.

For an SCF, $\mathbf{H}$, and therefore $\mathbf{F}$, can be calculated explicitly from the spatial derivatives of $\mathbf{v}$ using a theorem due to Noll (1962). In this case and with respect to the fixed orthogonal basis referred to above, the physical components of $\mathbf{H}$ are given by:

$$[\mathbf{H}] = \begin{bmatrix} 0 & 0 & 0 \\ a\chi & 0 & 0 \\ b\chi & 0 & 0 \end{bmatrix}$$

with $a = (e_2/e_1)$, $b = (e_3/e_1)$, and where the e_is are the magnitudes of the vectors of the natural basis and χ is the generalized shear rate

$$\chi = \{[(v')^2 e_2^2 + (w')^2 e_3^2]/e_1^2\}^{1/2}$$

(the primes denote differentiation with respect to x_1). For an SCF we have,

$$\mathbf{F} = \exp[t\mathbf{H}] = 1 + \mathbf{H}t, \qquad \mathbf{F}^{-1} = \exp[-t\mathbf{H}] = 1 - \mathbf{H}t,$$

and

$$\mathbf{C} = 1 + t[\mathbf{H} + \mathbf{H}^{\mathrm{T}}] + t^2 \mathbf{H}^{\mathrm{T}} \cdot \mathbf{H}, \qquad \mathbf{C}^{-1} = 1 - t[\mathbf{H} + \mathbf{H}^{\mathrm{T}}] + t^2 \mathbf{H}^{\mathrm{T}} \cdot \mathbf{H}.$$

Using the results of Chapter 2 and setting $\gamma = \chi t$, a generalized shear, after some algebraic manipulations, we get that

$$\lambda = [1 + 2M_1\gamma(aM_2 + bM_3) + M_1^2\gamma^2]^{1/2}$$

i.e., $\lambda \approx M_1\gamma$, for $\gamma \gg 1$, and except for the trivial case $M_1 = 0$ the stretch goes linearly with time, and since $(\mathbf{D}:\mathbf{D})^{1/2}$ is constant following a fluid particle, the efficiency decays as t^{-1}. This result is very important since examples of SCF are all rectilinear flow in pipes, Couette flow, helical flow, torsional flow, flow between coaxial cones or concentric spheres, etc.

Figure 4.4.1. Representation of a steady curvilineal flow.

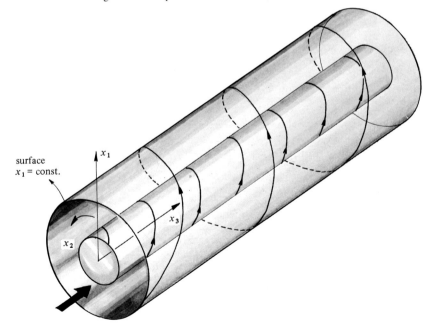

surface
x_1 = const.

Example 4.4.1

Consider the shear flow $v_1 = \dot{\gamma}x_2$, $v_2 = 0$, $v_3 = 0$. **F** can be written as

$$\mathbf{F} = 1 + t\mathbf{H}$$

where **H** has the representation

$$[\mathbf{H}] = \begin{bmatrix} 0 & 0 & 0 \\ \dot{\gamma} & 0 & 0 \\ 0 & 0 & 0 \end{bmatrix}.$$

Problem 4.4.1

Show that if $D(\nabla\mathbf{v})/Dt = 0$, then $\mathbf{F}(t)$ is given by

$$\mathbf{F}(t) = \exp[t\nabla\mathbf{v}]$$

The flow is called *viscometric* if $(\nabla\mathbf{v})^2 = 0$. Show that if $(\nabla\mathbf{v})^3 = 0$, $\lambda \approx t^2$ and therefore $D(\ln \lambda)/Dt \approx t^{-1}$ for long times.

Problem 4.4.2

Consider a velocity field such that

$$v_1 = 0, \qquad v_2 = \kappa x_1, \qquad v_3 = \varphi x_1 + \kappa x_2,$$

where κ and φ are constants. Show that **F** can be written as

$$\mathbf{F}(t) = \exp[t\mathbf{H}]$$

with $\mathbf{H}^3 = 0$ (from Noll, 1962).

4.5. Efficiency in linear three-dimensional flow

In this section we consider the stretching efficiency of the linear flow

$$\mathbf{v} = \mathbf{x} \cdot \nabla\mathbf{v}$$

which approximates the velocity field around a fluid particle.[9] In general, $\nabla\mathbf{v}$ will be a function of time but here we restrict ourselves to the case $D(\nabla\mathbf{v})/Dt = 0$ and $\nabla \cdot \mathbf{v} = 0$. The streamlines corresponding to two-dimensional flow were examined in Section 2.5.2.

From our studies with the linear two-dimensional flow (Example 4.2.3) we anticipate that the three-dimensional flow can give rise to either time-periodic or constant efficiencies and our analysis here will be restricted to the set of flows leading to constant efficiencies. We wish to display in a condensed form the numerical value of the asymptotic stretching efficiency in such a way as to put into evidence the roles of both vorticity and elongation. The basic idea is the following: First, we normalize the flow (i.e., $\nabla\mathbf{v}$) and second, we use the invariants of the normalized $\nabla\mathbf{v}$ to characterize the flow. Since one of the invariants happens to be zero $(\nabla \cdot \mathbf{v} = 0)$, two invariants are enough.

If we orient the axis along the eigenvectors of **D**, the most general expressions for **D** and Ω are:

$$[\mathbf{D}] = \begin{bmatrix} a & 0 & 0 \\ 0 & b & 0 \\ 0 & 0 & -(a+b) \end{bmatrix}$$

$$[\Omega] = \begin{bmatrix} 0 & j & -h \\ -j & 0 & g \\ h & -g & 0 \end{bmatrix}$$

where a, b, g, h, j are all constants. In order to evaluate e_∞ (recall that $e_{\lambda\infty} = e_{\eta\infty}$) we need to determine the eigenvalues of $\nabla\mathbf{v}$ normalized by $(\mathbf{D}:\mathbf{D})^{1/2}$. However, it turns out to be more convenient to calculate the eigenvalues of $\mathbf{L} \equiv \nabla\mathbf{v}/(\nabla\mathbf{v}:(\nabla\mathbf{v})^{\mathrm{T}})^{1/2}$, which gives e_∞^* (which is not invariant) and then use the relation between e_∞^* and e_∞ (Section 4.1) to convert the results to an invariant form. The invariants of **L** are:

$$I_{\mathbf{L}} = \mathrm{tr}\,\mathbf{L} = \nabla\cdot\mathbf{v}/(\nabla\mathbf{v}:(\nabla\mathbf{v})^{\mathrm{T}})^{1/2} = 0$$

$$II_{\mathbf{L}} = \tfrac{1}{2}[(\mathrm{tr}\,\mathbf{L})^2 - \mathrm{tr}\,\mathbf{L}^2] = -\tfrac{1}{2}\mathrm{tr}\,\mathbf{L}^2 = \tfrac{1}{2}(1 - W_{\mathrm{K}}^2)/(1 + W_{\mathrm{K}}^2)$$

$$III_{\mathbf{L}} = \det\mathbf{L} = (\nabla\mathbf{v}:(\nabla\mathbf{v})^{\mathrm{T}})^{1/2}\{ab(a+b) + [g^2(a+b) - j^2a - h^2b]\}$$

where

$$W_{\mathrm{K}}^2 = (g^2 + h^2 + j^2)/(a^2 + ab + b^2).$$

[See Olbricht, Rallison, and Leal (1982).] Thus, every linear three-dimensional flow can be represented by a point on a plane with coordinates $\mathrm{tr}(\mathbf{L}^2)$ and $\det(\mathbf{L})$ and assigned a unique value of e_∞^* or e_∞. The axis $\mathrm{tr}(\mathbf{L}^2)$ is related to the amount of vorticity in the flow whereas $\det(\mathbf{L})$ can be thought of as related to the three-dimensionality of the flow. The bounds for the region $\mathrm{tr}(\mathbf{L}^2)$, $\det(\mathbf{L})$ are:

$$-1 \leqslant \mathrm{tr}(\mathbf{L}^2) \leqslant 1 \tag{4.5.1}$$

and

$$-F(\mathrm{tr}(\mathbf{L}^2)) < \det(\mathbf{L}) < F(\mathrm{tr}(\mathbf{L}^2)) \tag{4.5.2}$$

where $F(\mathrm{tr}(\mathbf{L}^2))$ is given by

$$F(\mathrm{tr}(\mathbf{L}^2)) = (1/54^{1/2})\,\mathrm{sign}(b)(1 + 3W_{\mathrm{K}}^2)/(1 + W_{\mathrm{K}}^2)^{3/2}.$$

The region defined by these bounds is called the *accessible domain* and is plotted in Figure 4.5.1. The **L**s such that $\mathrm{tr}(\mathbf{L}^2) = 1$ correspond to irrotational flows whereas the value $\mathrm{tr}(\mathbf{L}^2) = -1$ corresponds to pure rotation ($a = b = 0$ in **D**). The upper bound of (4.5.2) corresponds to uniaxial extension ($a = -2b$), with the vorticity vector parallel to the extension axis, whereas the lower bound corresponds to biaxial extension

($a = b$) with the vorticity vector parallel to the compression axis. The det(L) axis corresponds to *generalized shear flows* ($W_K^2 = 1$). Inside the region $OADC$ all the eigenvalues of **L** are real; outside it there is one real eigenvalue and two complex conjugates. The level curves are iso-contours of the efficiency e_∞; i.e.,

$$e_\infty = f(\det(\mathbf{L}), \operatorname{tr}(\mathbf{L}^2)) = \text{constant.}$$

Note that the relation however, is not unique, and that more than one flow can lead to the same value of e_∞. Note also the following points:

(i) e_∞ does not decrease monotonically with increasing W_K^2; instead for any given value of det(L) one can find a value of $\operatorname{tr}(\mathbf{L}^2)$ that maximizes e_∞.

(ii) All flows along AB achieve the maximum value of e_∞ corresponding to three-dimensional flows $(= (2/3)^{1/2})$.

(iii) e_∞ is zero only at the point $\det(\mathbf{L}) = 0$, $\operatorname{tr}(\mathbf{L}^2) = 0$, corresponding to a simple two-dimensional shear flow.

Problem 4.5.1

Is the flow

$$\mathbf{v} = \mathbf{x} \cdot \nabla \mathbf{v}$$

a CSHM? Justify.

Figure 4.5.1. Accessible domain of a linear three-dimensional flow. (Reproduced with permission from Chella and Ottino (1985b).)

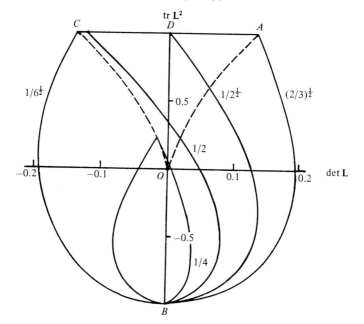

Problem 4.5.2
Obtain the pathlines of periodic points corresponding to the vector field
$$d\mathbf{x}/dt = \mathbf{x} \cdot \nabla \mathbf{v}.$$
Do all the flows that have closed orbits have poor efficiency? (*Hint*: Write $\nabla \mathbf{v}$ in Jordan form and expand the exponential.)

Problem 4.5.3
Identify the region in the accessible domain such that $\mathbf{v} = \text{const.}\ \omega$ (such flows are called Beltrami flows, see Section 8.7).

Problem 4.5.4
Construct the iso-helicity curves corresponding to Figure 4.5.1

4.6. The importance of reorientation; efficiencies in sequences of flows

It is apparent that for a wide class of flows (e.g., SCF) the efficiency decays as $1/t$. Some other flows (e.g., cavity flow) have partial restoration but, on the average, the efficiency decays also as $1/t$ for long times. A possible way to maintain a high average value of the efficiency is to design a sequence of flows (see Example 2.7.1) involving reorientation of material elements; another possibility is to 'reorient' the flow. We discuss several special cases below.

To start with the simplest case consider the flow
$$v_1 = Gx_2, \qquad v_2 = 0$$
and a line with initial orientation forming an angle θ_0, with the x_1 axis (Figure 4.6.1(a)). In this case we have
$$D(\ln \lambda)/Dt = G(Gt + \cot \theta_0)/[1 + (Gt + \cot \theta_0)^2]$$

Figure 4.6.1. Initial orientation of filament in a shear flow (a), and in a vortex flow (b).

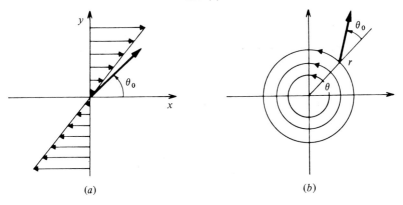

(a)   (b)

which is the behavior expected of SCFs. The average efficiency in the interval $0-\gamma$ ($\gamma \equiv Gt$), e_γ, is:

$$e_\gamma = (1/2^{1/2}) \ln(\gamma^2 \sin^2 \theta_0 + \gamma \sin 2\theta_0 + 1)/\gamma$$

and, if all initial orientations are equally likely, averaging over θ_0, we obtain:

$$\langle e \rangle_\gamma = (1/2^{1/2}) \ln[1 + \gamma^2/4]/\gamma. \tag{4.6.1}$$

The maximum value of $\langle e \rangle_\gamma$ is 0.284 which corresponds to a reorientation strain $\gamma = 3.98$. As might be expected, the same functional relationship holds for other flows. For example, for a point vortex

$$v_\theta = \omega/r, \qquad v_r = 0$$

and an infinitesimal material line oriented forming an angle θ_0, measured in the counter-clockwise direction with respect to r (see Figure 4.6.1(b)), we obtain

$$D(\ln \lambda)/Dt = \gamma(\gamma t \cos^2 \theta_0 - (\sin 2\theta_0)/2)/(1 + \gamma^2 t^2 \cos^2 \theta_0 - \gamma t \sin \theta_0)$$

where $\gamma \equiv 2\omega/r^2$. If all orientations are equally likely, we obtain, as before, the result (4.6.1).

Keeping the assumption of random angle reorientation, consider the effect of a distribution of strains, γ, and an infinite sequence of periods.[10] Consider first the case in which the distribution of time periods is random within an interval, i.e.,

$$f(\gamma) = \begin{cases} \frac{1}{2}\rho & \text{for } (\gamma_m - \rho) \leqslant \gamma \leqslant (\gamma_m + \rho) \\ 0 & \text{otherwise} \end{cases}.$$

In this case it is possible to obtain an exact expression for the efficiency of the sequence, $\langle e \rangle_{\gamma_m}$, which results in an infinite series which is evaluated numerically (Khakhar and Ottino, 1986a). The result is shown in Figure 4.6.2(a) as a function of the mean value, γ_m, and the width of the distribution, ρ. Note that the curves corresponding to $\rho = 0.05$ up to $\rho = 1$ almost coincide and that the best efficiency is achieved for $\rho = 0$ (uniform period).

Another case we might consider is a normal distribution of strains, i.e.,

$$f(\gamma) = (1/N) \exp[-(\gamma - \gamma_m)^2/2\rho^2]$$

Figure 4.6.2. Efficiency in shear flow and vortex flow as a function of shear γ_m: (a) random reorientation of the element after a strain γ. The strains are random in an interval 2ρ about a mean value γ_m. (b) Normal distribution of γ about a mean value γ_m and with standard deviation 2ρ. The broken line represents the maximum efficiency obtainable from the blinking vortex flow of Section 7.3. (Reproduced with permission from Khakhar and Ottino (1986a).)

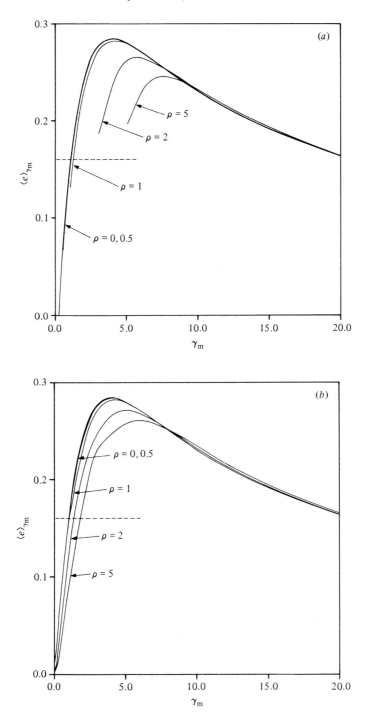

with

$$N = \int_0^\infty \exp\left(\frac{-(\gamma - \gamma_m)^2}{2\rho^2}\right) d\gamma.$$

In this case it is not possible to obtain an exact expression for $\langle e \rangle_{\gamma_m}$, but for ρ/γ_m small, $\langle e \rangle_{\gamma_m}$ is given by an expression similar to (4.6.1), i.e.,

$$\langle e \rangle_{\gamma_m} = K(\gamma_m, \rho) \ln[1 + \gamma_m^2/4]\gamma_m,$$

but where $K(\gamma_m, \rho)$ is a constant which is a complicated function of the variance (ρ) and the mean value (γ_m) of the distribution. A plot of $\langle e \rangle_{\gamma_m}$ as a function of γ_m for various values of the variance ρ is shown in Figure 4.6.2(b) (note similarity with 4.6.2(a)).

In these two examples we imagined that while the flow remains the same, the material elements are somehow reoriented. The idea has practical merit and finds application in the improvement of mixing in devices used to mix viscous fluids, such as extruders. In extruders, which can be modelled by the construction shown in Figure 4.6.3(a), sometimes one finds 'mixing sections' (e.g., protruding pins) that disrupt the flow and reorient material elements at the expense of only a small increase in power consumption. Chella and Ottino (1985) studied the case where material elements are assigned random reorientations when they reach designated planes. As expected, this mechanism greatly increases the mixing ability of the system (the length stretch becomes exponential with the number of mixing sections). A somewhat similar situation is to keep the orientation of the material element, but to somehow change the flow. One possibility is to 'join' flows as shown in Figure 4.6.3(b). In this particular case a fluid particle jumps from one streamsurface to another in a periodic manner, between an alternating sequence of flows produced by top and bottom moving walls. This idea is exploited further in Chapter 8, Section 8.2. As we shall see, it leads to very efficient mixing; indeed the mixing in such a device is chaotic. Several other possiblities might occur to the reader after studying the material of Sections 8.2 and 8.3.

Problem 4.6.1
Consider the family of irrotational flows given by the complex potential $w = z^n$. Show that these flows have an asymptotic efficiency equal to $1/2^{1/2}$, except for $n = 1$ (Franjione, 1986).

Problem 4.6.2
Generalize the ideas of these sections to sequences of Steady Curvilineal Flows.

Problem 4.6.3
Generalize these ideas to the linear two-dimensional flow.

4.7. Possible ways to improve mixing

It is apparent that there are very few velocity fields that can be integrated explicitly to obtain the stretching. Of those that can be integrated, hyperbolic flows produce the most significant stretching (and high values of $\langle e \rangle_\infty$) but they are *unbounded* and therefore not very practical to work with since they cannot be realized in the laboratory except in small regions of space. As we have seen in the previous section, time-periodic sequences of weak flows can achieve high efficiency and sequences of flows involving jumping from streamsurface to streamsurface hold similar promise. One might wonder if it is completely hopeless to try to achieve high efficiencies in steady *bounded* two-dimensional flows, and it is instructive to consider

Figure 4.6.3. (*a*) Extruder with mixing sections (generalization of Figure E4.2.4); (*b*) idealized flow produced by joining flows such as Figure E4.2.4, with upper and lower moving plates. A fluid particle jumps between streamsurfaces of adjacent sections.

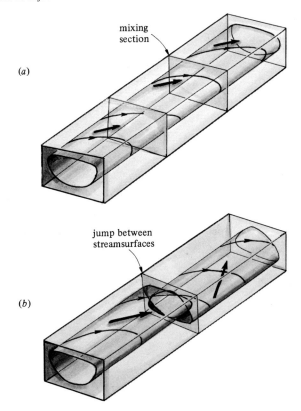

one possible way to increase mixing in such flows. One might start with hyperbolic flow (see Section 2.5) and try to improve the stretching by somehow closing the flow, i.e., by feeding the outflow into the inflow as indicated qualitatively in Figure 4.7.1. By continuity, nearby orbits have to close smoothly and form closed loops. The expectation might be that since the stretching in the neighborhood of the hyperbolic flow is exponential, then every time that a line segment passes by the hyperbolic region the stretches are compounded in such a way that the stretching is exponential.

However, this is illusory. As long as the flow is steady, isochoric, and two-dimensional, we can construct a streamfunction $\psi(x, y)$ such that $\mathbf{v} \cdot \nabla \psi = 0$. This implies that a material filament is trapped between curves of constant ψ. Another way of seeing this is to note that the topology of the streamlines in two-dimensional area preserving flows is composed of basically two building blocks: hyperbolic points (stagnation points) and elliptic points. It is obvious that if the flow is bounded and steady the streamlines join smoothly, and that an initially designated material filament can spiral between the two curves, ψ_1 and ψ_2 – very much in the same way as in a Couette flow – but that the length increase will grow linearly for long times (see Problem 4.7.1).[11] Indeed, as we shall see in Chapter 6, the long time value of the averaged stretching function

$$\alpha = \lim_{t \to \infty} \left\{ \frac{1}{t} \int_0^t \frac{D \ln \lambda(t')}{Dt'} \, dt' \right\}$$

is zero for these flows (termed 'integrable', a term defined in Chapter 6. Loosely speaking it means that it is possible to find the streamfunction[12]). The same argument applies to the combination of several hyperbolic points as shown in Figure 4.7.1. Apparently we face the conclusion that all two-dimensional flows are poor mixing flows. However, this need not be so, and as we shall see in Chapter 6 and demonstrate by means of examples in Chapter 7, if the flows become unsteady (e.g., time-periodic) the mixing can be excellent. In three dimensions the situation is more complicated. If the flow is three-dimensional (Chapter 8), combinations of inflows and outflows belonging to hyperbolic points can produce excellent mixing even if the flow is steady.

Problem 4.7.1

Consider a two-dimensional steady bounded flow containing a region of closed streamlines $\psi(x, y)$. Denote by $T(\psi)$ the time it takes for a fluid particle belonging to the streamline ψ to return to its original position.

Figure 4.7.1. Attempts at improving efficiencies in two-dimensional flows. By joining smoothly the outflow and the inflow of a hyperbolic point the streamlines form closed curves and the efficiency of mixing decays as t^{-1} (see Problem 4.7.1). In the top figure the broken line is a *homoclinic* trajectory; in the figure involving two hyperbolic points, A and B, the broken lines form *heteroclinic* trajectories.

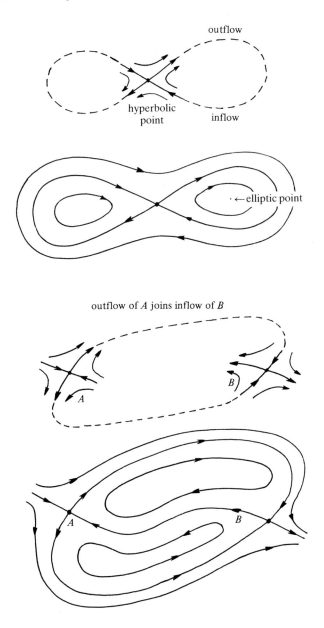

Show that a material filament $d\mathbf{x}$ at time t, $d\mathbf{x}(t)$, with initial orientation $\mathbf{m}_0$, is mapped to

$$d\mathbf{x}(t + T) = d\mathbf{x}(t) \cdot [1 - (dT/d\psi)(\nabla\psi)\mathbf{v}] + \text{higher order terms in } d\mathbf{x},$$

after a time T, and that the orientation $\mathbf{m}$ ($\equiv d\mathbf{x}/|d\mathbf{x}|$) after a time nT, where n is the number of cycles of the flow, is given by

$$\mathbf{m}_{t+T} = \mathbf{m}_0 \cdot [1 - (dT/d\psi)(\nabla\psi)\mathbf{v}]^n/\lambda.$$

Show also that, for $n \to \infty$, $\lim \mathbf{m}_{nT} \times \mathbf{v} \to 0$ (i.e., the filament becomes aligned with the streamlines) and that the stretching λ is linear in n (Franjione, 1987).

Bibliography

The foundations for the material discussed in this chapter can be found in continuum mechanics works. The most comprehensive treatment is given in Truesdell and Toupin (1960). A more accessible presentation can be found in Truesdell (1977). The original material on SCF is from Noll (1962). A particularly clear discussion can be found in Coleman, Markovitz, and Noll (1966). Major portions of this chapter can be found in Chella and Ottino (1985b). The example of the cavity flow can be found in Chella and Ottino (1985a). Section 4.5 is based on Khakhar and Ottino (1986a).

Notes

1 See Sections 56–61 in Truesdell (1954).

2 This flow allows the computation of many quantities of interest. The flow is a 'steady curvilinear flow' (see Section 4.4) and consequently all the mixing functions can be calculated analytically.

3 For example, mixing in many polymer processing applications. See for example, Chella and Ottino (1985a).

4 Such a flow forms the basis of a practical mixing device known as a single screw extruder. Fairly detailed analyses of the mixing are possible but numerical computations are necessary (see Chella and Ottino, 1985a).

5 These sections are based on R. Chella and J. M. Ottino (1985b). The notation is slightly different. Also we corrected several unfortunate typographical errors which appear in the paper.

6 This class of flows is relevant to low Reynolds number flows, such as some of those encountered in polymer processes, geophysics, etc.

7 The results of this section necessitate use of the concept of *relative motion*. However, since the concept is not used anywhere else in this work, we do not discuss it here even though we lose some rigor in not doing so (see Noll 1962).

8 A particularly clear and succinct discussion is given in Section 13 of Coleman, Markovitz, and Noll (1966).

9 For additional details the reader should consult Chella and Ottino (1985b).

10 This involves changing simultaneously the stretching time t and the location r if ω is regarded as constant in the vortex flow or the time t and the shear rate G in the shear flow.

11 There is substantially more to the analogy with the Couette flow. The Couette flow might be regarded as the prototype of an *integrable* flow (Chapter 6).

12 This does not imply that the streamfunction cannot be complicated. For example, Berker (1963), mentions three special cases studied by Oseen (1930) (*cf.* pp. 3 and 4). The streamfunctions studied can be interpreted as the superposition of two mutually orthogonal logarithmic spirals giving rise to $|\mathbf{v}| = \text{constant}$. In this case the portrait of the streamlines is complex enough to deserve the name of 'pseudo-turbulent' indicated by Berker.

Chaos in dynamical systems

In this chapter we consider the flow $x = \Phi_t(X)$ from a dynamical systems viewpoint. We study continuous and discrete dynamical systems, fixed and periodic points, invariant manifolds associated with hyperbolic points, and various signatures of chaos, such as homoclinic points and horseshoe maps.

5.1. Introduction

In Chapter 3 we considered classical ways of visualizing flows using 'classical' fluid mechanical tools such as streamlines, pathlines, and streaklines, which could be regarded as *global* in the sense that they give some information about the entire flow. The same can be said about instantaneous contour plots of viscous dissipation, vorticity, helicity, and various other quantities. On the other hand, in Chapter 4, we considered descriptors that have to do with stretching at *local* scales, e.g., specific length stretch, area stretch, and mixing efficiency. However, as we have seen, only in a few cases can the velocity field be integrated exactly to compute the stretching, and in those cases where it is indeed possible it does appear that the rate of stretching is mild. As we shall see, for most velocity fields with good mixing abilities, it is often impossible to obtain the flow. Consequently, in many cases of interest we are unable to calculate the local quantities described in Chapter 4.

In this chapter we reconsider the central question of our analysis, namely: What are the conditions under which a deterministic flow $x = \Phi_t(X)$ is able to stretch as efficiently as possible a material surface throughout the space occupied by the fluid?[1] Even though it is not possible to give a complete answer to this question, in this chapter we will focus on some relevant aspects of the general problem, to build intuition when we examine the prototypical model flows in Chapter 7 and 8 from a dynamical systems viewpoint. Therefore, in order to provide some background, we present here some of the main elements of the theory of dynamical systems.

In order to answer the general queston '*how does* $\mathbf{x} = \mathbf{\Phi}_t(\mathbf{X})$ *mix?*' we have to focus our attention on the periodic orbits and fixed points of flows (*continuous systems*) and the discrete systems derived from them (*discrete dynamical systems*). After this is done we will study the invariant sets associated with hyperbolic points. Due to its importance in practice, more than passing attention is given to the case of area preserving flows. Other general points should be made clear: In both Chapters 5 and 6 much of the treatment concentrates on time-periodic flows since this case is the easiest to analyze and gives rise to central concepts such as Poincaré sections and horseshoe maps. For convenience, the treatment of *flows* and *mappings* is carried out in parallel and most of the discussion is in terms of systems with low number of dimensions. This might require special alertness on the part of the reader. Note also that from a physical viewpoint we deal with *flows*, but the analysis is sometimes more conveniently carried out in terms of *maps*. For example, the case of time-periodic two-dimensional flow necessitates three-dimensions but in terms of mappings it can be reduced to just two.

The discussion is heuristic. For the mathematical foundations the reader should consult the books by Hirsch and Smale (1974) and Guckenheimer and Holmes (1983), and especially the article by Smale (1967) which gives a rigorous account of much of the material discussed here.

5.2. Dynamical systems

The starting point for our analysis is the flow or motion $\mathbf{x} = \mathbf{\Phi}_t(\mathbf{X})$. In practice the flow is generated by the Eulerian velocity field

$$d\mathbf{x}/dt = \mathbf{v}(\mathbf{x}, t),$$

with the initial condition $\mathbf{x} = \mathbf{X}$ at $t = 0$. Under fairly non-restrictive conditions solutions exist, at least locally, and they are unique with respect to the initial data if $\mathbf{v}(\mathbf{x})$ is Lipschitz (see Hirsch and Smale, 1974, Chap. 8).[2] The motion of fluid particles in an Eulerian velocity field is a dynamical system. In general we will consider that a continuous dynamical system is a system of differential equations

$$d\mathbf{x}/dt = \mathbf{f}(\mathbf{x}, t), \tag{5.2.1}$$

where the right hand side is arbitrary and $\mathbf{x} \in \mathbb{R}^n$. An important special case occurs when $\mathbf{f}$ is periodic in time, i.e.,

$$\mathbf{f}(\mathbf{x}, t) = \mathbf{f}(\mathbf{x}, t + T).$$

The space $\mathbf{x}$, with t as a parameter or with $\mathbf{x}$ augmented by t (space $\mathbb{R}^n \times \mathbb{R}$ if the time t is added as one of the axes), is called the *phase space* of the flow. Unless explicitly indicated we will deal exclusively with either autonomous systems or time-periodic systems.

If we denote $dV \equiv dX_1 \, dX_2 \dots dX_N$ as the volume of the initial conditions, and by $dv \equiv dx_1 \, dx_2 \dots dx_N$ as the volume of the initial conditions at time t, we have, as before,

$$dv = \det(D_{\mathbf{X}}\boldsymbol{\Phi}_t(\mathbf{X})) \, dV$$

where

$$[D_{\mathbf{X}}\boldsymbol{\Phi}_t(\mathbf{X})]_{ij} = \partial(\boldsymbol{\Phi}_t(\mathbf{X}))_i / \partial X_j$$

is the deformation gradient in N-dimensions.[3] In particular, setting $G \equiv$ const. in the transport theorem (Section 2.2), we obtain Liouville's theorem,

$$\frac{d}{dt} \int_{V(t)} dv = \int_{V(t)} \nabla \cdot \mathbf{f} \, dv = \int_{V(0)} \frac{D}{Dt} [\det D_{\mathbf{X}}(\boldsymbol{\Phi}_t(\mathbf{X}))] \, dV(0)$$

where $V(0)$ denotes the volume of the initial conditions, $\{\mathbf{X}\}$, and $V(t)$ denotes the volume of the initial conditions at time t, i.e., $\{\mathbf{x}\} = \boldsymbol{\Phi}_t\{\mathbf{X}\}$.[4]

If $\nabla \cdot \mathbf{f} < 0$ (see Equation (5.2.1)) the system is called *dissipative*. Dissipative systems contract volume in phase space. In Chapter 6 we will consider *Hamiltonian systems*, which conserve volume in phase space. These systems have the structure

$$dq_k/dt = \partial H/\partial p_k, \qquad dp_k/dt = -\partial H/\partial q_k,$$

where H is called the Hamiltonian and is a function of the p_ks and q_ks, and in some cases an explicit function of time. In particular, problems arising from Newton's law of motion can be cast into this structure.

Under fairly non-restrictive conditions the solution of the dynamical system (5.2.1) (see Section 2.4) gives rise to a flow $\mathbf{x} = \boldsymbol{\Phi}_t(\mathbf{X})$, i.e., the initial condition $\mathbf{X}$ is found at $\mathbf{x}$ at time t. Given any point $\mathbf{x}_0$ belonging to the phase space $\mathbb{R}^n \times \mathbb{R}$ at some time arbitrarily designated as zero, the *orbit* or *trajectory* based at $\mathbf{x}_0$ is given by $\boldsymbol{\Phi}_t(\mathbf{x}_0)$ for all times t. Thus, $\boldsymbol{\Phi}_{t>0}(\mathbf{x}_0)$ denotes the orbit of $\mathbf{x}_0$ for $t > 0$ (future times) and $\boldsymbol{\Phi}_{t<0}(\mathbf{x}_0)$ for all $t < 0$ (past times). In general, we refer to the system of ordinary differential equations (5.2.1) and its associated flow as a *continuous dynamical system*.

In some instances it is not necessary (or possible) to follow the trajectory of every initial condition continuously in time through the phase space and it is more convenient to record their positions at specified times. Consider for example that the position of the initial condition $\mathbf{X} = \mathbf{x}_0$ is recorded at $t = T$ and denoted $\mathbf{x}_1$, again at $t = 2T$ and denoted $\mathbf{x}_2$, and so on, i.e.,

$$\mathbf{x}_1 = \mathbf{\Phi}_T(\mathbf{x}_0)$$
$$\mathbf{x}_2 = \mathbf{\Phi}_T(\mathbf{\Phi}_T(\mathbf{x}_0))$$

$$\cdots$$

$$\mathbf{x}_n = \mathbf{\Phi}_T\{\mathbf{\Phi}_T[\ldots(\mathbf{x}_0)]\}$$

or in more compact notation,

$$\mathbf{x}_n = \mathbf{\Phi}_T^n(\mathbf{x}_0) \tag{5.2.2}$$

where $\mathbf{\Phi}_T^n(\cdot)$ denotes the composition of n mappings $\mathbf{\Phi}_T$. Alternative ways of expressing the same concept are

$$\mathbf{x}_n \to \mathbf{\Phi}_T(\mathbf{x}_n) \tag{5.2.3}$$

and

$$\mathbf{x}_{n+1} = \mathbf{\Phi}_T(\mathbf{x}_n). \tag{5.2.4}$$

Obviously, such a mapping contains some information about the original flow[5] and in some instances it is convenient to deal with such a map rather than with the entire flow. In general we refer to the mappings (5.2.2)—(5.2.4) as a *discrete dynamical system*.

5.3. Fixed points and periodic points

Given a flow $\mathbf{x} = \mathbf{\Phi}_t(\mathbf{X})$, $\mathbf{P}$ is a *fixed point of the flow* if

$$\mathbf{P} = \mathbf{\Phi}_t(\mathbf{P})$$

for all time t (i.e, the particle located at the position $\mathbf{P}$ stays at $\mathbf{P}$), as for example at a stagnation point in a velocity field. On the other hand $\mathbf{P}$ is a periodic point of period T (belonging to a periodic or closed orbit) if

$$\mathbf{P} = \mathbf{\Phi}_T(\mathbf{P})$$

i.e., the particle located at the position $\mathbf{P}$ with orbit $\mathbf{x} = \mathbf{\Phi}_t(\mathbf{P})$ returns to its initial position after a time T ($\mathbf{\Phi}_t(\mathbf{P}) \neq \mathbf{P}$ for any $t < T$), as for example in the orbits of circular Couette flow. For a discrete dynamical system

$$\mathbf{x}_{n+1} = \mathbf{f}(\mathbf{x}_n),$$

the definitions are similar (see Smale, 1967). A point $\mathbf{P}$ is a *fixed point of the mapping* $\mathbf{f}(\cdot)$ if

$$\mathbf{P} = \mathbf{f}^n(\mathbf{P})$$

for all n. Thus, a *fixed point of a flow* and its corresponding mapping are in general not the same. Generally, a fixed point of a mapping corresponds to a periodic point of the flow (see Figure 5.3.1).

We say that **P** is a *periodic point of order n* of the map $\mathbf{f}(\cdot)$ if

$$\mathbf{P} = \mathbf{f}^n(\mathbf{P})$$

i.e., **P** returns to its initial location after *exactly n* iterations ($\mathbf{f}^m(\mathbf{P}) \neq \mathbf{P}$ for any $m < n$). In this case we say that **P** is a *periodic point of order n*. Note that if we define a mapping

$$\mathbf{f}^n(\cdot) \equiv \mathbf{g}(\cdot),$$

then **P** is a fixed point of the map

$$\mathbf{x}_{n+1} = \mathbf{g}(\mathbf{x}_n).$$

The most common way (in this work) of generating mappings from flows is by means of the *Poincaré surface of section* which is introduced in Section 5.5. However, in this work this technique is used primarily in the context of Hamiltonian systems studied in Chapter 6.

5.4. Local stability and linearized maps

The behavior near fixed and periodic points is central to the understanding of dynamical systems. In this section we review a few of the most important points.

Figure 5.3.1. Fixed point of a map and periodic point of a flow. **P** is a periodic point of the flow $\mathbf{x} = \mathbf{\Phi}_t(\mathbf{X})$ and a fixed point of the mapping $\mathbf{x}_{n+1} = \mathbf{\Phi}_T(\mathbf{x}_n)$. At $t = 0$ the point is located at **P**, $\mathbf{\Phi}_{t_1}(\mathbf{P})$ denotes the position at t_1, $\mathbf{\Phi}_{t_2}(\mathbf{P})$ denotes the position at t_2.

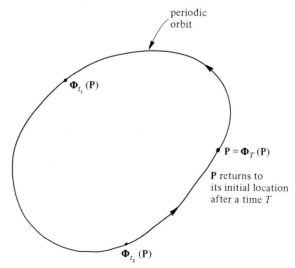

periodic orbit

$\mathbf{\Phi}_{t_1}(\mathbf{P})$

$\mathbf{P} = \mathbf{\Phi}_T(\mathbf{P})$

P returns to its initial location after a time T

$\mathbf{\Phi}_{t_2}(\mathbf{P})$

5.4.1. Definitions

Fixed and periodic points can be stable or unstable.[6] Consider first the definitions of stability (Guckenheimer and Holmes, 1983, p. 4; Hirsch and Smale, 1974, p. 185). These definitions are for local stability since they focus on the behavior near the points.[7] We discuss the definitions in terms of fixed points of flows; the definitions for mappings are similar (see Arnold, 1980, p. 115).

Liapunov stable

The point **P** is a stable equilibrium of the flow if for all neighborhoods U of **P** there exists a neighborhood U_1 of $\mathbf{P} \in U$ such that $\mathbf{x} = \mathbf{\Phi}_t(\mathbf{X})$ belongs to U for all times if **X** belongs to U_1 (see Figure 5.4.1(a)). This behavior is typified by centers; see Section 2.4.

Asymptotically stable

The point **P** is asymptotically stable if and only if there exists a neighborhood U of **P** such that for all **X** belonging to U we have

$$\lim_{t \to \infty} \mathbf{\Phi}_t(\mathbf{X}) = \mathbf{P}$$

and for $t > s$

$$\mathbf{\Phi}_t(U) \subset \mathbf{\Phi}_s(U).$$

See Figure 5.4.1(b). This behavior is typified by sinks; see Section 2.4.

5.4.2. Stability of area preserving two-dimensional maps

Consider the two-dimensional area preserving mapping,

$$\mathbf{x}_{n+1} = \mathbf{f}(\mathbf{x}_n)$$

with

$$\mathbf{P} = \mathbf{f}(\mathbf{P}).$$

Figure 5.4.1. (a) Representation of Liapunov stability, and (b) asymptotic stability.

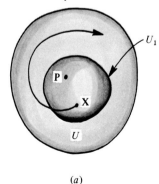
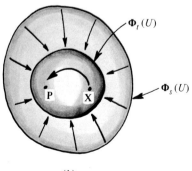

(a)   (b)

Such mappings are central to the understanding of fluid mixing in two-dimensional flows. The behavior near $\mathbf{P}$ is given by the linearized mapping

$$\xi_{n+1} = \mathbf{A} \cdot \xi_n$$

where $\mathbf{A}$ is 2×2 real matrix $(\mathbf{Df})$ and $\xi = \mathbf{x} - \mathbf{P}$. The eigenvalues of $\mathbf{A}$, λ_1, λ_2 are given by

$$\lambda^2 - \text{tr}[\mathbf{A}]\lambda + 1 = 0,$$

since $\det[\mathbf{A}] = 1$. In order to understand the dynamics of the mapping it is convenient to transform $\mathbf{A}$ into a Jordan form (i.e., there exists a matrix $\mathbf{R}$ such that $\mathbf{S} = \mathbf{R} \cdot \mathbf{A} \cdot \mathbf{R}^{-1}$ is of Jordan form, see Hirsch and Smale, 1974, Chap. 5. Note that $\det \mathbf{S} = \det \mathbf{A}$ and $\text{tr}\,\mathbf{S} = \text{tr}\,\mathbf{A}$. See Example 2.5.2). Thus, if $\mathbf{A}$ is

$$\mathbf{A} = \begin{bmatrix} a & b \\ c & d \end{bmatrix}$$

we have the following cases:

Hyperbolic $(\lambda_1, \lambda_2 \text{ real})$

$$\mathbf{R} = \begin{bmatrix} c & \lambda_1 - a \\ c & \lambda_2 - a \end{bmatrix} \qquad \mathbf{S} = \begin{bmatrix} \lambda_1 & 0 \\ 0 & \lambda_2 \end{bmatrix}.$$

Elliptic $(|\lambda_i| = 1 \text{ for } i = 1, 2, \text{ with } \lambda_{1,2} = \alpha \pm i\omega = \exp(\pm i\theta))$

$$\mathbf{R} = \begin{bmatrix} c & \alpha - a \\ 0 & \omega \end{bmatrix} \qquad \mathbf{S} = \begin{bmatrix} \cos\theta & -\sin\theta \\ \sin\theta & \cos\theta \end{bmatrix}.$$

Parabolic $(\lambda_i = \pm 1 \text{ for } i = 1, 2)$

$$\mathbf{R} = \begin{bmatrix} a - d & 2b \\ 2c & 0 \end{bmatrix} \qquad \mathbf{S} = \begin{bmatrix} \pm 1 & 0 \\ c & \pm 1 \end{bmatrix}.$$

The effects of the different linear mappings $\mathbf{S}$ on the square $(0, 1) \times (0, 1)$ are shown in Figure 5.4.2 (this analysis follows Percival and Richards, 1982, Chap. 2). The parabolic case corresponds to simple shear. Note that if the initial condition $\mathbf{x}_0$ belongs to one of the eigenspaces, successive mappings are given by $\mathbf{x}_n = \lambda_i^n \mathbf{x}_0$ $(i = 1, 2)$; therefore all the $\mathbf{x}_n$s remain in the eigenspace. It follows that if $|\text{tr}[\mathbf{A}]| > 2$ one of the eigenvalues has modulus greater than one, and $\mathbf{x}_n$ becomes unbounded. The mapping is said to be *unstable*. In general we say that the mapping is *hyperbolic* if none of its eigenvalues belongs to the unit circle; therefore hyperbolic mappings are unstable. On the other hand if $|\text{tr}[\mathbf{A}]| < 2$ the eigenvalues are complex conjugates and lie on the unit circle (Arnold, 1980, p. 116). In this case the system is called *elliptic* and the system is *stable*. The (degenerate) case $(\lambda_i = \pm 1)$ is called *parabolic*.

5.4.3. Families of periodic points

Consider a fixed point **P** of order k. Successive iterations produce the family

$$\mathbf{P} \to \mathbf{P}_1 \to \mathbf{P}_2 \to \cdots \to \mathbf{P}_k \qquad (5.4.1)$$

until reaching **P** with the kth iterate. The stability around each point is given by a linearized mapping as indicated earlier. If we want to relate the zeroth and nth iterates in the family (5.4.1) we compose the linearized mappings around each point to obtain

$$\mathbf{x}_n = \mathbf{J} \cdot \mathbf{x}_0$$

where **J** is the product of the Jacobian matrices evaluated at each point. The stability of the composition is given by the solution of the eigenvalue problem

$$\mathbf{J} \cdot \mathbf{x} = \lambda \mathbf{x}$$

or

$$\det(\mathbf{J} - \lambda \mathbf{1}) = 0$$

Problem 5.4.1
Prove that the eigenvalues, and consequently the stability, of all the members of the family (5.4.1) are the same (Lichtenberg and Lieberman, 1983, p. 186).

5.5. Poincaré sections

The Poincaré or surface section[8] method allows a systematic reduction in complexity of problems by means of a reduction in the number of dimensions (see Hirsch and Smale, 1974, Chap. 13) since it converts the

Figure 5.4.2. Effect of linear area preserving mappings on a square; (a) hyperbolic, (b) elliptic, (c) parabolic.

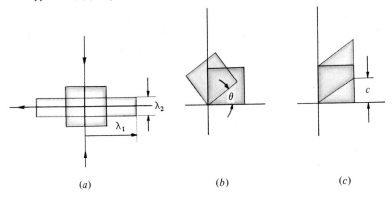

(a)   (b)   (c)

flow into a map. The concept was known to Poincaré and Birkhoff. As an example of the idea, consider an autonomous continuous dynamical system $d\mathbf{x}/dt = \mathbf{f}(\mathbf{x})$ with $\mathbf{x} \in \mathbb{R}^3$ and a periodic orbit with period T. Cut the orbit transversally with a (small) plane and denote the intersection as $\mathbf{P}$ (Figure 5.5.1). Consider the successive intersections of orbits originating from a point $\mathbf{x}_1$ near to $\mathbf{P}$. The successive intersections define a map $\mathbf{x}_{n+1} = \mathbf{\Phi}(\mathbf{x}_n)$, i.e., the point $\mathbf{x}_1$ is mapped to $\mathbf{x}_2 = \mathbf{\Phi}(\mathbf{x}_1)$, the point $\mathbf{x}_2$ is mapped to $\mathbf{x}_3 = \mathbf{\Phi}(\mathbf{x}_2) = \mathbf{\Phi}^2(\mathbf{x}_1)$, and so on. Note that the time between successive intersections need not be equal to T.[9] Figure 5.5.1(a) shows this idea graphically in three-dimensions. Thus the periodic trajectory in phase space corresponds to a fixed point of the mapping

$$\mathbf{P} = \mathbf{\Phi}(\mathbf{P}).$$

In the same fashion, a periodic trajectory of period-2 corresponds to a mapping such that

$$\mathbf{P}_2 = \mathbf{\Phi}(\mathbf{P}_1), \qquad \mathbf{P}_1 = \mathbf{\Phi}(\mathbf{P}_2)$$

or

$$\mathbf{P}_1 = \mathbf{\Phi}^2(\mathbf{P}_1)$$

and so on (see Figure 5.5.1(b)).

The ideas can be generalized; for example if $d\mathbf{x}/dt = \mathbf{f}(\mathbf{x})$ with $\mathbf{x} \in \mathbb{R}^N$ we consider a surface of dimension $N - 1$, $\sum(\mathbf{x}) = 0$, such that the flow is everywhere transverse to $\sum$, i.e.,

$$\nabla \sum(\mathbf{x}) \cdot \mathbf{f}(\mathbf{x}) \neq 0$$

Figure 5.5.1. Poincaré section iterates; (a) focuses on orbits near $\mathbf{P}$ and shows the intersections with a plane Σ orthogonal to the orbit passing by $\mathbf{P}$. A point close to $\mathbf{P}, \mathbf{x}_1$, is mapped to $\mathbf{x}_2 = \mathbf{\Phi}(\mathbf{x}_1)$, and then to $\mathbf{x}_3 = \mathbf{\Phi}(\mathbf{\Phi}(\mathbf{x}_1))$; (b) representation of a period-2 point in Poincaré section.

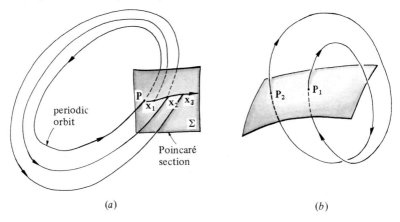

(a)   (b)

for all $\mathbf{x} \in \sum(\mathbf{x})$. This defines a map $\mathbb{R}^{N-1} \to \mathbb{R}^{N-1}$. Occasionally it is possible to define another surface of section and construct another mapping $\mathbb{R}^{N-2} \to \mathbb{R}^{N-2}$, and so on.

In this work Poincaré sections will be used in two main ways: globally, when the system is periodic in either space or time, or locally, near periodic or homoclinic orbits. Both ideas will be discussed and used in Chapters 6 through 8. In the case of time-periodic systems the technique amounts to taking stroboscopic pictures of initial conditions placed on $\sum(\mathbf{x})$ at intervals $T, 2T, 3T, \ldots$, etc. Thus Poincaré surfaces are a convenient way of displaying in a single plot *the character of the solutions belonging to all possible initial conditions.*[10] As we shall see some of the most important properties of dynamical systems are better understood with this viewpoint. In particular, Poincaré sections are of utmost importance in the case of periodic Hamiltonian systems.

5.6. Invariant subspaces: stable and unstable manifolds

Both the fixed (or periodic) hyperbolic points of discrete dynamical systems and the hyperbolic fixed points of continuous dynamical systems have associated with them invariant subspaces called *stable* and *unstable manifolds.*[11] To start with the simplest case, note that the linearization of discrete and continuous dynamical systems generates *linear* invariant subspaces.

Thus, if $d\xi/dt = D\mathbf{f}(\mathbf{P}) \cdot \xi$ is the linearization of $d\mathbf{x}/dt = f(\mathbf{x})$ around a point $\mathbf{P}$ such that $\mathbf{f}(\mathbf{P}) = 0$ with $\mathbf{x} \in \mathbb{R}^N$, the *stable, E^s, unstable, E^u,* and *center, E^c,* subspaces are defined in terms of the Jacobian $D\mathbf{f}(\mathbf{P})$ as:[12]

$E^s = \{space\ spanned\ by\ eigenvectors\ corresponding\ to\ eigenvalues\ whose$
real part $<0\}$

$E^u = \{space\ spanned\ by\ eigenvectors\ corresponding\ to\ eigenvalues\ whose$
real part $>0\}$

$E^c = \{space\ spanned\ by\ eigenvectors\ corresponding\ to\ eigenvalues\ whose$
real part $=0\}$.

It is clear that E^s, E^u, and E^c, are *invariant* sets; initial conditions remain trapped in the set. Examples are given in Figure 5.6.1.

Similarly, for the linearization of the mapping $\mathbf{x} \to \mathbf{G}(\mathbf{x})$ (or $\mathbf{x}_{n+1} = \mathbf{G}(\mathbf{x}_n)$) with $\mathbf{x} \in \mathbb{R}^N$, $\xi \to D\mathbf{G}(\mathbf{P})\xi$, around a fixed point $\mathbf{P}$, the *stable, E^s, unstable, E^u,* and *center, E^c,* subspaces are defined in terms of the Jacobian matrix $D\mathbf{G}(\mathbf{P})$ as:[13]

$E^s = \{$ *space spanned by eigenvectors corresponding to eigenvalues whose modulus* $< 1\}$

$E^u = \{$ *space spanned by eigenvectors corresponding to eigenvalues whose modulus* $> 1\}$

$E^c = \{$ *space spanned by eigenvectors corresponding to eigenvalues whose modulus* $= 1\}$.

These ideas can be extended to the non-linear system. For example, for the system $d\mathbf{x}/dt = \mathbf{f}(\mathbf{x})$ with a flow $\mathbf{x} = \mathbf{\Phi}_t(\mathbf{X})$, the *stable*, $W^s(\mathbf{P})$, and *unstable*, $W^u(\mathbf{P})$, *manifolds* associated with the hyperbolic point $\mathbf{P}$ are defined by:

$$W^s(\mathbf{P}) = \{\text{all } \mathbf{X} \in \mathbb{R}^N \text{ such that } \mathbf{\Phi}_t(\mathbf{X}) \to \mathbf{P} \text{ as } t \to \infty\}$$
$$W^u(\mathbf{P}) = \{\text{all } \mathbf{X} \in \mathbb{R}^N \text{ such that } \mathbf{\Phi}_t(\mathbf{X}) \to \mathbf{P} \text{ as } t \to -\infty\}.$$

Similarly, for mappings we have

$$W^s(\mathbf{P}) = \{\text{all } \mathbf{X} \in \mathbb{R}^N \text{ such that } \mathbf{G}^n(\mathbf{x}) \to \mathbf{P} \text{ as } n \to \infty\}$$
$$W^u(\mathbf{P}) = \{\text{all } \mathbf{X} \in \mathbb{R}^N \text{ such that } \mathbf{G}^n(\mathbf{x}) \to \mathbf{P} \text{ as } n \to -\infty\} \quad .[14]$$

An interpretation of $W^s(\mathbf{P})$ and $W^u(\mathbf{P})$ in the context of the Poincaré maps corresponding to a hyperbolic cycle is given in Figure 5.6.2.

Fluid dynamics provides a visual analog for periodic points and stable and unstable manifolds. Consider a two-dimensional time-periodic velocity field and focus on the behavior of a small fluid element near an elliptic cycle and a hyperbolic cycle. If a material point $\mathbf{P}$ belonging to the cycle is surrounded by a small blob S the elliptic cycle produces a net rotation

Figure 5.6.1. Examples of linear subspaces corresponding to the system $d\mathbf{x}/dt = \mathbf{f}(\mathbf{x})$ with $\mathbf{f}(\mathbf{P}) = \mathbf{0}$, $\mathbf{x} \in \mathbb{R}^3$.

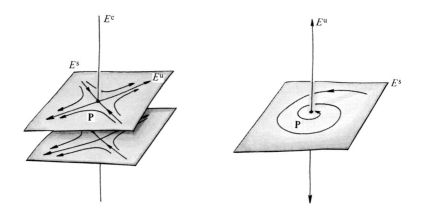

of the blob when **P** returns to its original position whereas the hyperbolic cycle contracts the blob in one direction and stretches it in another. In this case, the unstable manifold $W^u(\mathbf{P})$ corresponds to the region mapped out by the blob when time runs forward (see Figure 5.6.3). A seemingly minor complication in the blob experiment corresponds to the case when the blob S comes back folded upon its initial position, as shown in Figure 5.6.4. The complication, however, is profound. This construction, which will be studied in Section 5.8.3, is prototypical of chaotic flows.

Note that uniqueness of solutions places strong restrictions on the behavior of stable and unstable manifolds. For example, two unstable (or stable) manifolds belonging to two different periodic (or fixed) points cannot intersect; also $W^u(\mathbf{P})$ cannot intersect with itself, similarly, $W^s(\mathbf{P})$ cannot intersect with itself. However, intersections of stable and unstable manifolds belonging to the same or different points are permitted and are, in fact, responsible for much of the complex behavior of flows.

5.7. Structural stability

Suppose that $d\mathbf{x}/dt = \mathbf{f}(\mathbf{x})$ originates a flow $\mathbf{x} = \mathbf{\Phi}_t(\mathbf{X})$. An important question is: How 'different' is the flow generated by

$$d\mathbf{x}/dt = \mathbf{f}(\mathbf{x}) + \mu\mathbf{g}(\mathbf{x})$$

Figure 5.6.2. Interpretation of stable, W^s, and unstable, W^u, manifolds for a hyperbolic cycle and its corresponding map.

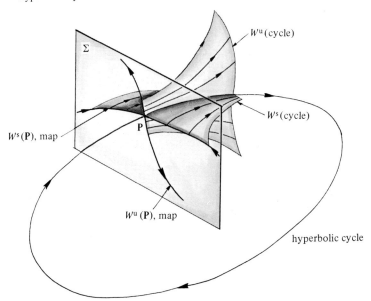

where $\mu\mathbf{g}(\mathbf{x})$ is some perturbation, $\|\mu\mathbf{g}(\mathbf{x})\| \ll \|\mathbf{f}(\mathbf{x})\|$?[15] In order to quantify this statement we give first (rather loosely) the idea of topological equivalence (see Smale, 1967).

Two flows are said to be *topologically equivalent* (or topologically conjugate) if there is a continuous invertible transformation and a time parametrization which maps one of the flows into the other. More precisely, two C^r flows, $\mathbf{x} = \mathbf{\Phi}_t(\mathbf{X})$ and $\mathbf{x} = \mathbf{\Psi}_t(\mathbf{X})$, are said to be C^k conjugate ($k < r$) if there exists a C^k diffeomorphism $\mathbf{h}$ such that

$$\mathbf{\Phi}_t = \mathbf{h}^{-1} \cdot \mathbf{\Psi}_t \cdot \mathbf{h}.$$

This makes the orbits of the flows coincide. C^0-equivalence (i.e., $\mathbf{h}$ is a homeomorphism) is called *topological equivalence*. This implies that fixed points of $\mathbf{\Phi}_t$ correspond to fixed points of $\mathbf{\Psi}_t$, unstable manifolds of $\mathbf{x} = \mathbf{\Phi}_t(\mathbf{X})$ to unstable manifolds of $\mathbf{\Psi}_t$, etc.

Figure 5.6.3. Visualization of manifolds $W^s(\mathbf{P})$ and $W^u(\mathbf{P})$ by means of a flow visualization (thought) experiment. The point $\mathbf{P}$ is a periodic point; a blob of tracer S encircles $\mathbf{P}$. The figure represents the state of the blob after one period $\mathbf{\Phi}_T(S)$ and two periods $\mathbf{\Phi}_T^2(S)$. Note that this picture is not possible in a steady flow since there is crossing of particle trajectories. The blob S is mapped to $\mathbf{\Phi}_T(S)$ after a time T and to $\mathbf{\Phi}_{2T}(S)$ after $2T$; $\mathbf{\Phi}_t(S)$ shows the position of the blob for $t < T$.

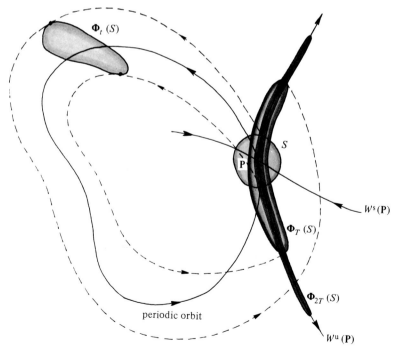

A vector field $d\mathbf{x}/dt = \mathbf{f}(\mathbf{x})$ (or its corresponding flow) has the property of *structural stability* if all μ-perturbations of it are topologically equivalent. More precisely: A vector field $d\mathbf{x}/dt = \mathbf{f}(\mathbf{x})$ (or flow $\mathbf{x} = \mathbf{\Phi}_t(\mathbf{X})$) is structurally stable if there exists $\mu > 0$ such that all C^1-μ-perturbations of $d\mathbf{x}/dt = \mathbf{f}(\mathbf{x})$ (or flow) are topologically equivalent to $d\mathbf{x}/dt = \mathbf{f}(\mathbf{x})$ (or flow). Similar definitions apply to maps. Pictorial examples are given in Figure 5.7.1 (based on Abraham and Shaw, 1985, Part 3, p. 37, where additional examples can be found).

The conditions under which structural stability is guaranteed are only known for the case of two-dimensional flows on orientable manifolds[16] and is the subject of Peixoto's theorem (Peixoto, 1962):

A C^k ($k \geqslant 1$) vector field $d\mathbf{x}/dt = \mathbf{f}(\mathbf{x})$ on an orientable two-dimensional manifold is structurally stable if and only if:
(i) *the number of fixed points and periodic orbits is finite and each is hyperbolic,*
(ii) *there are no trajectories joining saddle points,*
(iii) *the non-wandering set consists of fixed and periodic orbits only.*[17]

Figure 5.6.4. A complication on the case of Figure 5.6.3: the blob comes back folded ($t_1 < t_2 < T$).

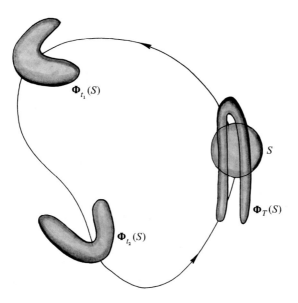

Chaos in dynamical systems

Problem 5.7.1

Study the structural stability of the one-dimensional equation $dx/dt = Cx^n$, where C is a constant, for $n = 1$, 2, and 3. Consider a perturbation of the form $\mu g(x)$. Take $g(x) = 1$.

Figure 5.7.1. Pictorial examples of topological equivalence, and structural stability and instability.

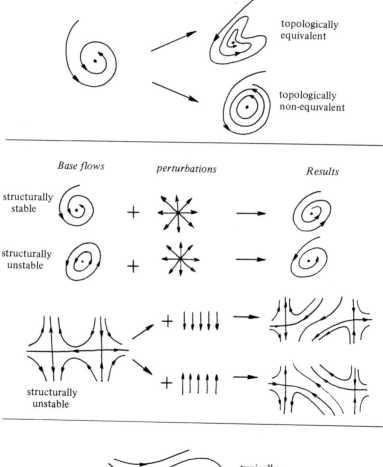

topologically equivalent

topologically non-equivalent

Base flows   *perturbations*   *Results*

structurally stable

structurally unstable

structurally unstable

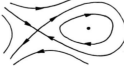

typically structurally unstable portrait

5.8. Signatures of chaos: homoclinic and heteroclinic points, Liapunov exponents, and horseshoe maps

5.8.1. *Homoclinic and heteroclinic connections and points*

If the unstable manifold of a hyperbolic cycle joins smoothly with the stable manifold of another cycle, the connection formed is called heteroclinic. If the unstable manifold of a cycle joins smoothly with the stable manifold

Figure 5.8.1. Homoclinic and heteroclinic connections. In (*a*) the stable and unstable manifolds of a hyperbolic cycle join smoothly forming a homoclinic connection; (*b*) shows two hyperbolic cycles before connecting, and in (*c*) the stable and unstable manifolds of the cycles join smoothly forming heteroclinic connections.

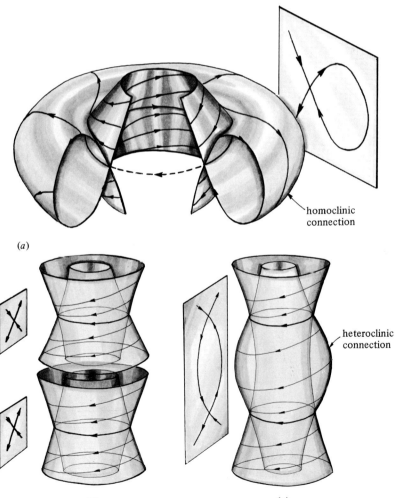

homoclinic
connection

(*a*)

heteroclinic
connection

(*b*)   (*c*)

of the same cycle, the connection is called homoclinic (see Figure 5.8.1). Similar definitions apply to maps. A point **y** is called *heteroclinic* (Smale, 1967; Guckenheimer and Holmes, 1983, p. 22, etc.) if it belongs simultaneously to both the stable and unstable manifolds of two different fixed (or periodic) points **p** and **q**. Thus, **y** is heteroclinic if

$$\mathbf{y} \in W^u(\mathbf{p}) \cap W^s(\mathbf{q}).$$

The point is called *homoclinic* if $\mathbf{p} = \mathbf{q}$. If the point of intersection of the manifolds at **y** is transversal, i.e., the manifolds intersect non-tangentially, then **y** is called a *transverse homoclinic point* (similarly for heteroclinic). Generally, when no confusion arises, a transverse homoclinic point is called a homoclinic point (similarly for heteroclinic). Figure 5.8.2 shows homoclinic and heteroclinic points for a map; only parts of the stable and unstable manifolds are shown. Such maps can be originated by the transverse intersection of one or two hyperbolic cycles, as shown in Figure 5.8.1.

By definition, homoclinic and heteroclinic points belong simultaneously to two invariant sets and therefore cannot escape from them.[18] One intersection implies infinitely many. In the case of an orientation preserving flow – as are all the systems originated from differential equations – the area A is mapped to A' (if the area A is mapped to B the map does not

Figure 5.8.2. Transverse homoclinic and heteroclinic connections for maps: (*a*) shows transverse homoclinic points (compare with 5.8.1(*a*) and 5.8.4(*a*)); (*b*) shows transverse heteroclinic points (compare with 5.8.1(*b*) and 5.8.4(*b*)). The region A is mapped to A', in turn A' is mapped into A''.

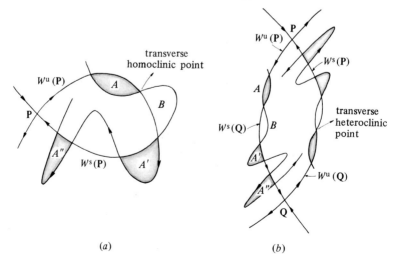

preserve orientation). Note that the intersection of invariant manifolds, as shown in Figure 5.8.2, is a projection onto the Poincaré section: the point belongs to the intersection at time T, $2T$, etc. However, what happens at intermediate times? Again a fluid flow visualization is useful. Figure 5.8.3 (which is a continuation of Figure 5.6.3) shows, schematically, intermediate

Figure 5.8.3. Visualization of $W^u(\mathbf{P})$ in a time-periodic system by means of a fluid mechanical experiment. The manifolds $W^s(\mathbf{P})$ and $W^u(\mathbf{P})$ intersect transversally.

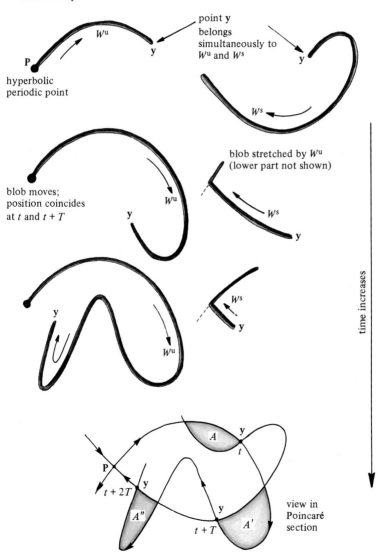

times by displaying the region of W^u already travelled by the homoclinic point **y** and the section of W^s to be travelled in future times. A tracer placed near the periodic point maps the unstable manifold (this is equivalent to the computational determination of W^s). By carefully placing a line of tracer on top of the stable manifold (this is extremely hard to do!) the line will form a blob in the neighborhood of the periodic point and is subsequently stretched by the unstable manifold. A construction using hyperbolic cycles serves to clarify why the area A is mapped to A' in Figure 5.8.3. Figure 5.8.4(*a*) shows the construction for a homoclinic connection whereas Figure 5.8.4(*b*) shows the construction for a heteroclinic connection.

The dynamics of the homoclinic points is extremely complex and is responsible for much of the behavior of chaotic systems. In order to state a few key results we need to introduce some additional terminology (the

Figure 5.8.4. Visualization of transverse (*a*) homoclinic and (*b*) heteroclinic connections corresponding to the maps of Figure 5.8.2. (*a*) indicates the mapping of the cross-hatched regions; the reader might compare this figure with 5.8.3.

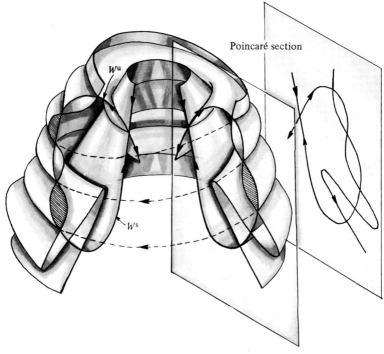

(*a*)

following concepts are due to Birkhoff, see Smale, 1967). A point **x** is *wandering* if there exists some neighborhood U of **x** such that

$$\bigcup_{t > t'} \Phi_t(U) \cap U = \phi$$

for some $t' > 0$. *Non-wandering* points are defined to be those points which are not wandering. For example, the points belonging to a closed orbit are non-wandering since portions of the neighborhood intersect the initial location. Non-wandering points do not imply periodic behavior. In fact, it is possible to show that homoclinic points are non-wandering (and they are clearly not periodic). Moreover, in every neighborhood of a homoclinic point there is a periodic point (see Smale, 1967).

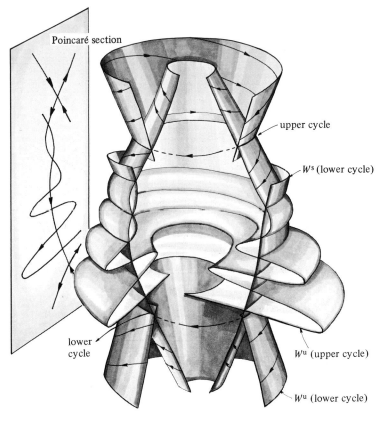

(*b*)

5.8.2. Sensitivity to initial conditions and Liapunov exponents

One of the manifestations of chaos most readily related to fluid mixing is the sensitivity to initial conditions.

The flow $\Phi_t(\mathbf{X})$ is said to be sensitive to initial conditions on a domain S if for all $\mathbf{X} \neq \mathbf{X}_0$, with $\mathbf{X}$ belonging to an ε-ball around $\mathbf{X}_0$ (i.e., the set of $\mathbf{X}$s such that $|\mathbf{X} - \mathbf{X}_0| < \varepsilon$, with ε small), there exists a time, $t < \infty$ such that $\Phi_t(\mathbf{X})$ lies outside the ε-ball for all $\mathbf{X}_0$ contained in S (see Figure 5.8.5). A similar definition applies to maps.[19]

A signature of chaotic systems is that they are characterized by a rapid divergence of initial conditions. Usually the divergence of initial conditions is quantified by means of numbers called Liapunov exponents which are related to the stretching of a filament of initial conditions. If $|d\mathbf{X}|$ represents the length of a vector of initial conditions $d\mathbf{X}$ around $\mathbf{X}$ with orientation $\mathbf{M} = d\mathbf{X}/|d\mathbf{X}|$, and its length at time t is $|d\mathbf{x}|$, the Liapunov exponent σ_i corresponding to a given orientation $\mathbf{M}_i$, is defined as (Lichtenberg and Lieberman, 1983, p. 262)[20]:

$$\sigma_i(\mathbf{X}, \mathbf{M}_i) \equiv \lim_{\substack{t \to \infty \\ |d\mathbf{X}| \to 0}} \left[\frac{1}{t} \ln\left(\frac{|d\mathbf{x}|}{|d\mathbf{X}|} \right) \right].$$

Another way of writing this, more suitable for computations, is

$$\sigma_i(\mathbf{X}, \mathbf{M}_i) \equiv \lim_{t \to \infty} \left(\frac{1}{t} \ln|D\Phi_t(\mathbf{X}) \cdot \mathbf{M}_i| \right),$$

since $\lambda = |(D\Phi_t(\mathbf{X}) \cdot \mathbf{M}_i|$.

An N-dimensional flow has at the most N, in general different, Liapunov exponents (i.e., there are N vectors $\mathbf{M}_i$, linearly independent, each of which gives, possibly, a different value). This is so since the eigenvalue problem

$$D\Phi_t(\mathbf{X}) \cdot \mathbf{M}_i = g_i \mathbf{M}_i$$

Figure 5.8.5. Sensitivity to initial conditions.

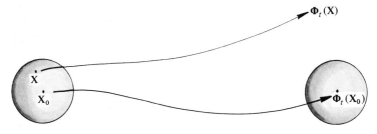

has at the most N different eigenvectors. In the cases of interest here, $\nabla \cdot \mathbf{v} = 0$, $\det(\mathbf{D}\Phi_t(\mathbf{X})) = 1$ and

$$\sum_{i=1}^{N} \sigma_i(\mathbf{X}, \mathbf{M}_i) = 0.$$

For dissipative systems this sum is negative.[21]

If one of the Liapunov exponents is non-zero, say σ_k, then

$$|d\mathbf{x}| \approx |d\mathbf{X}| \exp(\sigma_k t)$$

and the length of the filament grows exponentially with time.

The Liapunov exponents are related to the specific stretching rate and mixing efficiency discussed in Chapter 4. The relation to the stretching rate is

$$\sigma_i(\mathbf{X}, \mathbf{M}_i) \equiv \lim_{t \to \infty} \left\{ \frac{1}{t} \int_0^t \left(\frac{D \ln \lambda}{Dt} \right) dt' \right\} = \lim_{t \to \infty} \left[\frac{1}{t} \ln \lambda(\mathbf{X}, \mathbf{M}_i, t) \right]$$

i.e., the Liapunov exponent is the long time average of the specific rate of stretching, $D \ln \lambda/Dt$. Similarly, the average stretching efficiency

$$\langle e(\mathbf{X}, \mathbf{M}_i) \rangle \equiv \lim_{t \to \infty} \left[\frac{1}{t} \int_0^t e(\mathbf{X}, \mathbf{M}_i, t') \, dt' \right]$$

where

$$e(\mathbf{X}, \mathbf{M}_i, t) = \mathbf{D}{:}\mathbf{m}_i\mathbf{m}_i/(\mathbf{D}{:}\mathbf{D})^{1/2} = (D \ln \lambda/Dt)/(\mathbf{D}{:}\mathbf{D})^{1/2}$$

can be interpreted as a normalized Liapunov exponent (with respect to $(\mathbf{D}{:}\mathbf{D})^{1/2}$). The relationship between the (maximum) Liapunov exponent and the efficiency is not direct, unless $(\mathbf{D}{:}\mathbf{D})$ is constant over the pathlines. In most cases of interest $(\mathbf{D}{:}\mathbf{D})^{1/2}$ is a function of both $\mathbf{X}$ and t.

5.8.3. Horseshoe maps

Horseshoe maps occupy a central position in dynamical systems, and as we shall see they are very relevant to mixing. The existence of such a map indicates that the system is chaotic; in fact they can be regarded as the archetypical chaotic map (see equivalence among definitions of chaos in Section 5.9). Before going into their characteristics let us consider what we mean by 'mixing' from a mathematical standpoint. We will give two definitions. The difference between the two lies in the types of sets that are allowed to remain unmixed. Consider a flow region R and identify, arbitrarily, two others regions contained in R: a material volume A and another volume, fixed in space, B (see Figure 5.8.6). In a rough sense we say that the system mixes if there is a time T (or a number of mappings N), such that for any $t > T$, (or $n > N$), $\Phi_t(A) \cap B \neq \phi$ (or $\Phi_n(A) \cap B \neq \phi$) in such a way that this occurs for *most* As and Bs (similarly for mappings).

Mathematically, there are two main versions of this idea:

(i) We may take A and B to be the collection of all measurable sets in the region R. If the above property works for all sets except some of measure zero, we say that the system is *strongly measure-theoretic mixing*.

(ii) Another possibility is to take A and B to be all the sets with non-empty interior (i.e., we can fit an ε-ball inside). If the above property works,

Figure 5.8.6. Sketch for mathematical definitions of mixing. In (*a*) the blob A and the test region do not intersect, in (*b*) part of A lies within the region B. There is no widely accepted way of defining mixedness from a practical viewpoint. One possibility might be to divide the region R into cubes with side d and to specify that within each cube the volume fraction of the blob A has to be within $\phi - \varepsilon$ and $\phi + \varepsilon$ (ϕ is the overall volume fraction of A in R). The dimensions of the cube specify the resolution required and ε indicates the desired uniformity. If the materials are miscible, the time scale necessary to achieve uniformity at a length scale d is proportional to d^2. However, large regions of poorly mixed material might persist for long times and can be reduced only by further fluid mechanical action (see Figure E9.2.1).

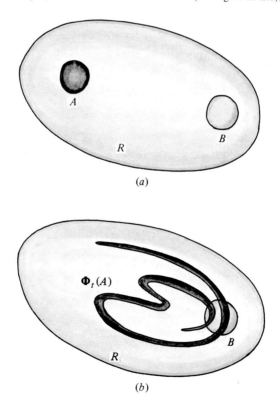

(*a*)

(*b*)

we say that the system is *strongly topologically mixing*. Measure-theoretic mixing implies topological mixing.[22]

Measure-theoretic mixing is the best achievable mixing. A physical feeling for measure-theoretic mixing can be obtained by means of a deceptively looking simple transformation which is called the 'baker's transformation', so named by the analogy of rolling and cutting dough. In the simplest case we cut a square and then we stack the pieces as indicated in Figure 5.8.7(*a*). A theorem asserts that a flow is measure-theoretic mixing if there exists a homeomorphism (i.e., a C^0 diffeomorphism) of the region R to a square under which the flow becomes a baker's transformation.[23] Since measure-theoretic mixing is the best possible mixing then the baker's transformation is the best mixing device. However, generally we are dealing with continuous flows which preserve connectedness and do not allow for cutting and welding.[24]

The closest that we can come to a 'baker's transformation' is stretching and folding as shown in Figure 5.8.7(*b*) and hope that not much material is 'lost' in the transformation (i.e., it leaves the quadrilateral). Such a construction is probably the simplest example of the so called *Smale Horseshoe Map* (Smale, 1967). The forward mapping is denoted $f(S)$; the inverse map is denoted $f^{-1}(S)$; forward iterates are denoted as $f^n(S)$ and backward iterates as $f^{-n}(S)$. Both maps are defined geometrically in Figure 5.8.7 (the reader should check that the mapping of all points $A, B,$ to $A', B',$ etc. in the forward and backward transformations is indeed correct). Note that the horseshoe map can approximate a baker's transformation by restricting the amount of material that goes out of the quadrilateral.[25]

In the case of the figure area elements in both horizontal striations (H_0

Figure 5.8.7. (*a*) Baker's transformation; (*b*) horseshoe map.

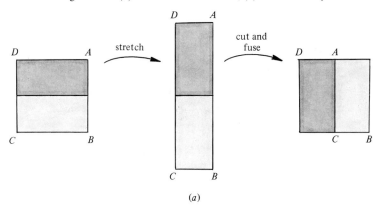

(*a*)

Figure 5.8.7 *continued*

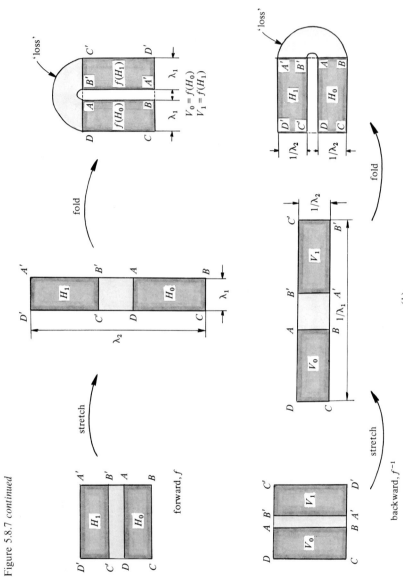

(b)

and H_1) are stretched by the same amount in the vertical direction by a factor λ_1 and contracted in the horizontal direction by a factor λ_2. Thus, the Jacobian everywhere in H_0, is

$$\begin{bmatrix} \lambda_1 & 0 \\ 0 & \lambda_2 \end{bmatrix}$$

whereas the Jacobian in H_1 is

$$\begin{bmatrix} -1 & 0 \\ 0 & -1 \end{bmatrix} \begin{bmatrix} \lambda_1 & 0 \\ 0 & \lambda_2 \end{bmatrix}$$

since it involves a $180°$ rotation. The values of λ_1 and λ_2 are selected in such a way that $0 < \lambda_1 < \frac{1}{2}$ and $\lambda_2 > 2$ so that the vertical striations are disjoint. Thus, both f restricted to $f(S) \cap S$ and f^{-1} restricted to $f^{-1}(S) \cap S$ are linear; the non-linearity comes in due to the folding. It is clear that n-applications of the forward mapping create 2^n vertical striations occupying an area $2^n \lambda_1^n$ (Figure 5.8.8 shows $f^2(S)$). Since $0 < \lambda_1 < \frac{1}{2}$ the total area goes to zero as n tends to infinity. Something similar occurs for the backward striations (area $2^n \lambda_2^n$). Note that if the map is area preserving $\lambda_1 \lambda_2 = 1$.

Figure 5.8.8 shows the 'loss' of material that does not get mixed by the transformation (i.e., it goes somewhere else in the flow region where it may or may not get mixed by another horseshoe map) and, in fact we might think of the horseshoe map as a baker's transformation with loss (Rising and Ottino, 1985). The set remaining in the original square, denoted I,

$$I \equiv \lim_{n \to \infty} \left[\bigcap_{-n}^{+n} f^n(S) \right]$$

Figure 5.8.8. (a) $f^2(S)$ and (b) $f^n(S) \cap f^{-n}(S) \cap S$ for $n = 4$.

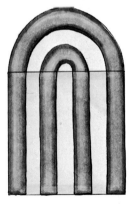

(a)   (b)

is in fact completely equivalent, in terms of its dynamics, to a baker's transformation (Figure 5.8.8 shows $f^n(S) \cap f^{-n}(S) \cap S$ for $n = 4$). In fact, the set I has several other properties: I is an invariant (Cantor) set consisting of an uncountable number of points which contains an infinite but countable number of periodic hyperbolic points of arbitrarily long periods and an uncountable set of bounded non-periodic motions. Furthermore, if $f(\cdot)$ is restricted to I it is equivalent to a 'Bernoulli shift' (see Schuster, 1984, Chap. 2) and it is structurally stable. This last point is very significant and indicates that these properties are not exclusive to the special horseshoe defined geometrically in Figure 5.8.7(b). There are many other types of horseshoes, some of which with their respective inverses are shown in Figure 5.8.9 (see Smale, 1967). In particular the striations need not be perfectly horizontal or vertical, the Jacobian need not be uniform, the striation thicknesses can be different, etc. Homoclinic/ heteroclinic points and horseshoes are intimately related. In fact the existence of one implies the other. There are several qualitative ways of seeing this. Figure 5.8.10 shows one possible way. Figure 5.8.10(a) shows the forward and backward transformations due to the presence of a homoclinic point. Figure 5.8.10(b) is just a deformed version of 5.8.10(b)

Figure 5.8.9. Various types of horseshoes with corresponding inverses.

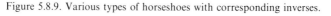

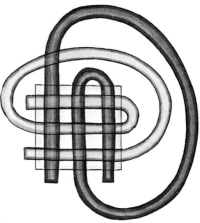

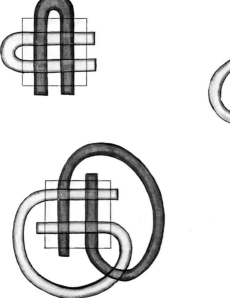

which clearly displays the intersections of forward and backward transformations characteristic of horseshoes (see Abraham and Shaw, 1985, Part 3, Chap. 5). Thus, we might visualize the horseshoe as a sort of 'black box' mixing device: material enters and comes out partially mixed. It is then natural to think of improving the mixing characteristics of the system by feeding back its output; another possibility is to 'connect' several horseshoes (Rising, 1989).

Figure 5.8.10. (*a*) Forward and backward transformations due to homoclinic point associated with a fixed point **P**, (*b*) a deformed version of (*a*).

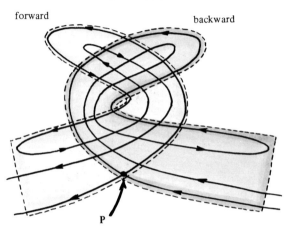

(*a*)

(*b*)

Problem 5.8.1

Verify that the Jacobian corresponding to H_1 has the form indicated in the text.

Problem 5.8.2

Verify the inverses of Figure 5.8.7(b), and Figure 5.8.9.

5.9. Summary of definitions of chaos

Throughout this work we will say that a system displays chaos when it satisfies one of the following conditions:

(i) *There is an invariant set S (i.e., $\Phi_t(S) = S$) and the flow is sensitive to initial conditions on S.*[26]

(ii) *The flow has homoclinic and/or heteroclinic points.*

(iii) *The flow produces horseshoe maps.*

We note that there are other possible definitions[27] and that, in a strict sense, definitions (1), (2), and (3), are not equivalent. (In 'general' (1) implies (2) and (3); (3) implies (1); and (2) and (3) are equivalent.) In principle, these three definitions are amenable to mathematical proof. However, in practice, given a flow, (1) is extremely hard to prove analytically (although Liapunov exponents are routinely reported via numerical computations); (2) can be proved by means of the Melnikov method,[28] and (3) is accessible from both an analytical as well as experimental viewpoint (see Chapter 7).

5.10. Possibilities in higher dimensions

Most of the concepts discussed in the previous section were explained in terms of time-periodic two-dimensional systems. The central ideas – horseshoes, homoclinic/heteroclinic intersections, etc. – can be generalized to higher dimensions. For example, intersections can involve manifolds of different dimensions.[29] Since the implications to fluid mixing are so obvious we present a few visual examples here (the reader interested in a more complete account should consult Abraham and Shaw, 1985, Part 3). It is clear that all these possibilities can occur in a system of the form

$$dx/dt = f(x) \qquad \text{with } \nabla \cdot f(x) = 0. \qquad (5.10.1)$$

Figure 5.10.1 shows a transverse heteroclinic trajectory produced by the intersection between a two-dimensional unstable manifold and a two-dimensonal stable manifold belonging to two different hyperbolic points.

Figure 5.10.2 shows a heteroclinic connection between the unstable manifold of a hyperbolic point of the saddle type and the two-dimensional stable manifold of a hyperbolic (saddle) cycle. Finally, note that Figure 5.8.3 can be interpreted as a cycle to cycle connection.[30]

Bibliography

The origin of most of the material presented in this chapter is mathematical but much of the motivation can be found in physics and astronomy. In a very general sense there are two kinds of chaotic systems: *dissipative systems* – with one-dimensional linear mappings being the most pervasive example – and *Hamiltonian systems*, which will be treated in Chapter 6. Historically, the earliest studies focused on Hamiltonian systems and were concerned with celestial mechanics; e.g., stability of the solar system. The earliest work can be traced back to Poincaré (1892, 1893, 1899); the

Figure 5.10.1. Connection between two hyperbolic points, P_1 and P_2, in three-dimensions.

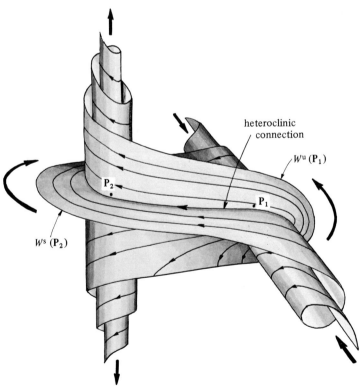

concept of intersections of manifolds is already present in his work. Another work originating in the celestial mechanics literature and reaching into the mathematics literature is Moser's (1973) *Stable and random motion in dynamical systems*; it contains a thorough and fairly readable discussion of horseshoe maps. Major portions of this chapter are contained in Smale's (1967) 'Differentiable dynamical systems'. The reader interested in the more mathematical aspects of this material is encouraged to consult this work.

Much of the recent emphasis in volume contracting flows can be traced to a pioneering paper by Lorenz (1963), 'Deterministic nonperiodic flow' (here is where strange attractors find their home). Lorenz's work remained

Figure 5.10.2. Connection between a hyperbolic point (**P**) and a hyperbolic cycle in three-dimensions.

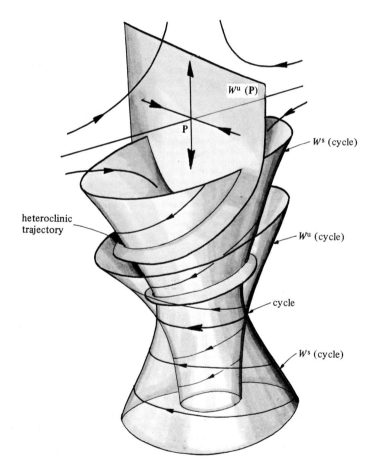

buried in the meterological literature until it was rediscovered in the physics literature in the last fifteen years or so. The applications of these ideas in fluid mechanics are many. For leads the reader should consult Landford (1982) and Guckenheimer (1986). Fluid mechanical and other applications are reviewed by Swinney (1985) (the Appendix of this paper was published in 1983 in *Physica*, **7D**, 3–15).

The literature on one-dimensional non-linear mappings and period doubling is abundant. One of the earliest general references is May (1976). One of the most influential works in one-dimensional mappings is Feigenbaum (1980), 'Universal behavior in non-linear systems', where references to many of the original works can be found. A simple book describing some of these matters is Schuster (1984), *Deterministic chaos: an introduction* (the discussion on Hamiltonian systems is very brief). A more mathematical treatment is given by Devaney (1986).

The matters discussed in this chapter are closely connected with stability and bifurcations (see for example, Ioos and Joseph, 1980), the mathematical foundations of mechanics (e.g., Arnold, 1980; highly recommended, Thirring, 1978), and the qualitative theory of differential equations (e.g., Arnold, 1985; Hirsch and Smale, 1974; Arnold, 1983). The reader unfamiliar with classical mechanics should consult the textbook *Introduction to dynamics* by Percival and Richards (1982). This work focuses on the systems with one- and two-degrees of freedom and it covers Hamiltonian systems and transformation theory.

There are a large number of works dealing with the matters discussed in this chapter and it is not possible to do justice to the subject in such a short space. Fortunately most of the above matters have been condensed in the book *Nonlinear oscillations, dynamical systems, and bifurcation of vector fields*, by Guckenheimer and Holmes (1983). This book focuses on systems with a few degrees of freedom and has extensive discussions on the Smale horseshoe maps, homoclinic and heteroclinic bifurcations, and a large number of examples. The visually oriented reader should consult the series by Abraham and Shaw (1985). Another book, probably more accessible for the non-mathematical reader, is the work by Lichtenberg and Lieberman (1983). This book is particularly useful for Hamiltonian systems.

The reader is warned that this chapter is a rather myopic view of dynamical systems; many important topics, concepts, and methods, are not covered. Missing are symbolic dynamics, normal forms, symplectic manifolds, and presentations of well studied flows such as the logistic equation, Hénon's map, the standard map, and Lorenz's equations, etc.

Notes

1 Thus, the goal is not to be able to integrate the flow exactly but rather to find the general conditions for efficient stretching in *any* flow.

2 Recall that if $\mathbf{v} \in \mathbb{R}^n$ the non-autonomous systems can be converted into an autonomous system by defining $x_{n+1} = t$. See also Section 2.4.

3 Recall that, with the exception of vorticity, most of our results carry from $\mathbb{R}^3$ to $\mathbb{R}^N$.

4 The relationship between the Reynolds theorem and the Liouville theorem is mentioned by Synge (1960, p. 174). The date of Liouville theorem is 1838; the date of Reynolds (or transport) theorem is 1903.

5 For example if the flow is smooth, say k-times continuously differentiable with bounded Jacobian, then the mapping is smooth and has a smooth inverse.

6 We restrict the discussion to autonomous systems.

7 Stability can be defined also in a global sense, but it is usually much harder to prove (see Hirsch and Smale, 1974, Section 9.3.).

8 Sometimes called cross-section.

9 In some cases, such as time-periodic systems, we can select surfaces in such a way that the time between successive intersections is the same.

10 It sould be mentioned that there is a converse construction which is also of importance. A *suspension* allows the reconstruction of the flow starting with the map (Smale, 1967).

11 A set S is called an *invariant* set of the flow $\mathbf{x} = \mathbf{\Phi}_t(\mathbf{X})$ on a manifold M ($S \subset M$) if $\mathbf{\Phi}_t(\mathbf{X}) \in S$ for all $\mathbf{X} \in S$, for all time t (for our purposes we can regard a manifold as a smooth surface). Thus, if a point belongs to an invariant set, then when acted on by the flow, it remains in the set. The definition for mappings is analogous.

12 For additional details see Section 2.4.

13 See Section 5.4.2 for the two-dimensional case.

14 According to the stable manifold theorem near the fixed point the stable and unstable manifolds are tangents to the linear eigenspaces E^s and E^u (for references see Smale, 1967, also Guckenheimer and Holmes, 1983, p. 18). This theorem was known to Poincaré and Birkhoff.

15 For conditions on $\mathbf{g}(\mathbf{x})$ see Andronov, Vitt, and Khaiken (1966). The concept of structural stability is largely due to Andronov. A similar question is addressed in Section 6.10.

16 That is, the surface has two 'sides'.

17 Note that in many cases this theorem is of rather limited use to us since we are generally interested in area (or volume) preserving perturbations and not just any perturbation.

18 The implications of this invariance are extremely important when the flow is area preserving as in the case of Hamiltonian systems (see Chapter 6).

19 Note that sensitivity to initial conditions is compatible with continuity with respect to initial conditions.

20 A similar definition, with time t replaced by n, applies to mappings (Lichtenberg and Lieberman, 1983, p. 267).

21 In general the numerical calculation of *all* Liapunov exponents is complicated. Straightforward application of the definition produces the value of the maximum Liapunov exponent for almost all initial $\mathbf{M}_i$s (Lichtenberg and Lieberman, 1983, p. 280). See also Greene and Kim, 1987.

22 The term mixing has a precise meaning in ergodic theory. The term has been used in a fluid mechanical sense throughout this work (stretching plus diffusion). This is the only section in which it is used, unavoidably, in a mathematical sense. The terms 'recurrence', 'wandering', 'mixing' (measure-theoretic), and 'ergodic', are obviously related but their relationship will not be discussed in detail here. For example, ergodicity implies recurrence, but the converse is not true. Both recurrence and ergodicity do not imply mixing.

23 The relationship between fluid mixing and a baker's transformaton was pointed out in the 1950s by Spencer and Wiley (1951) but its full mathematical implications were apparently not realized.

24 Static mixers provide, possibly, the closest experimental and practical approximation to a baker's transformation in the context of mixing (see Section 8.1). These devices are used in the polymer processing industry to mix viscous liquids (see Middleman, 1977). The most popular is the Kenics" mixer which consists of a tube with internal helical subdivisions, a twisted plane, of alternating right hand and left hand pitches. Each subdivision is called an element. Ideally, after each element the streams fed into the mixer are subdivided into two and after n elements the striation thickness decreases as 2^{-n}. Experiments carried out by the Kenics corporation show indeed a layered structure. Photographs are reproduced in Middleman's book (*op. cit*). Some of these matters are discussed in the context of the 'partitioned-pipe mixer' described in Chapter 8.

25 Many other types of horseshoes are possible; the one depicted in Figure 5.8.7(*b*) is the simplest and most popular one.

26 In dissipative systems a system is called chaotic if it possesses a *strange attractor* (an attractor is called strange if it is an attractor, and it is sensitive to initial conditions, i.e., it possesses at least one positive Liapunov exponent). Attractors are impossible in Hamiltonian systems. For definitions see Guckenheimer and Holmes, 1983. For applications of strange attractors concepts in fluid mechanics, see Landford (1982) and Guckenheimer (1986).

27 Other possible definitions of (temporal) chaos are to observe a signal $x(t)$ as a function of time and to compute the power spectrum of $x(t)$. An indication of chaos is a continuous spectrum. Another possibility is to compute the correlation function, $c(\tau)$, of the signal $x(t)$, i.e.,

$$c(\tau) = \lim_{t \to \infty} \left[\frac{1}{t} \int_0^T x'(t) x'(t + \tau) \, dt \right]$$

where $x'(t)$ is the difference between $x(t)$ and the time average value of $x(t)$. If $c(\tau) \to 0$ as $T \to \infty$ the system is considered chaotic. Sometimes, the visual appearance of numerically computed Poincaré sections is taken as evidence of chaos. If possible, we prefer to designate as chaotic any system which satisfies any of the conditions (1)–(3) above, since they are, in principle mathematically based and amenable to proof. If these cannot be proved we then might consider any of these alternative definitions. In any case we will clearly specify which criterion is being used.

28 Since this technique will be used mostly in the context of Hamiltonian systems the discussion is reserved until Section 6.10.

29 Sometimes the word 'index' is used in this context. The index of W^u is the number of dimensions of W^u.

30 The situations shown here are common since hyperbolic linear volume preserving flows are dense and open among all volume preserving flows (Smale, 1967). However, it is an unsolved problem whether or not this is true when the right hand side of the evolution equation $d\mathbf{x}/dt = \mathbf{v}(\mathbf{x}, t)$ is constrained to satisfy the Navier–Stokes equations or the Euler equations.

6

Chaos in Hamiltonian systems

The equations describing the trajectory of a fluid particle in a two-dimensional isochoric flow are a Hamiltonian system, a special case of a volume preserving dynamical system. The objective of this chapter is to discuss the general structure of chaotic Hamiltonian systems, with an emphasis on periodic systems with one-degree of freedom and time-periodic Hamiltonians. Central to the understanding of these systems is the study of the flow near hyperbolic and elliptic fixed and periodic points. Hamiltonian systems conserve volume in phase space and in Poincaré sections. As we shall see, this restriction has important implications in unravelling the general structure of these systems, especially near elliptic points, and is the most important point of departure from the previous chapter.

6.1. Introduction

It seems at first contradictory to focus on Hamiltonian systems in the context of mixing of viscous fluids. However, the connection is purely kinematic and involves no approximations. As seen in Chapter 3, the equations describing the trajectory of a fluid particle in an isochoric two-dimensional velocity field are

$$dx_1/dt = \partial\psi/\partial x_2, \qquad dx_2/dt = -\partial\psi/\partial x_1,$$

where ψ is the streamfunction. Such a system of equations, regardless of the form of ψ is a Hamiltonian system (Aref, 1984) and therefore it is profitable to study fluid mixing from such a viewpoint to exploit the substantial body of theory focusing on these systems. The system is said to have one degree of freedom if the flow is steady, $\psi = \psi(x_1, x_2)$, and two if the flow is unsteady, $\psi = \psi(x_1, x_2, t)$. If ψ is time-periodic we say that the system has 'one and a half' degrees of freedom. As a confirmation of the ideas of Section 4.7 we will prove that two-dimensional steady flows are poor mixing flows and that time-periodic flows are, most likely, effective mixing flows at least in some region of space.

In the following sections we discuss Hamiltonian systems from a general viewpoint. Even though most of our applications will be to systems with one and a half degrees of freedom, some of the results will be given for systems with N-degrees of freedom. The objective is to show the underlying features shared by all Hamiltonian systems. We shall see that Hamiltonian systems with one-degree of freedom are *integrable* and hence they cannot be *chaotic*. On the other hand, Hamiltonian systems with one and a half degrees or two-degrees of freedom stand a very good chance of being non-integrable and chaotic (non-integrability is a necessary but not sufficient condition for chaos).

The implications of the results presented in this chapter in the context of mixing are mainly two: (i) all steady bounded two-dimensional flows have zero asymptotic mixing efficiencies; (ii) many time-periodic two-dimensional flows have positive mixing efficiencies, at least in some part of the flow.

Some of the theorems and methods sketched in this chapter describe what happens under *small* perturbations of the integrable case. It should be noted that we are primarily interested in the behavior for *large* perturbations, since it is where the best mixing occurs; however, much less is known for this case from a mathematical viewpoint and standard analytical techniques do not work (see Section 6.9). Nevertheless, the concepts presented here should constitute a reasonable basis on which to launch such an analysis. In Chapters 7 and 8 we discuss specific examples which serve to clarify some of the points discussed here. It should be clear that as long as the system is Hamiltonian, the general features will appear even in the simplest representatives of the class.

6.2. Hamilton's equations

As seen in Chapter 5 a Hamiltonian system has the structure given by the $2N$ first order differential equations

$$dq_k/dt = \partial H/\partial p_k, \qquad dp_k/dt = -\partial H/\partial q_k, \qquad (6.2.1a,b)$$

where the q_ks are the components of a vector $\mathbf{q} = (q_1, \ldots, q_N)$, the '*position*', and the p_ks, the components of, $\mathbf{p} = (p_1, \ldots, p_N)$, the '*momentum*', and H is the Hamiltonian, which is a scalar function of $\mathbf{p}$ and $\mathbf{q}$ and, in some cases, as we shall consider here, an explicit function of time.

In mechanical systems $\mathbf{q}$ can actually represent the position and $\mathbf{p}$ the momentum understood in the usual way.[1] In other applications $\mathbf{p}$ and $\mathbf{q}$ take different physical significance.[2]

If $\mathbf{p}$ and $\mathbf{q}$ have N components each, the Hamiltonian system is said to have N-degrees of freedom. If H is an *explicit* function of time the system has an additional degree of freedom (i.e., $N+1$). In the case of time-periodic Hamiltonians, time is regarded as an additional $\frac{1}{2}$ degree of freedom.

6.3. Integrability of Hamiltonian systems

Central to the discussion of chaos in Hamiltonian systems is the concept of integrability. By 'integrability' we do not mean the computation of the solution in terms of known functions but rather the ability of finding sufficient number of constants of the motion so as to be able to predict qualitatively the motion in phase space. As we shall see, if a system is *integrable* it cannot be chaotic. There are two equivalent ways of defining integrability; the first one requires the use of Poisson brackets; the second action-angle variables.

The Poisson bracket between the two functions $u = u(\mathbf{p}, \mathbf{q})$ and $v = v(\mathbf{p}, \mathbf{q})$ is defined as (Arnold, 1980):

$$[u, v] = \sum_k \left\{ \frac{\partial u}{\partial q_k} \frac{\partial v}{\partial p_k} - \frac{\partial v}{\partial q_k} \frac{\partial u}{\partial p_k} \right\}. \qquad (6.3.1)$$

With this definition Hamilton's equations can be written as:

$$dq_i/dt = [q_i, H] \qquad (6.3.2a)$$
$$dp_i/dt = [p_i, H]. \qquad (6.3.2b)$$

A Hamiltonian H, with N degrees of freedom, is called *integrable* if there exist N different functions, to be determined, $\{F_i\}, i = 1, \ldots, N$, such that $[F_i, H] = 0$. The condition $[F_i, H] = 0$ implies $dF_i/dt = 0$ over a surface of constant H. In the case of a fluid flow system they can be thought of as a streamsurface.[3]

Example 6.3.1

Consider a one-degree of freedom system, i.e., $H = H(p, q)$. In this case we need to find only one F_i. In this case the necessary function is H itself since $[H, H] = 0$. Therefore all steady two-dimensional flow fields are integrable. In this case material is trapped by streamlines and the system is a poor mixing flow (see Problem 4.7.1).

In the case of a two-degrees of freedom system, $H = H(p_1, p_2, q_1, q_2)$ or $H = H(p, q, t)$, we need to find two functions, F_1 and F_2. The first one is easy to find since $[H, H] = 0$, and we can take $F_1 = H$. Then we have to find $F_2 \neq aH \neq H + b$, where a and b are constants, such that $[F_2, H] = 0$. However, F_2 need not exist! As Helleman (1977) points out, most

Hamiltonian systems with two-degrees of freedom are *non-integrable*, i.e., it is in general not possible to find F_1 and F_2 such that $dF_1/dt = 0$ and $dF_2/dt = 0$ along a trajectory.

Another way of defining integrability is by means of *action-angle variables*, which is a special case of a canonical transformation. In general a transformation of variables

$$\mathbf{p}, \mathbf{q} \to \boldsymbol{\theta}, \mathbf{I}$$

$\boldsymbol{\theta} = \boldsymbol{\theta}(\theta_1, \ldots, \theta_N)$, $\mathbf{I} = \mathbf{I}(I_1, \ldots, I_N)$, with $I_i = I_i(\mathbf{p}, \mathbf{q}, t)$, $\theta_i = \theta_i(\mathbf{p}, \mathbf{q}, t)$, is called canonical if the structure of the system in the $\mathbf{I}, \boldsymbol{\theta}$ variables has the form

$$dI_k/dt = \partial H/\partial \theta_k, \qquad d\theta_k/dt = -\partial H/\partial I_k, \qquad (6.3.3a,b)$$

where $H = H(\mathbf{I}, \boldsymbol{\theta})$. That is, a transformation is called canonical if the Hamiltonian structure is preserved (the appearance of (6.3.3a,b) is identical to (6.2.1a,b)).[4]

A very useful subset of canonical transformations is the *action-angle* transformation $\mathbf{p}, \mathbf{q} \to \boldsymbol{\theta}, \mathbf{I}$ the Hamiltonian in the $\boldsymbol{\theta}, \mathbf{I}$ co-ordinates is not a function of $\boldsymbol{\theta}$. (This is indeed one of the definitions of *cyclic* variable. Conversely, it can be stated that the action-angle transformation is possible only in cyclic systems.) Under these conditions the system becomes

$$d\mathbf{I}_i/dt = (\partial H(\mathbf{I})/\partial \theta_i) = 0, \qquad d\theta_i/dt = -(\partial H(\mathbf{I})/\partial r_i) \equiv \omega(\mathbf{I})$$

which can be readily integrated to

$$\mathbf{I}(t) = \mathbf{I}(0), \qquad \theta_i(t) = \omega_i(\mathbf{I})t + \theta_i(0). \qquad (6.3.4a,b)$$

Thus, the action-angle transformation allows us to trivially integrate the Hamiltonian system (hence the justification of the name integrable). Conversely, if the action-angle transformation is possible the system is called integrable.[5] The mapping (6.3.4a,b) is called an integrable *twist mapping*, the *action* r_i remains constant while the *angle* θ_i rotates with speed $\omega_i(\mathbf{I})$. *By definition, all integrable Hamiltonians with bounded orbits*[6] *can be reduced to integrable twist mappings.*

6.4. General structure of integrable systems

Consider the simplest case of a Hamiltonian system, namely

$$dq/dt = \partial H/\partial p, \qquad dp/dt = -\partial H/\partial q, \qquad (6.4.1a,b)$$

with $H = H(p, q)$, i.e., a system with one-degree of freedom. The linearization of the system around a fixed point, $dq/dt = dp/dt \equiv 0$, shows that the system admits only three kinds of fixed points: saddle or hyperbolic points, centers or elliptic points, and the degenerate case of parabolic points (see Example 2.5.2, two-dimensional area preserving case). In the most general case the unstable and stable manifolds of hyperbolic points join smoothly by a curve of constant H. The essence of any integrable system is embodied in Figure 6.4.1. To see why this is so, note that all integrable systems with N-degrees of freedom can be transformed into N-uncoupled one-degree of freedom systems, i.e., the system

$$dI_i/dt = 0, \qquad d\theta_i/dt \equiv \omega(\mathbf{I})$$

yields N uncoupled linear equations for θ_i as a function of t. Since this is generic to all integrable systems, *all* integrable systems can be non-linearly transformed into each other, and are in this sense equivalent. Since the prototype of a one-degree of freedom is a simple pendulum, the essence of an integrable system with N-degrees of freedom is embodied in a system of N non-interacting pendula. Thus, we expect the phase plane of any integrable system to be some non-linearly deformed version of the phase portrait of Figure 6.4.1 or any of the streamfunction portraits of Figure 4.7.1.[7]

Figure 6.4.1. General picture of an integrable Hamiltonian system. The prototype is a pendulum without friction; in this case **q** represents the angle from the vertical $(-\pi, \pi)$ and **p** is proportional to the angular momentum. The level curves correspond to $H(\mathbf{p}, \mathbf{q}) = $ constant.

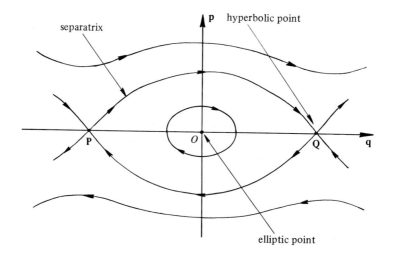

6.5. Phase space of Hamiltonian systems

As in Chapter 5 the state of a Hamiltonian system can be represented in the 2N-dimensional phase space by a vector $\mathbf{x}$,

$$\mathbf{x} = (p_1, \ldots, p_N, q_1, \ldots, q_N)$$

with time t being a parameter. The initial condition is represented by $\mathbf{X}$,

$$\mathbf{X} = (p_1^0, \ldots, p_N^0, q_1^0, \ldots, q_N^0)$$

such that the solution to the system (6.2.1a,b) can be represented by the flow

$$\mathbf{x} = \boldsymbol{\Phi}_t(\mathbf{X}) \qquad \text{with } \mathbf{X} = \boldsymbol{\Phi}_{t=0}(\mathbf{X}).$$

The velocity of a point $\mathbf{x}$ in phase space is given by

$$\mathbf{v} = d\mathbf{x}/dt = (dp_1/dt, \ldots, dp_N/dt, dq_1/dt, \ldots, dq_N/dt).$$

If $H = H(\mathbf{p}, \mathbf{q}, t)$ and ∇H is defined as

$$\nabla H = (\partial H/\partial p_1, \ldots, \partial H/\partial p_N, \partial H/\partial q_1, \ldots, \partial H/\partial q_N),$$

then by the chain rule,

$$\frac{dH}{dt} = \sum_i \left\{ \frac{\partial H}{\partial q_i} \frac{dq_i}{dt} + \frac{\partial H}{\partial p_i} \frac{dp_i}{dt} \right\} + \frac{\partial H}{\partial t}$$

or

$$\frac{dH}{dt} = \mathbf{v} \cdot \nabla H + \frac{\partial H}{\partial t} \tag{6.5.1}$$

which is interpreted as the rate of change of H (energy in the usual case) following the flow. For Hamiltonian systems $\mathbf{v} \cdot \nabla H \equiv 0$, which implies that $\mathbf{v}$ belongs to surfaces in phase space of constant H. Furthermore, if H is not an *explicit* function of time H remains constant for a given initial condition $\mathbf{X}$.

Another way of seeing this, and at the same time adding some physical meaning to the Poisson brackets, is to note that (6.5.1) can be written as

$$\frac{dH}{dt} = [H, H] + \frac{\partial H}{\partial t}.$$

Thus, the result $[H, H] = 0$ corresponds to the statement $(\nabla H) \cdot \mathbf{v} = 0$. The Hamiltonian phase space has also several other properties: If $V(0)$ denotes the volume of a set of initial conditions, according to Liouville's theorem (Chapter 5), $V(t)$ evolves according to

$$\frac{dV(t)}{dt} = \int_{V(t)} (\nabla \cdot \mathbf{v}) \, dv.$$

For Hamiltonian systems $\nabla \cdot \mathbf{v} = 0$, by substitution, and $V(t) = V(0)$. Hence the system conserves volume in phase space.

The Liouville theorem has many implications in Hamiltonian mechanics, here we will mention only two:

(i) *Poincaré's recurrence theorem* (see Arnold, 1980, p. 71): Consider a region $\{D\}$ in phase space such that $\mathbf{\Phi}_t\{D\} = \{D\}$ for all t. Then any trajectory in D returns infinitely close, infinitely often, to its initial location (i.e., all the points are non-wandering). The reason for this is easy to see. Consider an ε-ball around $\mathbf{X}$, denoted $B_\varepsilon(\mathbf{X})$. Since the initial conditions are trapped in $\{D\}$, conservation of volume requires that after a finite time t, $\mathbf{\Phi}_t(B_\varepsilon(\mathbf{X})) \cap B_\varepsilon(\mathbf{X}) \neq \phi$.

(ii) A Hamiltonian system cannot have asymptotic equilibrium positions and asymptotically stable limit cycles in phase space (for asymptotically stable equilibrium and limit cycles see Hirsch and Smale, 1974).

6.6. Phase space in periodic Hamiltonian flows: Poincaré sections and tori

Consider a system with a time-periodic Hamiltonian with period T. In this case it is traditional to represent the flow of the system on a *torus* with the time being a cyclic coordinate around the torus (Figure 6.6.1(*a*)). A surface transversal to the flow gives a good idea about the behavior of the trajectories for long times. For example, consider a surface of section (or Poincaré section) Σ such that

$$\Sigma = \{(\mathbf{x}, t) \in \mathbb{R}^{2N} \times t \in [nT, n = 0, 1, 2, \ldots]\}.$$

Such a section defines a return map in the plane Σ, $\mathbf{x}_{n+1} = \mathbf{G}(\mathbf{x}_n)$. Note that all trajectories originating from Σ return to Σ after the period T. Note also that area is conserved in the Poincaré section.

In the case of an integrable system, the cross-section of the torus can be regarded as an integrable twist mapping (Equation (6.3.4a,b)). An initial trajectory starting on Σ wraps around the torus as indicated in Figure 6.6.1(*b*). The trajectory may or may not join with itself depending on the value of $\omega_i(\mathbf{r})$. If the trajectory comes back to its initial position $\mathbf{p}$ in the Poincaré section after going around the torus m-times, then $\mathbf{p}$ is a periodic point of order m.

As seen in Chapter 5 the stability of fixed and periodic points corresponds to the solution of the eigenvalue problem

$$\mathbf{DG}(\mathbf{p}) \cdot \boldsymbol{\xi} = \lambda \boldsymbol{\xi}$$

where $\mathbf{DG}(\mathbf{p})$ is the Jacobian matrix of the mapping evaluated at the point in question. In a Hamiltonian system with N degrees of freedom $DG(\mathbf{p})$ is a $2N \times 2N$ matrix. In general, in Hamiltonian systems the eigenvalues appear as four-tuples

$$\lambda, \bar{\lambda}, 1/\lambda, 1/\bar{\lambda}$$

where the overbars represent the complex conjugate (see Lichtenberg and Lieberman, 1983, p. 181). For $N = 1$ ($\mathbf{x} \in \mathbb{R}^2$) the only possibilities for fixed points of the map $\mathbf{G}(\mathbf{x}_n) = \mathbf{x}_{n+1}$ are (Moser, 1973, p. 54, see Section 5.4.2):

Hyperbolic $|\lambda_1| > 1 > |\lambda_2|, \quad \lambda_1 \lambda_2 = 1$

Elliptic $|\lambda_i| = 1 \quad (i = 1, 2)$ but $\lambda_i \neq 1$

Parabolic $\lambda_i = \pm 1 \quad (i = 1, 2)$.

Figure 6.6.1. (*a*) Flow on a torus and corresponding map on Poincaré section; (*b*) rational trajectory.

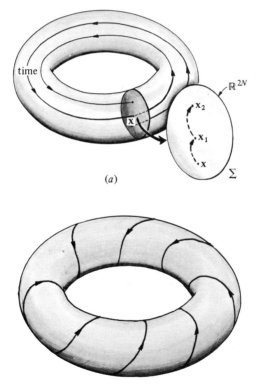

(*a*)

(*b*)

Problem 6.6.1
Prove that area is conserved in the Poincaré section of a Hamiltonian system.

6.7. Liapunov exponents

As was seen in Chapter 5, chaotic systems are characterized by a rapid divergence of initial conditions and usually, the divergence of initial conditions is quantified by means of Liapunov exponents. We know that a $2N$-dimensional flow has at the most $2N$, in general different, Liapunov exponents (i.e., there are $2N$ vectors $\mathbf{M}_i$, linearly independent, each of which gives, possibly, a different value). In Hamiltonian systems, due to volume conservation,

$$\sum_i^{2N} \sigma_i(\mathbf{X}, \mathbf{M}_i) = 0.$$

As seen in Chapter 5, for dissipative systems this sum is negative.

Integrable systems have all σ_is equal to zero. To see why this is so recall that all integrable systems can be reduced to an integrable twist mapping. From our discussion in Chapter 4 it is easy to see that in a twist mapping a filament of initial conditions $|d\mathbf{x}|$ grows linearly with time and the Liapunov exponents are zero (a twist mapping is essentially a Couette flow which is also a SCF). Furthermore, since chaotic systems have to have at least one positive Liapunov exponent, *integrable systems cannot be chaotic*. An important consequence is that all steady bounded two-dimensional flows are poor mixing flows.

Example 6.7.1
Consider a flow given by the streamfunction

$$\psi(x_1, x_2, t) = Ux_2 + [\Delta Uh/2] \ln\{\cosh(x_2/h) + A \cos[(x_1 - Ut)/h]\}$$

representing a train of vortices moving in the direction x_1 at speed U. The non-dimensional parameter, A, represents the concentration of vorticity. The value $A = 0$ corresponds to parallel flow with the classical hyperbolic tangent velocity profile. The meaning of the other terms is the following: ΔU is the velocity difference across the layer and h is proportional to the vortex spacing (for details see Stuart, 1967). This expression was used by Roberts (1985) to model the stretching of material lines and the roll-up of streaklines in shear layers. The evolution of streaklines and material lines corresponds to the solution of

$$\frac{dx_1}{dt} = \frac{\partial \psi}{\partial x_2}, \qquad \frac{dx_2}{dt} = -\frac{\partial \psi}{\partial x_1}.$$

In a moving frame $F'(x_1', x_2')$ moving at speed U in the direction $x_1 > 0$ the streamfunction becomes

$$\psi'(x_1', x_2') = (\Delta Uh/2) \ln[\cosh(x_2'/h) + A \cos(x_1'/h)]$$

and the flow is *autonomous* in this frame, and therefore integrable, with a streamline portrait similar to Figure 6.4.1 (this flow is usually referred to as 'Kelvin cat eyes', see Lamb, 1932, p. 225). If

$$\psi'(x_1', x_2')/(\Delta Uh/2) < \ln(1 + A)$$

the orbits are closed and periodic. Since the flow is integrable we can conclude that a material line marked in the flow will stretch linearly. Also, every segment of a streakline fed into the flow will stretch linearly. However, as we have seen in similar examples (Example 2.5.1) the streaklines can be quite complicated. This flow is reconsidered in Chapter 8, Section 8.3.

Example 6.7.2

In the simplest case, the basic equations for the motion of point vortices in the plane form a Hamiltonian dynamical system (Batchelor, 1967, p. 530). Thus if we consider point vortices of strengths $\kappa_1, \kappa_2, \ldots, \kappa_n$, with positions $(x_1, y_1), (x_2, y_2), \ldots, (x_n, y_n)$ the instantaneous value of the streamfunction is

$$\psi(x, y) = \left(\frac{-1}{4\pi}\right) \sum_i \kappa_i \ln[(x - x_i)^2 + (y - y_i)^2].$$

Since there is no self induced motion, the movement of the vortex of strength κ_j is equal to the velocity of the fluid at the point (x_j, y_j) due to all the other vortices. The result is

$$\kappa_j \frac{dx_j}{dt} = \frac{\partial W}{\partial y_j}, \qquad \kappa_j \frac{dy_j}{dt} = -\frac{\partial W}{\partial x_j}$$

where

$$W = \left(\frac{-1}{4\pi}\right) \sum_i \sum_{\substack{j \\ (i \neq j)}} [\kappa_i \kappa_j \ln(r_{ij})]$$

with

$$r_{ij} = [(x_j - x_i)^2 + (y_j - y_i)^2]^{1/2}.$$

Thus, the system is Hamiltonian with a number of degrees of freedom equal to the number of vortices. Note, however, that in this case the value of W (Hamiltonian) associated with a fluid particle is, in general, a function of time since it depends on the instantaneous positions of all the vortices (contrast with previous example) and it is not obviously clear that invariants should exist. Clearly, nothing interesting happens in the case of just one vortex (one-degree of freedom) and that the case of two vortices in the plane can be integrated has been known for a long time. In this case, for example, the distance between the two vortices remains constant, and details can be found in Batchelor (1967, p. 530), and various other places. However, it was only recently that the case of three vortices in the plane was shown to be integrable, and that a system of *four*, or more than four, vortices in an unbounded two-dimensional region is in general chaotic (Aref and Pomphrey, 1982; for a review see Aref, 1983).

6.8. Homoclinic and heteroclinic points in Hamiltonian systems

As seen in Chapter 5 a point **y** is called *heteroclinic* if it belongs simultaneously to both the stable and unstable manifolds of two different fixed or periodic points. In Hamiltonian systems we have volume conservation and this fact has important consequences in the behavior of homoclinic and heteroclinic points (see Figure 6.8.1). Since the manifolds are invariant sets (see Section 5.6) mappings of the heteroclinic point **y** belong to the intersection of manifolds. The only difference now is that since area is conserved, the shadowed regions of Figure 6.8.1 transform as indicated (Area A = Area A'); a point in $W^s(\mathbf{Q})$ moves asymptotically slowly as it approaches the stable fixed point $\mathbf{Q}^8$; the trajectory must

Figure 6.8.1. System of Figure 6.4.1 after a perturbation showing a transverse heteroclinic point; the region A is mapped to A', the point **y** is mapped to **y'**. The area of A is equal to the area of A'. Some curves survive the perturbation (KAM curves, see Section 6.10.2).

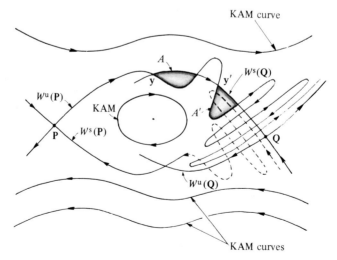

wander greater and greater distances normal to the stable manifold of **Q**. Similar behavior arises from the stable manifold of **Q**. Similar behavior arises from the stable manifold of **P** in the neighborhood of **Q**. Multiple intersections occur and other heteroclinic points appear.[9] As seen in Chapter 5 the *presence of transverse homoclinic and/or heteroclinic points is one of the accepted definitions of chaos.* Furthermore, transverse homoclinic points imply horseshoes (Smale–Birkhoff theorem, Guckenheimer and Holmes, 1983, p. 252).

6.9. Perturbations of Hamiltonian systems: Melnikov's method

The method of Melnikov (1963) provides one of the few analytical ways to determine the existence of intersections of stable and unstable manifolds. In this section we present the method in its simplest possible form (for a more complete treatment the reader should consult Guckenheimer and Holmes, 1983, Section 4.5; generalization to systems with n-degree of freedom are possible).

Consider a system of the form

$$d\mathbf{x}/dt = \mathbf{f}(\mathbf{x}) + \varepsilon\mathbf{g}(\mathbf{x}, t), \qquad \varepsilon \text{ small}$$

with $\mathbf{x} = (x_1, x_2)$, $\mathbf{f} = (f_1, f_2)$ such that $\mathbf{f}$ is Hamiltonian

$$f_1 = \partial H/\partial x_2, \qquad f_2 = -\partial H/\partial x_1,$$

and $\mathbf{g} = (g_1, g_2)$ time-periodic but not necessarily Hamiltonian. Both $\mathbf{f}$ and $\mathbf{g}$ are smooth. In this work we will identify $\mathbf{f}$ with an Eulerian velocity field and H with the streamfunction. Consider that the unperturbed system ($\varepsilon \equiv 0$) has a hyperbolic saddle point **H** with a homoclinic orbit, $\mathbf{q}^{(0)}(x_1, x_2)$ $=$ const. or $\mathbf{q}^{(0)}(t)$, and with stable, W^s, and unstable, W^u, manifolds surrounding an elliptic point **E** as shown in Figure 6.9.1(*a*). Assume also that the interior of the homoclinic orbit $\mathbf{q}^{(0)}$ is filled with elliptic orbits $\mathbf{q}^{(\alpha)}$, where α is a parameter ranging from 0 to 1. Denote by $T(\mathbf{q}^{(\alpha)})$ the period corresponding to the orbit and assume also that

$$T \to \infty \qquad \text{as } \mathbf{q}^{(\alpha)} \to \mathbf{q}^{(0)}.$$

Consider now the behavior of the perturbed system using a Poincaré section with the period of $\mathbf{g}$ (see Figure 6.9.1(*b*)). It can be shown that the perturbed system has a unique hyperbolic saddle point with stable and unstable manifolds close to those of the unperturbed system. If t_0 denotes a parameter which measures length along the unperturbed orbit, the distance in the Poincaré section between the stable, W_ε^s, and the unstable manifold, W_ε^u, *of the perturbed system* is given by

$$d(t_0) = \frac{\varepsilon M(t_0)}{|\mathbf{f}[\mathbf{q}^{(0)}(t)]|} + O(\varepsilon^2)$$

where $M(t_0)$ is the so-called Melnikov's integral and is given by

$$M(t_0) = \int_{-\infty}^{+\infty} \left\{ \mathbf{f}(\mathbf{q}^{(0)}(t)) \wedge \mathbf{g}(\mathbf{q}^{(0)}(t), t + t_0) \right\} dt$$

Figure 6.9.1. (*a*) Homoclinic orbit before perturbation; (*b*) after perturbation.

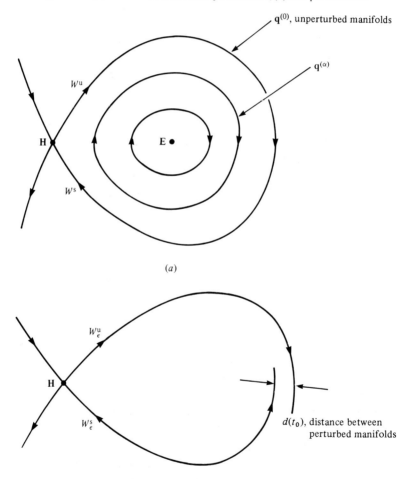

$\mathbf{q}^{(0)}$, unperturbed manifolds

$\mathbf{q}^{(\alpha)}$

W^u

H

$E \bullet$

W^s

(*a*)

W_ε^u

H

W_ε^s

$d(t_0)$, distance between perturbed manifolds

(*b*)

or

$$M(t_0) = \int_{-\infty}^{+\infty} \left\{ \mathbf{f}[\mathbf{q}^{(0)}(t-t_0)] \wedge \mathbf{g}[\mathbf{q}^{(0)}(t-t_0), t] \right\} dt$$

where ' $\wedge$ ' is the wedge product, defined by

$$\mathbf{f} \wedge \mathbf{g} = f_1 g_2 - f_2 g_1.$$

Therefore the manifolds cross when $M(t_0) = 0$. If the perturbation $\mathbf{g}$ is Hamiltonian,

$$g_1 = \partial G / \partial x_2, \quad g_2 = -\partial G / \partial x_1,$$

the Melnikov integral can be written as

$$M(t_0) = \int_{-\infty}^{+\infty} [H(\mathbf{q}^{(0)}(t-t_0)), G(\mathbf{q}^{(0)}(t-t_0), t)] dt$$

where [ ] denotes the Poisson bracket (Section 6.3).

The applications of this method are numerous. An example is discussed in Section 7.2; problems are suggested in Sections 7.3, 8.3, and 8.4 (for recent developments see Wiggins, 1988a,b).

The relation between the information provided by the Melnikov function and Poincaré sections is not trivial. A measure of the 'extent of chaos' is provided by the area A (Figure 6.8.1) which is the integral of the distance between manifolds. The dynamics of A governs the transport in the system. Such an approach was adopted by Leonard, Rom-Kedar, and Wiggins (1987) and extended by Rom-Kedar, Leonard, and Wiggins (1990).

6.10. Behavior near elliptic points

Consider an integrable twist mapping in the neighborhood of an elliptic point in the Poincaré section of a time-periodic Hamiltonian system, i.e.,

$$r_{n+1} = r_n \tag{6.10.1a}$$
$$\theta_{n+1} = \theta_n + 2\pi\sigma(r_n). \tag{6.10.1b}$$

(Defining the *winding number* or *rotation number* σ as $\omega_i = 2\pi\sigma$, see 6.3.4a,b). According to the value of r, $\sigma(r_n)$ can be a rational or irrational number. If $\sigma(r_n)$ is irrational the trajectories wrap densely (in the mathematical sense) around a torus, never intersecting themselves (see Figure 6.10.1). On the other hand, if $\sigma(r_n)$ is rational $(=m/n)$ the trajectory returns exactly to its initial position after m turns around the torus. In the Poincaré section (Figure 6.6.1(b)) the rational orbits (periodic) produce a finite number of intersections with a circle (according to the value of n). The intersections of the irrational orbits, also called quasi-periodic, fill the circle densely.

The central question is what happens to the twist map under a perturbation of strength μ such that

$$r_{n+1} = r_n + \mu f(r_n, \theta_n) \tag{6.10.2a}$$

$$\theta_{n+1} = \theta_n + 2\pi\sigma(r_n) + \mu g(r_n, \theta_n). \tag{6.10.2b}$$

The perturbations f and g are such that they are 2π-periodic, area preserving, and vanish faster than r as r tends to the origin.

Over the years several theorems have been discovered that produce a fairly complete picture of the events that occur after the perturbation. Three theorems will be described here and in chronological order: the Poincaré–Birkhoff theorem, the Kolmogorov–Arnold–Moser theorem, and Moser's twist theorem. The discussion is mostly qualitative.

6.10.1. Poincaré–Birkhoff theorem

This theorem describes the fate of the *rational* tori upon perturbations of the twist mapping. A necessary condition is that $\sigma' \neq 0$ (the prime denotes the derivative with respect to r). Consider the integrable twist mapping of Equations (6.10.2a,b). Select a radius r_0 such that $\sigma(r_0) = m/n$ and an initial angle θ_0. After n mappings we get

$$\theta_n = \theta_0 + n2\pi\sigma(r_0) = \theta_0 + n2\pi(m/n) \equiv \theta_0 \text{ (up to } 2\pi) \qquad \text{and} \qquad r_0 = r_n.$$

On the other hand, in the non-integrable case we get, in general,

$$\theta_n = \theta_0 + n[2\pi\sigma(r_0) + \mu h(r_0, \theta_0)] \neq \theta_0, \qquad \text{and} \qquad r_n \neq r_0$$

Figure 6.10.1. Integrable twist mapping.

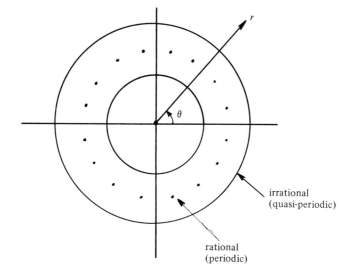

irrational
(quasi-periodic)

rational
(periodic)

where μ is a perturbation and $h(r, \theta_0)$ is some complicated but calculable function that can be obtained as a function of the perturbations f and g in the non-integrable mapping (Equation (6.10.2a,b)). In general the point θ_0, r_0 will return neither to its initial radial position nor to its initial angular position. However, if both the functions $h(r, \theta_0)$ and $\sigma(r)$ have a Taylor series expansion in r (condition $\sigma' \neq 0$) we can find a point θ_0, ρ_0 near the initial θ_0, r_0 such that

$$2\pi\sigma(\rho_0) + \mu h(\rho_0, \theta_0) = 2\pi m/n.$$

Thus, for this value of ρ_0 we do get $\theta_0 = \theta_n$, *but* $\rho_0 \neq \rho_n$. Since the same argument holds for any θ_0 we obtain a curve $\rho_0(\theta)$ of points that are mapped *radially*. Since this happens after exactly n mappings and the winding number is monotonic in the neighborhood of r_0, we can envision that the circle corresponding to r_0 does not rotate and that the neighboring orbits ($r > r_0$ and $r < r_0$) rotate in opposite directions, as shown in Figure 6.10.2 (see Helleman, 1980, p. 185). Also, since the map preserves area, not all the points belonging to the curve $\rho_0(\theta)$ can move inward or outward; some have to move inward and some have to move outward. Also since the mapping is continuous some points do not move at all (i.e., they cross the circle r_0 or in other words, they are fixed points) and due to periodicity and conservation of area the number of such points has to be even, say $2k$. Examination of the flow in the neighborhood of the fixed points shows that k points are hyperbolic and k are elliptic (see Figure 6.10.2). Thus the Poincaré–Birkhoff theorem (Birkhoff, 1935) tells us under an μ-perturbation the rational orbits break into a collection of k-hyperbolic

Figure 6.10.2. Sketch used in the description of the Poincaré–Birkhoff theorem.

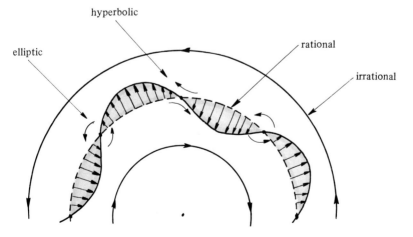

and k-elliptic points. Such a prediction is confirmed by numerous computational studies. Note that this theorem does not say anything about the irrational orbits.

6.10.2. The Kolmogorov–Arnold–Moser theorem (the KAM theorem)

The local picture of what happens to the *irrational* orbits in the neighborhood of an elliptic point is the context of the celebrated KAM theorem. It says basically that most of the invariant irrational tori of the integrable system are conserved in the perturbed system. This theorem was outlined by Kolmogorov (1954a,b) and proved by Arnold in a series of papers starting in 1961. A detailed exposition is given in a very long paper written in 1963 (Arnold, 1963). This paper gives probably the best account since it gives background, examples, a non-rigorous heuristic derivation, and concludes with the rigorous proofs. The results were generalized by Moser in 1962 and subsequently by Rüssman in 1970. As indicated by Arnold, the proof is rather cumbersome and it is based on a very large number of inequalities. The technical aspects of the proof go back to methods developed by astronomers in the nineteenth century and, as indicated by Kolmogorov, on earlier work by Birkhoff (1927) and a method of successive approximations developed by Kolmogorov himself. Keeping in line with the previous sections we will not attempt to reproduce the theorem in detail here but rather give an outline of some of the central ideas involved in the arguments.

Consider for simplicity an integrable Hamiltonian system with two-degrees of freedom i.e. the phase space is four dimensional. The Hamiltonian is of the form

$$H(\mathbf{p}, \mathbf{q}) = H^0(\mathbf{p})$$

with $\mathbf{p} = (p_1, p_2)$ and $\mathbf{q} = (q_1, q_2)$ [the results can be generalized to a system with N-degrees of freedom]. In this case the equations of motion take the form

$$dq_k/dt = \partial H^0(\mathbf{p})/\partial p_k, \qquad dp_k/dt = 0 \qquad \text{(with } k = 1, 2\text{)},$$

i.e., $\mathbf{p} = $ constant, and

$$q_k = (\partial H^0(\mathbf{p})/\partial p_k)t + q_k^0 \qquad \text{(with } k = 1, 2\text{)},$$

i.e., the four-dimensional phase space contains an invariant two-dimensional torus which is characterized by two frequencies $\omega_1 = \partial H^0(\mathbf{p})/\partial p_1$ and $\omega_2 = \partial H^0(\mathbf{p})/\partial p_2$ which depend on $\mathbf{p}$. As we have seen, if the frequencies are incommensurable, the trajectories never return to the initial position on the torus (the motion is called quasi-periodic or conditionally periodic). Let us suppose, for the moment, that indeed the frequencies ω_1 and ω_2

are incommensurable, i.e., it is not possible to find integers k_1, k_2 such that

$$k_1\omega_1 + k_2\omega_2 = \mathbf{k}\cdot\boldsymbol{\omega} = 0,$$

where $\mathbf{k} = (k_1, k_2)$ and $\boldsymbol{\omega} = (\omega_1, \omega_2)$. In order to avoid the possibility $\mathbf{k}\cdot\boldsymbol{\omega} = 0$ it is sufficient to require that no relationship of the form $f(\omega_1(\mathbf{p}), \omega_2(\mathbf{p})) = 0$ exists for arbitrary $\mathbf{p}$. Taking the differential of the function $f(\omega_1(\mathbf{p}), \omega_2(\mathbf{p}))$ it is easy to see that a *sufficient* condition for incommensurability is to require

$$\det(\partial\omega_i/\partial p_j) = \det(\partial^2 H^0(\mathbf{p})/\partial p_i p_j) \neq 0.$$

Also to avoid problems at the origin we require that $\partial\omega_i/\partial p_j \neq 0$ at $p_j = 0$. Consider now a perturbation of the form (near integrable system),

$$H(\mathbf{p}, \mathbf{q}, \mu) = H^0(\mathbf{p}) + \mu H^1(\mathbf{p}, \mathbf{q})$$

where H^0 and H^1 are analytical (Moser's contribution was to relax this assumption by requiring a finite number of derivatives).

If the above conditions are met the phase flow in the four-dimensional space is stable with respect to small changes in H. It can be proved that given any $\mu > 0$ there exists $\delta > 0$ such that if the strength of the perturbation is such that $\mu < \delta$, the phase space of the perturbed system consists entirely of invariant two-dimensional tori, *except for a set of measure less than* μ. Arnold's proof is constructive and requires an infinite number of steps. One of the major difficulties is how to quantify the influence of rationally related frequencies since terms of the form $\mathbf{k}\cdot\boldsymbol{\omega} = 0$ appear as denominators in series solutions of the equations of motion of the perturbed Hamiltonian system (this is the problem of the 'vanishing denominators'). However, the central question is how important they are to the overall picture. The answer turns out to be that they are not very important and that the previous claim holds; the original tori of the integrable system are not destroyed but merely displaced. The conditions for the abundance of terms $\mathbf{k}\cdot\boldsymbol{\omega} = 0$ can be quantified by means of theorems of Diophantine approximations. It turns out that for *almost all*[10] randomly selected frequencies $\boldsymbol{\omega} = (\omega_1, \ldots, \omega_n)$, with ω_i real, the components are incommensurable for all ks, $\mathbf{k} = (k_1, \ldots, k_n)$, with k_i integer. Almost every ω satisfies the inequality

$$|(\mathbf{k}\cdot\boldsymbol{\omega})| \geq K|k|^{-\upsilon} \tag{6.10.3}$$

where $|k| = |k_1| + \cdots + |k_n|$, $\upsilon = n + 1$, for all $\mathbf{k}$, where K is a function of ω (the quantity $K|k|^{-\upsilon}$ can be interpreted as a measure of the 'irrationality' of the frequencies ω). Using these arguments it is possible to estimate the measure of the denominators that vanish.

The conclusion is that when μ is sufficiently small, all tori corresponding to frequencies satisfying the inequality (6.10.3) do survive.[11] The great majority of the initial conditions in the four-dimensional space are in fact conditionally periodic. Note that in systems with two degrees of freedom the presence of closed invariant curves precludes ergodicity.[12]

The general picture obtained by means of uncountable computer simulations (see Chapters 7 and 8) is that in fact tori survive for *large* perturbations although they gradually disappear (and eventually reappear) until no closed curves survive near the original ones.

6.10.3. The twist theorem

The KAM theorem tells us that there are many invariant tori (indeed a set of *positive Lebesgue measure*) that *do not* disappear (if they satisfy a given set of conditions they do not disappear). The Poincaré–Birkhoff theorems tells us what happens to the rational orbits that disappear. The twist theorem (Moser, 1973) is, in a loose sense, the reverse of the KAM theorem.

Consider that $\sigma(r)$ is C^s $(s > 5)$,[13] $|\sigma'(r)| > 0$, on an annulus of radius r, $a < r < b$, and that the perturbation is C^s close (in the sense of a C^s norm. The C^s norm is defined as the maximum of the absolute values of the kth partial derivatives, $0 < k < s$). *Then:*

  (i) a torus survives in $a < r < b$
  (ii) the perturbation is a twist mapping on the perturbed torus
  (iii) the radius of the torus satisfies

$$(\sigma(r) - (m/n)) > Cn^{-2.5} \qquad (6.10.4)$$

for all integers m, n (C is a constant).

Note that the inequality (6.10.4) is a *consequence* in Moser's twist theorem and an *assumption* in the KAM theorem. In the KAM theorem we select a specific perturbed orbit that satisfies some conditions (e.g. the inequality (6.10.3)) and conclude that it survives. In the twist theorem we focus on an annulus where the perturbation is small and conclude that there is some torus which satisfies inequality (6.10.4).

6.11. General qualitative picture of near integrable chaotic Hamiltonian systems

We have discussed what happens to the general picture of an integrable Hamiltonian system (Figure 6.4.1) containing two hyperbolic points connected smoothly by their stable and unstable manifolds encircling one stable elliptic point. Under perturbations the stable and unstable manifolds intersect transversally and a complex picture appears near the hyperbolic points. We have also seen what happens under perturbations near the elliptic points. Some orbits disappear, the 'most rational' tori first, each giving rise to a string of elliptic and hyperbolic points with their own stable and unstable manifolds. These points produce in turn a picture similar to the general integrable system (Figure 6.4.1) and under perturbation a picture similar to Figure 6.8.1. Simultaneously, within the elliptic points the picture is as Figure 6.10.2 and repeats itself in the newly formed elliptic points. KAM curves break up into islands exhibiting the same structure at all length scales (self-similarity) suggesting the applicability of renormalization methods.[14] A qualitative picture of the phase space is shown in Figure 6.11.1. *We expect this picture to be characteristic of all near-integrable chaotic Hamiltonian systems.* Such a prediction is confirmed by numerous computational studies, regardless of the formal details of the models.

The Poincaré section is a display of the behavior of the system to all possible initial conditions. The picture represents phenomena with multiple time and length scales (e.g., islands corresponding to high periods take a long time to form in a computer experiment). It is important to recognize that not all the phenomena might be displayed in the time scales of interest (for example, in mixing we are interested in low period events).[15]

Figure 6.11.1. General picture of a near-integrable system.

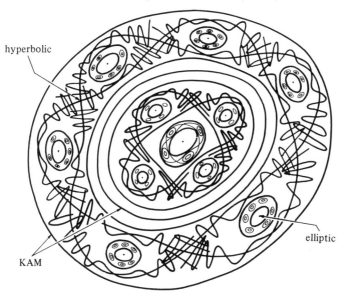

hyperbolic

elliptic

KAM

In this context it is interesting to anticipate the effect of such a phase portrait on a line (Berry *et al.*, 1979). Figure 6.11.2(*a*) shows part of the manifold structure of a Hamiltonian system. A line is initially placed horizontally passing by H_1, E, and H_2. After some time t the line is

Figure 6.11.2. Sketch of formation of tendrils and whorls; (*a*) shows part of the manifold structure and orbits near the elliptic point E. (*b*) shows the deformation of a line with an initial condition passing by H_1, H_2, and E (broken lines). In general, the stretching produced by tendrils is much more noticable than the whorl produced by the elliptic point.

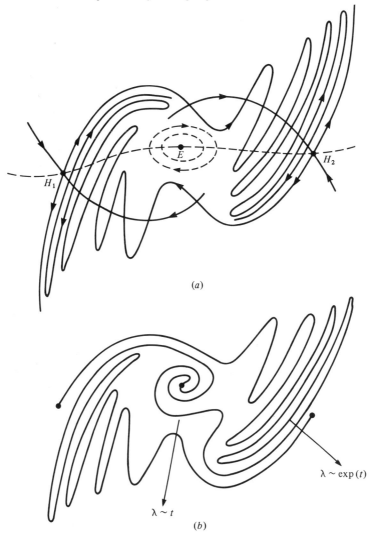

deformed as indicated in Figure 6.11.2(*b*) forming both folded and 'vortical' structures (named 'tendrils' and 'whorls' respectively by Berry *et al.*).[16] Note that if the 'experiment' is continued for a longer time other higher period points will deform the line as well (for example, the elliptic points, at a rate dependent upon both the winding numbers and the period of the point). These effects will be felt at smaller scales and we can anticipate whorls inside folded structures, whorls inside whorls, folds within folds, and so on. This action is reminiscent of a turbulent flow.

Problem 6.11.1
Elaborate on the differences between the stretching of a line in an actual turbulent flow and Figure 6.11.2(*b*).

Problem 6.11.2
Criticize the Figure P6.11.2 representing the evolution of a line in a Hamiltonian system after the breakup of KAM surfaces.

Figure P6.11.2.

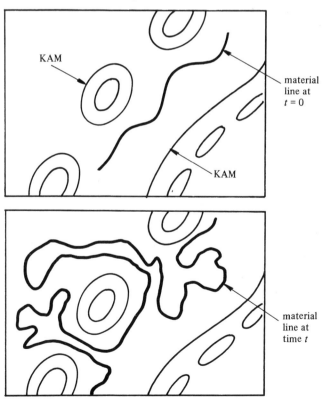

Bibliography

Much of the material of this chapter, especially in the first sections, makes reference to classical mechanics and theory of differential equations and dynamical systems. For an introduction to classical mechanics we suggest Percival and Richards (1982). An excellent but more advanced treatment is given by Arnold (1980). A modern encyclopedic treatment covering quantitative dynamics and the three-body problem is given by Abraham and Marsden (1985). Traditional treatments are given by Goldstein (1950) and by Synge (1960). For the general theory of dynamical systems we recommend the books by Hirsch and Smale (1974) and any of the books by Arnold (Arnold, 1983; 1985) and the article by Birkhoff (1920), which is a long but accessible account of the foundations of dynamical systems with two-degrees of freedom written in the language of classical mechanics.

Much of the material on chaos reviewed in this section can be found in the recent books by Lichtenberg and Lieberman (1983), Guckenheimer and Holmes (1983) and in the review article by Helleman (1980). The book by Lichtenberg and Lieberman (1983) addresses primarily Hamiltonian systems and is especially useful. Other general articles containing material on dynamical systems and chaos are: Moser (1973) which is the main reference for the twist theorem, and Arnold (1963) which is the main reference for the KAM theorem. This last paper is especially recommended in spite of its difficulty and length. A useful collection of classical papers addressing chaos in Hamiltonian systems is presented by MacKay and Meiss (1987). The Melnikov method is analyzed in detail by Wiggins (1988b).

Notes

1 Newton's second law for a particle of unit mass is:

$$d^2\mathbf{x}/dt^2 = \mathbf{F}$$

where $\mathbf{x}$ represents the position of the particle and $\mathbf{F}$ the sum of the forces acting on the particle. Using standard techniques (see Hirsch and Smale, 1974) the second order system can be transformed into a first order system by defining $q_i = x_i$ and $p_i = dx_i/dt$. If in addition we define the Hamiltonian $H(\mathbf{p}, \mathbf{q})$

$$H(\mathbf{p}, \mathbf{q}) = \tfrac{1}{2}|\mathbf{p}|^2 + V(\mathbf{q}),$$

which corresponds to total energy of a particle moving in a potential V, Newton's equations can be written as

$$dq_k/dt = \partial H/\partial p_k, \qquad dp_k/dt = -\partial H/\partial q_k,$$

which are known as Hamilton's equations (Goldstein, 1950). From a more general viewpoint, in classical mechanics it is shown that Hamilton's equations can be derived from Lagrange's equations which in turn are an extension of Newton's equations (see Goldstein, 1950, for a complete discussion). It should be noted that although there is no new physics involved, Hamiltonian theory often gives a much more powerful insight into the structure of problems in classical mechanics. Hamiltonian systems with the hamiltonian even if the momenta and position are called *reversible*, i.e., $t \to -t$ if $\mathbf{p} \to -\mathbf{p}$.

2 For example, in an oscillating circuit with no resistance but containing a saturated core, inductance can be described in terms of Hamilton's equations (see Minorsky, 1962, p. 62).

3 Note that in a steady two-dimensional isochoric velocity field the streamfunctions satisfies $D\psi/Dt = 0$.

4 A necessary condition is that the Jacobian $\partial(\mathbf{p}, \mathbf{q})/\partial(\mathbf{I}, \boldsymbol{\theta})$ be equal to one.

5 Since finding the action-angle variables (often referred to as normal coordinates) is a way of solving problems in Hamiltonian mechanics, much of the initial effort in this area was focused on finding such transformations, *assuming* their existence. The general possible ways to accomplish this task are the subject of the Hamilton–Jacobi theory (Goldstein, 1950, Chaps. 8 and 9). However, as it became clear in recent years it is almost always impossible to find the action-angle variables corresponding to a given Hamiltonian.

6 Note that the action-angle variable transformation does not work on the separatrix (see Figure 6.4.1). For example a hyperbolic system $dx/dt = x$, $dy/dt = -y$ cannot be transformed into a twist map.

7 According to Peixoto's theorem these phase portraits are structurally unstable in the class of *all* systems (for example if one allows a small amount of dissipation; i.e., volume contraction). Note, however, that this takes us out from the class of Hamiltonian systems.

8 This has to happen even in dissipative systems and is the content of the 'λ-lemma'. See Guckenheimer and Holmes, 1983, Section 5.2; Palis, 1969.

9 The 'λ-lemma' implies homoclinic tangles. The implications of infinitely many intersections of the stable and unstable manifolds in Hamiltonian systems were pointed out by Poincaré in 1899 (see Arnold, 1963, p. 178).

10 Meaning all, except for a set of Lebesgue measure zero.

11 Distinguish clearly the assumptions from the conclusions; compare with the twist theorem.

12 The effects of surviving KAM surfaces on transport in Hamiltonian systems are discussed by MacKay, Meiss, and Percival (1984). A tutorial presentation of these and other matters is given by Salam, Marsden, and Varaiya (1983).

13 The condition C^s ($s > 5$) is not sharp (see Moser, 1973). This theorem was proved by Moser in 1962 with $s \geqslant 333$ and later improved by Rüssman to $s \geqslant 5$; with additional conditions it can be reduced to $s \geqslant 3$.

14 A substantial review is given by Escande (1985). An application example is discussed by Greene (1986).

15 Also, it should be emphasized that according to the values of the parameters, the region of chaos might occupy a very small region of the phase space. In most Hamiltonian systems there is no sharp 'transition to chaos' and chaotic islands and bands surrounding islands of instability, sometimes too small to be detected by computational means, may exist for any non-zero value of the perturbation. Examples are given in Chapters 7 and 8. Sometimes the term 'transition to global chaos' is used to refer to the breakup of the last KAM torus.

16 Note that the whorls can be produced in flows without vorticity and even in flows without circulation (Jones and Aref, 1988).

Figure 1.3.4. Mixing of two immiscible polymers. The image-processed Fourier filtered transmission electron micrograph shows poly(butadiene) (black and light blue) domains in a poly(styrene) matrix (light grey). The black domains correspond to the domains not connected to the boundary. The volume fraction of poly(butadiene) is 0.312 and the average cluster size is 4.8 μm. (Reproduced with permission from Sax (1985).)

Figure 1.3.5. Mixing below the mixing transition in a mixing layer, visualization by laser induced fluorescence [the term 'mixing transition' refers to the onset of small scale three-dimensional motion]. The experimental conditions correspond to a speed ratio $U_2/U_1 = 0.45$, and a Reynolds number based on the local thickness of 1,750. A fluorescent dye is pre-mixed with the low speed free stream. Laser induced fluorescence allows the measurement of the local dye concentration, (a) single vortex, (b) pairing vortices. (Reproduced with permission from Koochesfahani and Dimotakis (1986).)

Figure 7.3.10. Manifolds corresponding to various values of μ, (a) $\mu = 0.3$, and (b) $\mu = 0.5$. (c) Magnified view of the stable and unstable manifolds of two period-1 points for $\mu = 0.5$ (see 7.3.10(a)).The unstable manifolds of the central point are shown in red, the stable manifold in yellow, the unstable manifolds of the outer point are shown in green, the stable manifold in light blue. The figure shows the results for 15 iterations. More iterations would make the picture hard to visualize; the number of points in the manifolds of the central point is 4,000, the number of points in the manifolds of the outer point is 2,000. (d) Magnified view of the manifolds near the transition to global chaos (the figure corresponds to $\mu = 0.38$). Note that the manifolds of the central and outer points intersect slightly. Compare with Figure 7.3.2. (Reproduced with permission from Khakhar, Rising, and Ottino (1987).)

Figure 7.4.4. Comparison of Poincaré sections and experiments ($\theta_{out} = 360°$): (a) Poincaré section corresponding to 8 initial conditions and 1,000 iterations; the symmetry of the problem produces a total of 16,000 points. (b) Experiments corresponding to the stretching of a blob initially located in the neighborhood of the period-1 hyperbolic point for 15 periods.

Figure 7.4.5. Similar comparison to that of Figure 7.4.4 except that $\theta_{out} = 180°$: (a) Poincaré section; colored points reveal that the initial conditions 'do not mix' even for a large number of periods, e.g., see magenta points; (b) corresponding experimental result for 10 periods (blob initially located in the neighborhood of the period-1 hyperbolic point), in this case, the blob does not invade the entire chaotic region.

Figure 7.4.6. Comparison of two Poincaré sections identical in all respects, except that the angular displacement θ_{out} of the outer cylinder in case (a) is 165° whereas in case (b) it is 166°. The value of θ_{in} is such that $\Omega_{in}/\Omega_{out} = -2$. The initial placement of colored points is the same in both cases.

Figure 7.4.7. (a) Location of periodic points, and (b) corresponding Poincaré section ($\theta_{out} = 180°$). The crosses represent hyperbolic points, the circles elliptic points. The period of the point is indicated by the color; green = 1, red = 2, blue = 3, orange = 4, yellow = 5, ... White results due to 'superposition' of nearby points of different order.

Figure 7.4.10. Poincaré sections corresponding to the angular histories of Figure 7.4.9: (*a*) square ($\theta_{out} = 180°$), (*b*) $\sin^2$, (*c*) sawtooth, and (*d*) $|\sin|$). The initial placement of colored points is the same in all cases. Note that whereas the large scale features are remarkably similar the distribution of colors reveals some local differences.

Figure 7.4.11. Stretching map corresponding to square history and 10 periods. The figures were constructed by placing two vectors per pixel and averaging over 10^2 initial orientations. In (*a*) $\theta_{out} = 180°$ and the white regions correspond to stretching of greater than 50; in (*b*) $\theta_{out} = 360°$ and the cut-off value is 5,000. Figure (*a*) should be compared with 7.4.8 and 7.4.5, (*b*) with 7.4.4(*a,b*).

Figure 7.5.7. An experiment similar to that of Figure 7.5.5 showing an island at the point of bifurcation ($T = 48.2$ s, displacement 970 cm, $Re = 1.2$, $Sr = 0.08$). This structure is characteristic of a golden mean rotation speed.

Figure 7.5.8. Partial structure of periodic points corresponding to the system of Figure 7.5.2(*d*), A = hyperbolic period-1; B = elliptic period-2; C = hyperbolic period-2; D = elliptic period-4; and E represents a hole of period-1. Note that the circles represent elliptic points and the squares hyperbolic points. The full white lines connect two elliptic points to a central hyperbolic point; all three points move as a unit. The broken lines connect two period-2 hyperbolic points to a period-1 hyperbolic point; the two period-2 hyperbolic points were born from the period-1 hyperbolic point. The two period-2 hyperbolic points interchange their positions after one period.

Figure 7.5.9. Illustration of reversibility in regular and chaotic regions. The mixing protocol corresponds to Equation (7.5.1a,b) with a period of 30 s. (*a*) is the initial condition at which the vertical line is placed in the chaotic region, while the oblique line is placed in the regular region. (*b*) shows the lines after two periods. Note that the stretching and bending of the line placed in the regular region is very small, and that the line is merely translated. On the other hand, the line placed in the chaotic regions suffers significant stretching. (*c*) shows the state of the system after being reversed for two periods. Clearly, the line in the regular region managed to return to its initial location successfully, however, the line in the chaotic region loses its identity due to the magnification of errors in the experimental set-up.

Fig. 1.3.4

Fig. 1.3.5

(a)   (b)

Fig. 7.3.10

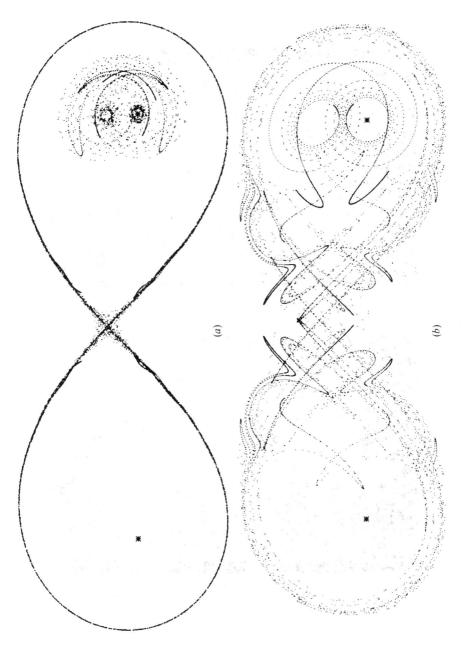

(a)

(b)

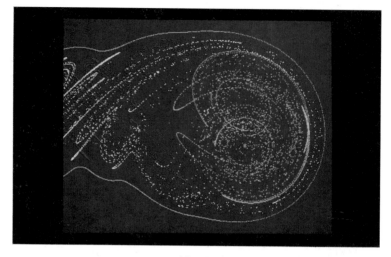

(c)

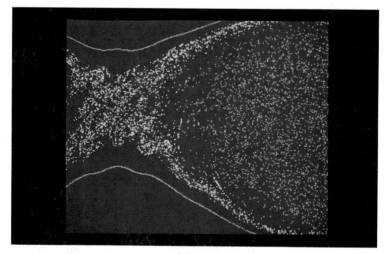

(d)

Fig. 7.4.4

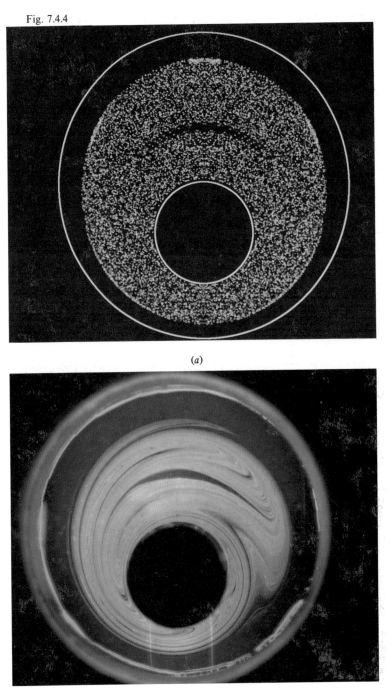

(a)

(b)

Fig. 7.4.5

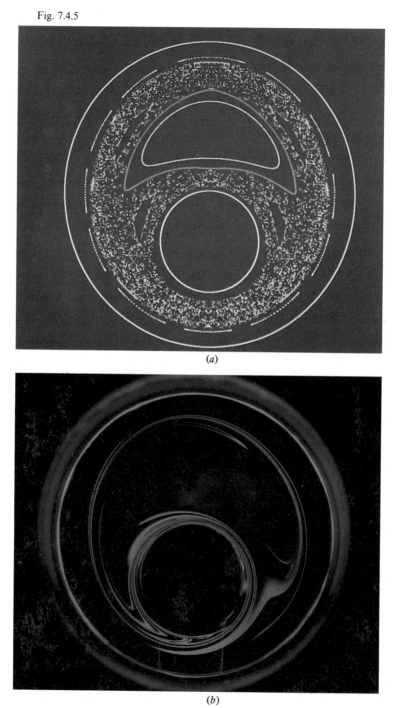

(a)

(b)

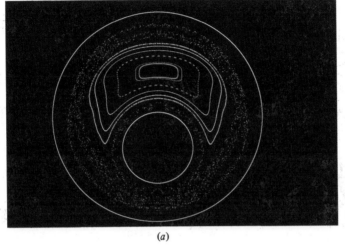

(a)

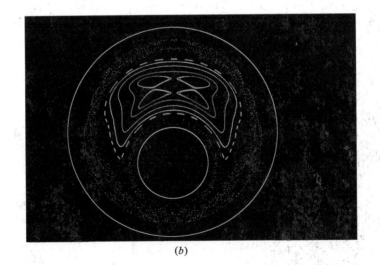

(b)

Fig. 7.4.7

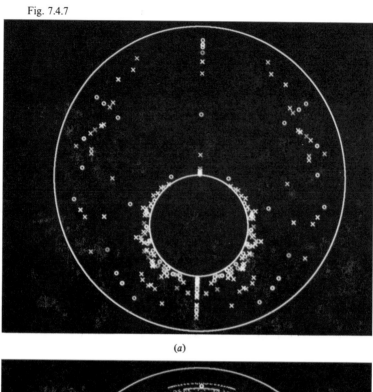

(a)

(b)

Fig. 7.4.10

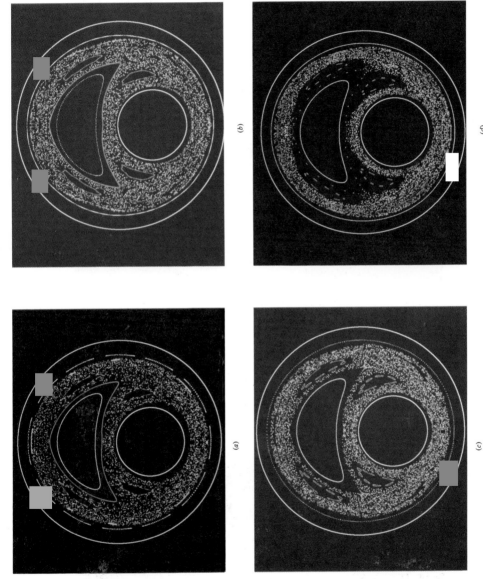

(b)

(d)

(a)

(c)

Fig. 7.4.11

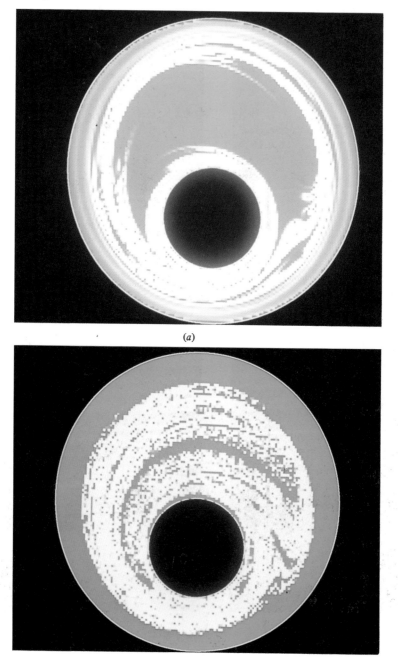

(a)

(b)

Fig. 7.5.7

Fig. 7.5.8

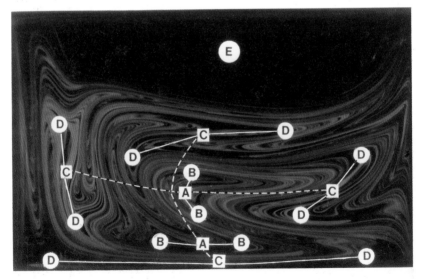

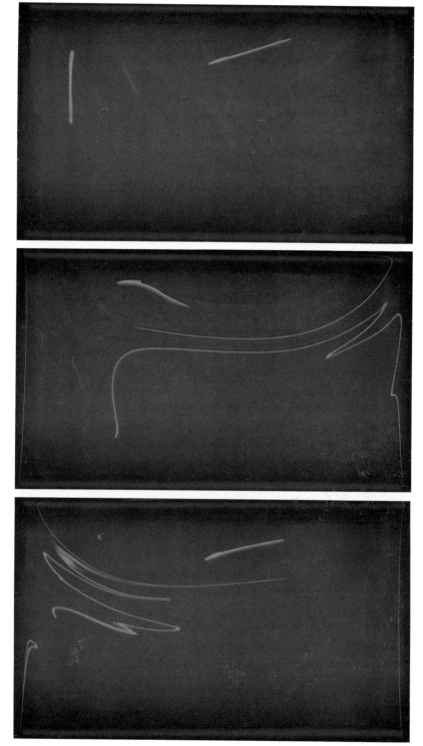

7

Mixing and chaos in two-dimensional time-periodic flows

In this chapter we study fluid mixing in several two-dimensional time-periodic flows. The first two flows – the tendril–whorl flow and the blinking vortex flow – are somewhat idealized and admit analytical treatment; the last two examples – the journal bearing flow and the cavity flow – can be simulated experimentally.

7.1. Introduction

This chapter consists of two parts. In the first part (Sections 7.2 and 7.3) we study in some detail two idealized periodic mappings, which can be regarded, in an approximate sense, as building blocks for complex velocity fields: (1) the tendril–whorl mapping (TW), and (2) the blinking vortex flow (BV).[1] Both flows can be regarded as prototypical local flow representations of more complex flows and they are simple enough so as to admit some detailed analytical treatment. The analysis concentrates on: (i) the structure of periodic points and their local bifurcations, and (ii) the global bifurcations, or interactions among the stable and unstable manifolds belonging to hyperbolic points. It is found that these simple systems possess a rich and complex mixing behavior. Both mappings are chaotic according to *mathematically accepted* definitions. One of the lessons to be extracted from the complexities encountered in the analysis of these mappings is that there are practical limits to the level of probing and understanding of the details of mixing in complex systems.

In the second part (Sections 7.4 and 7.5) we focus on two flows which do not admit such a detailed treatment: (3) a Stokes's time-modulated flow in a journal bearing, where we focus on the comparison between experiments and computations, as well as the need for various computational and analytical tools; and (4) several kinds of low Reynolds number cavity flows – in this case we compare the performance of steady flows with that of periodic flows and study the evolution and bifurcations of coherent structures. Undoubtedly, many other examples and variations will occur to the reader.

It should be clear that even though the examples presented here are governed by the theory of Chapters 5 and 6, our problem is not simply one of adaptation and much work remains. Possibly the most important difference between conventional dynamical systems studies and the work presented here is that we are interested in the *rate* at which phenomena occur. Therefore attention is focused on low period phenomena since we want to mix both efficiently and quickly. There is also the matter of the area occupied by chaotic and 'non-chaotic' or regular regions. Presently, there are no predictions of the size of the mixed and unmixed regions, no thorough studies regarding the stability of the results with respect to geometrical changes, and no indepth studies of the relationship of the flow to the morphology produced by the mixing.[2]

An important question to keep in mind throughout this chapter is: What constitutes a complete analysis or understanding of a mixing flow? In the first two examples we follow a program such as the one outlined in Table 7.1. A loose definition as to what constitutes *complete understanding* is the following: A system can be regarded as 'completely understood', from a practical viewpoint, when the answer to the $n + 1$ question in a program, such as the one sketched in the table, can be qualitatively predicted from the previous n answers. In the first two examples, since we have an explicit expression for the mapping, we can accomplish a substantial portion of the program. In the case of the journal bearing flow we have an exact solution to the Navier–Stokes equations, but the Eulerian velocity is not nearly as convenient as having the mapping itself. However, there are number of avenues left for analysis and we exploit many of these. In the case of the cavity flow we do not have a solution for the velocity field and the comparison with computations and analysis has to be of a less direct nature.

7.2. The tendril–whorl flow

Flows in two dimensions increase length by forming two basic kinds of structures: *tendrils* and *whorls* (see Figure 7.2.1)[3] and their combinations. In complex two-dimensional fluid flows we can encounter tendrils within tendrils, whorls within whorls, and all other possible combinations. The mapping we study in this section seems to be the simplest one capable of displaying this kind of behavior. The *tendril–whorl flow* (TW) introduced by Khakhar, Rising, and Ottino (1987) is a discontinuous succession of extensional flows and twist maps. In the simplest case all the flows are identical and the period of alternation is also constant. As we shall see

Table 7.1

	Analytical	Computational	Experimental	Comments
Study of symmetries	Entirely analytical	Convenient to produce symmetric Poincaré sections	Can be used to verify experimental accuracy	Can be exploited to reduce computational time (e.g., search of periodic points)
Location of periodic points Order one, or fixed points, Order two ... Order n	Only in very simple cases	Order one given by conjugate lines. Higher order obtained by minimization/symmetry methods		Lowest order points most important in rapid mixing
Local analysis Stability of periodic points. Bifurcation as parameters are changed	Possible in simple cases	Possible (often only method to obtain eigenvalues)		Regular islands surrounding elliptic points (obstruction to mixing)
Poincaré sections	Only in very simple cases	Relatively easy		Long time behavior of the system for all initial conditions
Global analysis Stable and unstable manifolds associated with hyperbolic points (order one, two, n).	Only possible in very simple cases	Can be done by placing a blob encircling the point	Hard, purely accidental	Speed along manifolds is proportional to magnitude of eigenvalues/period. Degree of manifold overlapping controls overall rate of dispersion.
Interactions between manifolds Melnikov method	Possible in simple cases	Relatively easy		Generally valid for small perturbations only; convenient if analytical expression for homoclinic/heteroclinic trajectory is available
Liapunov exponents		Relatively easy		
Stretching of material lines and blobs		Hard for large stretchings	Relatively easy	
Formations of horseshoes	Possible in simple cases	Resolution might be a problem but possible	Possible, especially in systems with suitable symmetries	
Pathlines Streamlines Streaklines	Impossible in chaotic flows	Streaklines hardest to obtain		Relatively little information about mixing (except streaklines)
Maps of constant properties Stretching Efficiency Striation thickness, etc.		Best route	Possible via image analysis	Useful to make indirect connections

the simplest case is complex enough and can be considered as the point of departure for several generalizations (smooth variation, distribution of time periods, etc.).

The physical motivation for this flow is that, locally, a velocity field can be decomposed into extension and rotation (see Section 2.8). Alternatively, according to the polar decomposition theorem, a local deformation can be decomposed into stretching and rotation (see Section 2.7).

In the simplest case the velocity field over a single period is given by

$$v_x = -\varepsilon x, \qquad v_y = \varepsilon y, \qquad \text{for } 0 < t < T_{\text{ext}}, \text{ extensional part} \qquad (7.2.1)$$

$$v_r = 0, \qquad v_\theta = -\omega(r) \qquad \text{for } T_{\text{ext}} < t < T_{\text{ext}} + T_{\text{rot}}, \text{ rotational part} \qquad (7.2.2)$$

where T_{ext} denotes the duration of the extensional component and T_{rot} the duration of the rotational component. Thus, the mapping consists of vortices producing whorls which are periodically squeezed by the hyperbolic flow leading to the formation of tendrils, and so on. The function $\omega(r)$ is a positive quantity that specifies the rate of rotation. Its form is fairly arbitrary and the results are, in a qualitative sense, fairly independent of this choice. The most important aspect is that $\omega(r)$ has a maximum, i.e. $d(\omega(r))/dr = 0$ for some r. Nevertheless, independently of the form of $\omega(r)$, we can integrate the velocity fields (7.2.1) and (7.2.2) over one period to give

$$\mathbf{f}_{\text{ext}}(x, y) = (x/\alpha, \alpha y), \qquad \mathbf{f}_{\text{rot}}(r, \theta) = (r, \theta + \Delta\theta),$$

where $\alpha = \exp(\varepsilon T_{\text{ext}})$ and $\Delta\theta = -\omega(r)T_{\text{rot}}/r$, i.e., the point (x, y) goes to $(x/\alpha, \alpha y)$ due to the extensional part, the point (r, θ), which corresponds to $(x/\alpha, \alpha y)$, goes to $(r, \theta + \Delta\theta)$ due to the rotational part, and so on.

Figure 7.2.1. Basic structures produced by mixing in two-dimensional flows; (a) whorl, (b) tendril.

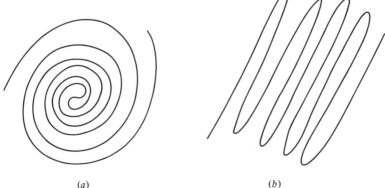

(a)   (b)

We consider a rotation given by

$$\Delta\theta = -Br\exp(-r).$$

We will investigate only a few aspects of this flow. A more thorough analysis is given by Khakhar, Rising, and Ottino (1987), and especially in the Ph.D. thesis of Khakhar (1986). As we shall see, due to the simplicity of the TW map, we can calculate, analytically, up to periodic points of order 2. A similar analysis can be carried out, in theory, for periodic points of order n. The analysis is however, extremely difficult.

7.2.1. Local analysis: location and stability of period-1 and period-2 periodic points

The qualitative idea of period-1 (or fixed) and period-2 periodic points is indicated in Figure 7.2.2. The composition of the extensional and rotational maps produces, in polar coordinates,

$$\mathbf{f}(r, \theta) = (r', \tan^{-1}(\alpha^2 \tan \theta) + \Delta\theta(r'))$$

where $\mathbf{f}$ denotes the composition of (7.2.1) and (7.2.2), $\mathbf{f} = \mathbf{f}_{\text{ext}} \cdot \mathbf{f}_{\text{rot}}$, and where r' is given by

$$r' = r\left[\left(\frac{1}{\alpha^2}\right)\cos^2\theta + \alpha^2\sin^2\theta\right]^{1/2}$$

At the period-1 periodic point (see Figure 7.2.2) we have

$$\mathbf{f}(r, \theta) = (r, \theta).$$

Figure 7.2.2. Pictorial representation of period-1 and period-2 points.

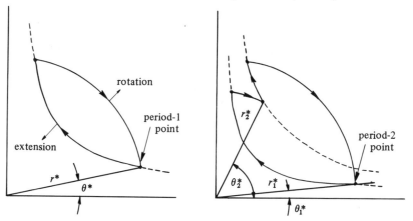

Let us denote the periodic point as $\mathbf{p}^* = (r^*, \theta^*)$. Since r stays constant we require

$$\left(\frac{1}{\alpha^2}\right) \cos^2 \theta^* + \alpha^2 \sin^2 \theta^* = 1,$$

which implies

$$\theta^* = \tan^{-1}(1/\alpha).$$

Also, since the initial and final angles coincide (up to an integer number of rotations, $2\pi n$),

$$\theta^* = \underset{\text{extension}}{\tan^{-1}(\alpha^2 \tan \theta^*)} + \underset{\text{rotation}}{\Delta\theta(r^*)} + 2\pi n.$$

Simplifying, we obtain

$$Br^* \exp(-r^*) = \tan^{-1}[(\alpha - 1/\alpha)/2] + 2\pi n \qquad n = 0, 1, 2, \ldots, M$$

so that depending on the values of α and M we obtain a number of different solutions. We consider for convenience only the case of $M = 0$, and make B dimensionless by defining

$$\beta = \frac{Be}{\tan^{-1}[(\alpha - 1/\alpha)/2]}$$

so that the radius and angular position of the period-1 periodic points are given by the equations:

$$\theta^* = \tan^{-1}(1/\alpha)$$
$$r^* \exp(1 - r^*) = 1/\beta$$

Thus, the TW mapping is characterized by two parameters: α and β. We notice that the angular position depends only on α, the radial position only on β. It is also easy to see that the origin is a fixed point for all parameter values. For $M = 0$ (and $\beta < 1 + 4\pi$) there can be 0 ($\beta < 1$), 1 ($\beta = 0, 1$) or 2 ($\beta > 1$) periodic points. A graphical interpretation of the equations is given in Figure 7.2.3. If $M > 0$ additional periodic points might appear.

The local behavior of the period-1 fixed points is given by the eigenvalues of the linearized mapping evaluated at the point $\mathbf{p}^* = (r^*, \theta^*)$,

$$\lambda_{1,2} = \tfrac{1}{2} \operatorname{tr}(\mathbf{Df}) \pm \{[\tfrac{1}{2} \operatorname{tr}(\mathbf{Df})]^2 - 1\}^{1/2}.$$

The mapping is

$$\mathbf{f}(x, y) = ((x/\alpha) \cos \Delta\theta - \alpha y \sin \Delta\theta, (x/\alpha) \sin \Delta\theta + \alpha y \cos \Delta\theta)$$

where

$$\Delta\theta = \Delta\theta(r) \qquad \text{and} \qquad r = (x^2/\alpha^2 + \alpha^2 y^2)^{1/2}$$

and

$$\mathbf{Df} = \begin{bmatrix} \dfrac{\partial}{\partial x}\left[\left(\dfrac{x}{\alpha}\right)\cos\Delta\theta - \alpha y\sin\Delta\theta\right] & \dfrac{\partial}{\partial y}\left[\left(\dfrac{x}{\alpha}\right)\cos\Delta\theta - \alpha y\sin\Delta\theta\right] \\[3mm] \dfrac{\partial}{\partial x}\left[\left(\dfrac{x}{\alpha}\right)\sin\Delta\theta + \alpha y\cos\Delta\theta\right] & \dfrac{\partial}{\partial y}\left[\left(\dfrac{x}{\alpha}\right)\sin\Delta\theta + \alpha y\cos\Delta\theta\right] \end{bmatrix}$$

After some manipulations we get

$$\text{tr}[\mathbf{Df(p^*)}] = 2 + (\alpha - 1/\alpha)r^*\left(\frac{d\Delta\theta}{dr}\right)_{r^*}.$$

This expression is valid for all choices of $\Delta\theta(r)$. For our choice, we obtain

$$\text{tr}[\mathbf{Df(p^*)}] = 2 + G \tag{7.2.3}$$

where G is defined as

$$G \equiv (\alpha - 1/\alpha)\tan^{-1}\{(\alpha - 1/\alpha)/2\}(r^* - 1)$$
$$G \equiv g(\alpha)(r^* - 1).$$

Thus, the character of the eigenvalues depends on the value of G. We have:

$$\begin{aligned} G > 0 &\quad\rightarrow \text{hyperbolic,} \\ G = 0 &\quad\rightarrow \text{parabolic,} \\ 0 > G > -4 &\quad\rightarrow \text{elliptic,} \\ G = -4 &\quad\rightarrow \text{parabolic,} \\ -4 > G &\quad\rightarrow \text{hyperbolic.} \end{aligned}$$

Based on the above analysis we can now study the sequence of (local) bifurcations that take place as β is increased from zero for a fixed value

Figure 7.2.3. Graphical interpretation of the equation for location of period-1 points.

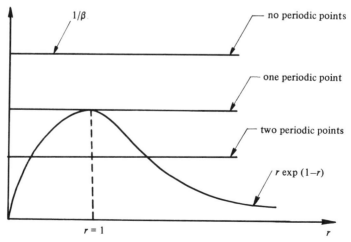

of α (recall that the position of r^* depends on β). This is explained graphically in Figure 7.2.4. If β exceeds $1 + 4\pi$ the situation is like Figure 7.2.5.

At $\beta = 1$ two period-1 points are formed at $r^* = 1$ in the first and third quadrant as shown earlier. Both points are parabolic since $G = 0$ in this case. When β becomes greater than 1, each fixed point splits into two;

Figure 7.2.4. Graphical representation of bifurcation sequences; the periodic points are located at an angle $\theta = \tan^{-1}(1/\alpha)$; the line is mapped by the elongational flow and twisted clockwise. The intersection(s) correspond to the periodic points.

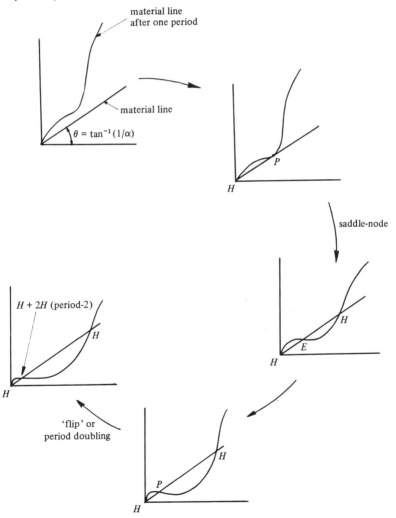

the two at radial distance $r^* > 1$ are hyperbolic, while those at $r^* < 1$ are initially elliptic. Thus, at $\beta = 1$, $r^* = 1$ there is a 'saddle-node' bifurcation (Guckenheimer and Holmes, 1983, p. 156).[4] As β is increased further, the inner fixed point, which is initially elliptic, moves closer to the origin and G becomes more negative. If $g(\alpha) > 4$, for β large enough, $G = -4$ and a second bifurcation takes place in which the elliptic point becomes parabolic and then hyperbolic. In addition there are two additional period-2 elliptic points. This bifurcation is called 'flip' or 'period-doubling'.

Assuming that α and β are large enough to have period-2 fixed points we have

$$\mathbf{f}^2(\mathbf{p}^*) = \mathbf{p}^* \qquad (7.2.4)$$

i.e., the following relations hold:

$$\mathbf{f}(r_1^*, \theta_1^*) = (r_2^*, \theta_2^*)$$
$$\mathbf{f}(r_2^*, \theta_2^*) = (r_1^*, \theta_1^*)$$

where (r_1^*, θ_1^*) and (r_2^*, θ_2^*) are the two period-2 points formed near r^*, θ^*

Figure 7.2.5. Creation of additional periodic points; the whorl formed by rotation intersects the line $\theta = \tan^{-1}(1/\alpha)$ more than once in the quadrant $x > 0$, $y > 0$ if the rotational strength β is greater than $1 + 4\pi$.

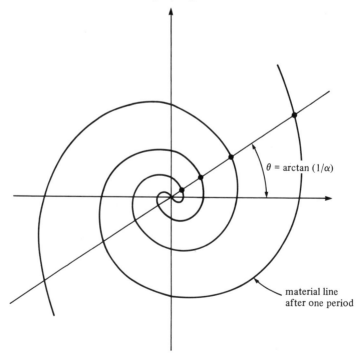

$\theta = \arctan(1/\alpha)$

material line
after one period

(by symmetry we can easily obtain the results, r^*, $\theta^* + \pi$). On substituting in the mapping, we obtain:

$$r_2^* = r_1^*(\cos^2 \theta_1^*/\alpha^2 + \alpha^2 \sin^2 \theta_1^*)^{1/2}$$
$$r_1^* = r_2^*(\cos^2 \theta_2^*/\alpha^2 + \alpha^2 \sin^2 \theta_2^*)^{1/2}$$
$$\theta_2^* = \tan^{-1}(\alpha^2 \tan \theta_1^*) + \Delta\theta(r_2^*)$$
$$\theta_1^* = \tan^{-1}(\alpha^2 \tan \theta_2^*) + \Delta\theta(r_1^*)$$

which can be simplified to give:

$$\tan \theta_1^* \tan \theta_2^* = 1/\alpha^2$$
$$\Delta\theta_1^* = -(\pi/2 - 2\theta_1^*)$$
$$\Delta\theta_2^* = -(\pi/2 - 2\theta_2^*)$$
$$r_2^* = \gamma r_1^*$$

where

$$\Delta\theta_i^* = -Br_i \exp(-r_i), \qquad i = 1, 2,$$
$$\gamma = \left(\frac{\alpha_1(1 + \alpha_2^2)}{\alpha_2(1 + \alpha_1^2)}\right)^{1/2} \qquad (7.2.5)$$

and $\alpha_1 = 1/\tan \theta_1^*$, $\alpha_2 = 1/\tan \theta_2^*$.

The stability of the period-2 periodic points depends on the eigenvalues of the Jacobian of $f^2(\cdot)$. For period-2 periodic points we have

$$\mathbf{x}_1 \to \mathbf{x}_2 \to \mathbf{x}_1,$$
$$\mathbf{x}_1 = \mathbf{f}^2(\mathbf{x}_1) = \mathbf{f}(\mathbf{x}_2)$$

and the Jacobian is calculated as

$$D\mathbf{f}^2(\mathbf{x}_1) = D\mathbf{f}(\mathbf{x}_2)D\mathbf{f}(\mathbf{x}_1)$$

or $\qquad\qquad\qquad\qquad\qquad\qquad\qquad\qquad\qquad (7.2.6)$

$$D\mathbf{f}^2(\mathbf{x}_2) = D\mathbf{f}(\mathbf{x}_1)D\mathbf{f}(\mathbf{x}_2).$$

After considerable simplification, we obtain:

$$\text{tr}[D\mathbf{f}(\mathbf{x}_1)D\mathbf{f}(\mathbf{x}_2)] = 2 + G_1 G_2 + 2(G_1 + G_2) \equiv 2 + G' \qquad (7.2.7)$$

where

$$G_1 = \frac{\Delta\theta_1^*(\alpha^4 - 1)(1 - r_1^*)}{\alpha_1(\alpha_2^2 + 1)}$$
$$G_2 = \frac{\Delta\theta_2^*(\alpha^4 - 1)(1 - r_2^*)}{\alpha_2(\alpha_1^2 + 1)}.$$

At the birth of the period-2 fixed points $r_1^* = r_2^*$, $\alpha_1 = \alpha_2 = \alpha$ so that $G_1 = G_2 = G$, and

$$G' = G^2 + 4G.$$

At the flip bifurcation, $G = -4$ which implies that $G' = 0$ so that the period-2 points are parabolic at their point of formation. For slightly

Figure 7.2.6. (*a*) Rotational strength for transition of the periodic point from elliptic to hyperbolic, β_{e-h}, as a function of the strength of the extensional flow, α, for period-1 and period-2 points. (Reproduced with permission from Khakhar, Rising, and Ottino (1986).) (*b*) Magnification of region in (*a*). The broken line corresponds to $\alpha = 10$.

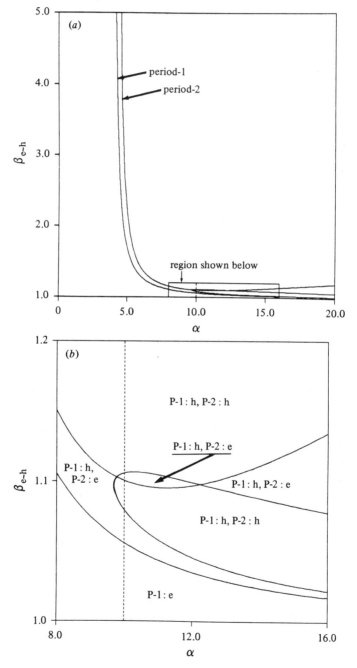

larger values of β we find that $G' < 0$ and the period-2 points become elliptic. As β is increased further, depending on the value of α, a second flip may occur in which the period-2 points become hyperbolic and two period-4 elliptic points are formed for each period-2 point. Let us now consider a few aspects of the problem which are discussed in more detail in Khakhar, Rising, and Ottino (1986). Some aspects of the bifurcation behavior are summarized in Figure 7.2.6, where we have plotted flip bifurcation values (i.e., the value of β at which the elliptic point becomes hyperbolic) versus α for the period-1 and period-2 elliptic points. Figure 7.2.6(b) shows a magnified region with the broken line corresponding to $\alpha = 10$ (as is apparent from this figure the period-1 elliptic points *do not* exhibit a period doubling cascade of bifurcations). A similar study, much harder from an algebraic viewpoint, can be carried out for period-3 periodic points, etc.

7.2.2. Global analysis and interactions between manifolds

Here we consider some of the global bifurcations of the TW mapping by studying the interactions of the manifolds of the hyperbolic period-1 periodic points. Obviously, a similar study can be carried out for period-2 periodic points, etc.

From the point of view of mixing, the stable and unstable manifolds provide us with broad features of the flow and its ability to mix. As indicated in Chapter 5 we obtain the locus of the manifolds by surrounding the hyperbolic points by a circle of small radius made up of a large number of points and convecting them by the flow. Forward mappings give the unstable manifolds, backward mappings the stable manifolds (actually, the result is a thin filament to the resolution of the graphics output device which encases the manifolds). Obviously, in the case of homoclinic or heteroclinic behavior larger and larger portions of the manifold become apparent with the number of iterations. However, given the limit of resolution, and the fact that we can work only with a finite number of points, a very large number of iterations is extremely hard to interpret. The manifolds will not appear as continuous lines, which they are, but as a series of disconnected dots.[5]

Far away from the origin ($r \to \infty$) the twist mapping acts as a solid body rotation. The interesting behavior occurs for $r < O(1)$; the chaotic behavior is confined to a well defined region shown schematically in Figure 7.2.7. The flow drags outside material which enters via conduits around the stable manifolds and leaves via conduits formed around the unstable manifolds. For $\beta < 1$ the mixing zone disappears and the flow is regular (mostly) everywhere.[6]

Figure 7.2.8 shows the manifolds of the hyperbolic period-1 periodic points at $r^* > 1$ ($\mathbf{P}_1$ and $\mathbf{P}'_1$ in Figure 7.2.7) and the origin, $\mathbf{O}$, for $\alpha = 1.5$ and two different values of β. The manifolds appear to join smoothly[7] and the system has the classical appearance of an integrable Hamiltonian system. At $\beta = 1.075$ the stable and unstable manifolds of $\mathbf{P}_1$ join smoothly with each other (similarly for $\mathbf{P}'_1$) whereas the manifolds of $\mathbf{O}$ fly-off to infinity. However, when $\beta = 1.099$ the situation is somehow reversed: the stable and unstable manifolds of $\mathbf{O}$ join each other and are confined by the region of the stable and unstable manifolds of $\mathbf{P}_1$ and $\mathbf{P}'_1$, which also join smoothly with each other.

Figure 7.2.9 shows the system at $\alpha = 5.000$ and increasing values of β. In Figure 7.2.9(*a*) the mixing zone consists of two compartments which are isolated from each other. We expect that KAM surfaces surround the elliptic points in the center of the islands. The picture changes in Figure 7.2.9(*b*), in which good mixing is expected to take place, due to the presence

Figure 7.2.7. Schematic view of mixing zone. $\mathbf{P}'_1$ and $\mathbf{P}_1$ are the outer period-1 points, $\mathbf{P}'_2$ and $\mathbf{P}_2$ are the inner period-1 points. The large arrows show the direction of transport along the stable and unstable manifolds.

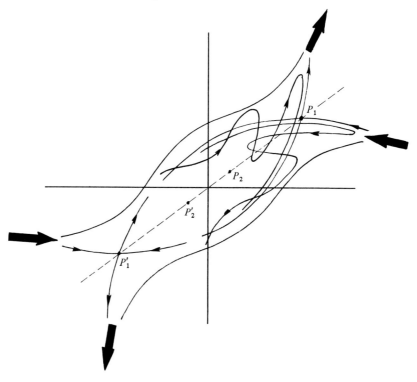

Figure 7.2.8. Manifolds associated with the central hyperbolic fixed point and period-1 points for various values of α and β. (a) $\alpha = 1.500$ and $\beta = 1.075$, (b) $\alpha = 1.500$ and $\beta = 1.099$. Note the change between figures (a) and (b; in (a) $\mathbf{P}_1$ forms a homoclinic connection; in (b) the central point forms a homoclinic connection. (Reproduced with permission from Khakhar, Rising, and Ottino (1986).)

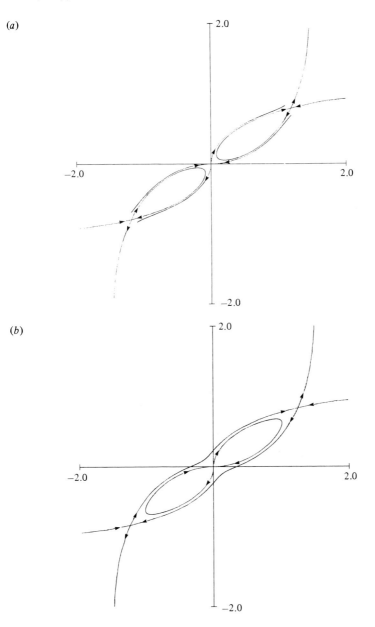

Figure 7.2.9. Same as Figure 7.2.8, but at larger values of α. In (a) $\alpha = 5.000$ and $\beta = 1.082$, (b) $\alpha = 5.000$ and $\beta = 1.595$. Elliptic islands are expected surrounding $\mathbf{P}_2'$ and $\mathbf{P}_2$ in case (a). In (b) the islands have been reduced in size (note change of scale). (Reproduced with permission from Khakhar, Rising, and Ottino (1986).)

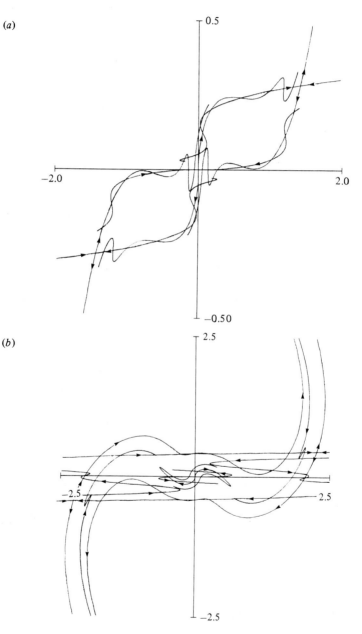

of homoclinic and heteroclinic points (the manifolds have been only partially drawn to make the picture understandable).

In summary, from the above analysis (which is not complete since the analysis pertains to only period-1 periodic points, and even more, those restricted to $M = 0$) we anticipate that if the extensional flow is weak (small α) the mixing will be poor for all β (strength of the rotation). As α is increased and for a large enough β there are intersections of the manifolds belonging to the outer periodic points and those of the origin resulting in a single mixing zone. However, based on the results of Chapter 6, we know that the mixing zone is not truly homogeneous, and is expected to contain islands around each elliptic point. Of these the largest are those corresponding to the period-1 elliptic points. The size of the islands decreases with increasing α and β.

7.2.3. *Formation of horseshoe maps in the TW map*

The above discussion indicates that the TW system is capable of homoclinic/heteroclinic behavior. We know that a homoclinic intersection implies the existence of horseshoe maps. It is possible to prove that the TW system is capable of producing horseshoe maps of period-1. An analytical proof of existence of horseshoes constitutes a *proof* of chaos. We can show the existence of period-1 horseshoes by the construction sketched in Figure 7.2.10 (Rising and Ottino, 1985). We select a rectangle with sides in a ratio α:1 and with two corners being periodic points of period-1. By selecting suitable values of β, and mapping forward, i.e., first the extensional flow with y being the stretching axis and then the twist mapping acting clockwise, we can produce vertical stripes. Then, by the inverse mapping, i.e., first the twist mapping counter-clockwise and then the extensional flow with x being the axis of extension, we can produce horizontal stripes. Actual examples are given by Khakhar, Rising, and Ottino (1986).

Problem 7.2.1
Show that $\mathbf{f}_{ext}(x, y) = (x/\alpha, \alpha y)$ in rectangular coordinates, and
$$\mathbf{f}_{ext}(r, \theta) = (r[(\cos^2 \theta/\alpha^2) + \alpha^2 \sin^2 \theta]^{1/2}, \tan^{-1}(\alpha^2 \tan \theta))$$
in polar coordinates. Show that $\mathbf{f}_{rot}(r, \theta) \to (r, \theta + \Delta\theta)$ in polar coordinates, and
$$\mathbf{f}_{rot}(r, \theta) = (x \cos \Delta\theta - y \sin \Delta\theta, x \sin \Delta\theta + y \cos \Delta\theta)$$
in rectangular coordinates.

Problem 7.2.2
Prove that the point at $r = 0$ is hyperbolic.

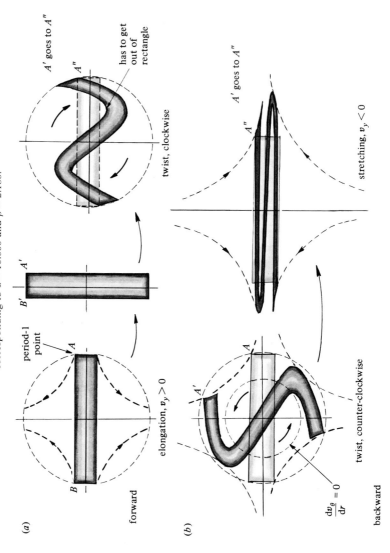

Figure 7.2.10. Horseshoe formation in the tendril–whorl flow. In (a) the elongational flow acts first, followed by a twist clockwise. In (b) the twist acts first, counter-clockwise, followed by the elongational flow. (c) Actual example corresponding to $\alpha = 10.000$ and $\beta = 2.180$.

Problem 7.2.3*
Inject particles in one of the conduits and study the distribution of residence times in the 'chaotic' region. Interpret in terms of the behavior of manifolds.

Problem 7.2.4
Consider the flow $v_1 = Gx_2$, $v_2 = KGx_1$. Can a function $K = K(t)$ be specified such that this system will give horseshoes?

7.3. The blinking vortex flow (BV)

This flow, introduced by Aref (1984), consists of two co-rotating point vortices, separated by a fixed distance $2a$, that blink on and off periodically with a constant period T (Figure 7.3.1).[8] At any given time only one of the vortices is on, so that the motion is made up of consecutive twist maps about different centers and is qualitatively similar to the cavity flows studied in Section 7.5.[9] When the two vortices act simultaneously, the system is integrable and the streamlines are similar to Figure 7.3.2(a).

Figure 7.2.10 *continued*

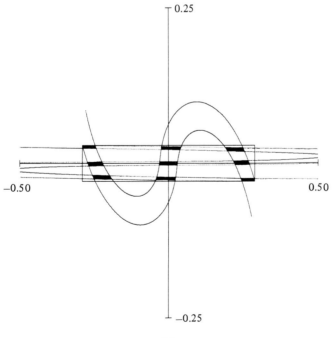

(c)

The velocity field due to a single point vortex at the origin is given by:

$$v_r = 0, \qquad v_\theta = \Gamma/2\pi r$$

where Γ is the strength of the vortex. The mapping consists of two parts; each of the form

$$\mathbf{G}(r, \theta) = (r, \theta + \Delta\theta)$$

where r is measured with respect to the center of the vortex and where the angle $\Delta\theta$ is given by $\Delta\theta = \Gamma T/2\pi r^2$. Placing two vortices at distances $(-a, 0)$ and $(a, 0)$ in a cartesian co-ordinate system (Figure 7.3.1), the complete mapping, in dimensionless form, is given by:

$$\mathbf{f}_i(x, y) = (\xi_i + (x - \xi_i) \cos \Delta\theta - y \sin \Delta\theta, (x - \xi_i) \sin \Delta\theta + y \cos \Delta\theta)$$

where ξ_i denotes the position of the vortex i ($i = A, B$), $\Delta\theta = \mu/r^2$, with $\mu = \Gamma T/2\pi a^2$, and $r = ((x - \xi_i)^2 + y^2)^{1/2}$ (distances are made dimensionless with respect to a so that the vortices A and B are placed at $\xi_i = \pm 1$). In the following analysis we assume that the vortex at $\xi_{i=A} = 1$ is switched on first and that rotation is counter-clockwise.

In this case perturbations are introduced by varying the period of the flow, i.e., by starting from the integrable case, $\mu = 0$, and increasing the value of μ. As we shall see there are several important differences with the TW mapping: (i) in the BV both flows are weak[10] (i.e., the length of filaments increases linearly in time) instead of having a sequence of strong and weak flows as in the TW; and (ii) the BV flow is bounded, i.e., material is not attracted into the flow from large distances; and (iii) the flow is a one-parameter system rather than two as for the TW map.[11]

7.3.1. Poincaré sections

Let us start the analysis by considering Poincaré sections of the flow. In the process of doing so, the virtues and limitations of such an approach will become apparent: Poincaré sections provide *some* of the information about the limits of possible mixing, but in order to truly understand the inner workings of chaotic systems, we are forced to examine stable and

Figure 7.3.1. Schematic diagram of the blinking vortex system.

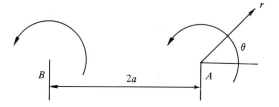

unstable manifolds and to investigate the existence of horseshoes, etc. (see Table 7.1).

Figure 7.3.2 shows Poincaré sections, $t = nT$ with $n = 1, 2, 3, \ldots$, for different values of the perturbation μ (Doherty and Ottino, 1988). For $\mu = 0$, the system is integrable (see Example 7.3.1). As μ is increased, regions of chaos form first near the vortices, then in the center region, until for $\mu = 0.5$ they occupy the entire region. Figure 7.3.3 shows a magnified view of the Poincaré section characterizing the motion near the boundary for $\mu = 0.38$, for a number of different particles starting at different positions. Near the boundary between the 'regular' and 'chaotic' regions we can observe a chain of islands of regular flow, each island

Figure 7.3.2. Poincaré sections corresponding to 12 different initial conditions for various values of the flow strength μ: (a) $\mu = 0.01$, (b) $\mu = 0.15$, (c) $\mu = 0.25$, (d) $\mu = 0.3$, (e) $\mu = 0.4$, and (f) $\mu = 0.5$. A transition to global chaos occurs approximately at $\mu = 0.36$. (Reproduced with permission from Doherty and Ottino (1988).)

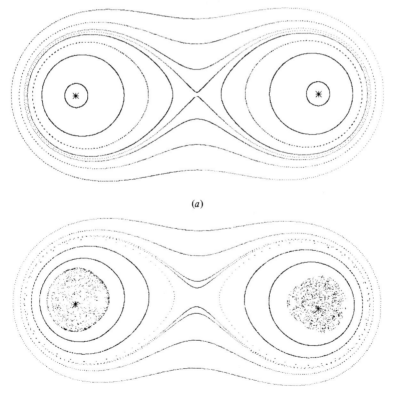

(a)

(b)

containing an elliptic periodic point. Between islands there is a hyperbolic point, of the same period as the elliptic points, whose manifolds generate heteroclinic behavior. In addition, in each island the elliptic point is surrounded by a chain of higher period islands and, in theory, the picture repeats itself *ad infinitum*. The restriction to mixing imposed by the small

Figure 7.3.2 *continued*

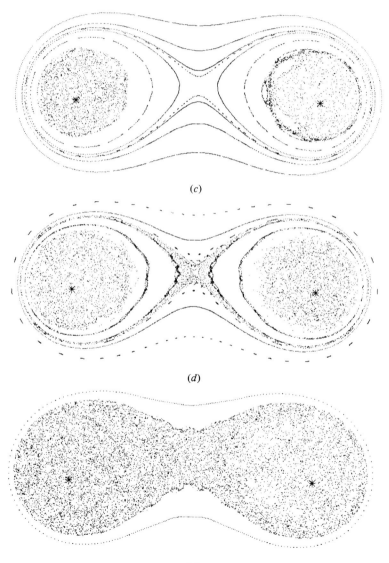

(*c*)

(*d*)

(*e*)

islands appears only for long times ($O(10^4)$ iterations) since they are of high period.

With this as a basis, we explore some of the above behavior by focusing on aspects of local and global bifurcations. The analysis is augmented by computations of Liapunov exponents, 'irreversibility', and mixing

Figure 7.3.2 *continued*

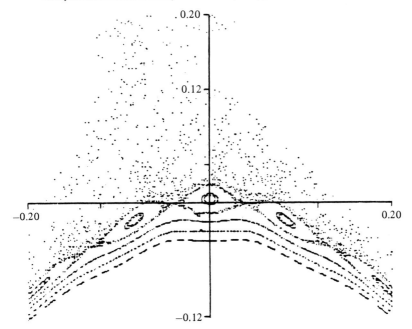

(*f*)

Figure 7.3.3. Magnified view of boundary region for $\mu = 0.38$. (Reproduced with permission from Doherty and Ottino (1988).)

efficiencies. In the last part of the analysis we use the mapping to compute the evolution of the intensity of segregation which might be used to mimic mixing times. Several other possible numerical experiments are suggested and many others might occur to the reader.

7.3.2. Stability of period-1 periodic points and conjugate lines

Figure 7.3.4 shows two candidates for period-1 periodic points (or fixed points). The simplest one is A'. First, vortex A moves A' to B' and then vortex B moves B' back to A'. There are many other ones: For example, the point A'' corresponds to more than one rotation due to vortex A, to position B'', and then less than one rotation due to vortex B back to position A''. It is clear that if (x^*, y^*) is a period-1 periodic point of the BV flow, then it is necessary that

$$\mathbf{f}_A(x^*, y^*) = (x^*, -y^*)$$

$$\mathbf{f}_B(x^*, -y^*) = (x^*, y^*).$$

To find the period-1 points, first we map the x-axis by vortex A *clockwise and with half the angle*. Then we do the same thing with vortex B but *counter-clockwise*, i.e., one line is the mirror image of the other about $y = 0$. The points at the intersection of the mapped lines, *conjugate lines*,

Figure 7.3.4. Schematic representation of period-1 points.

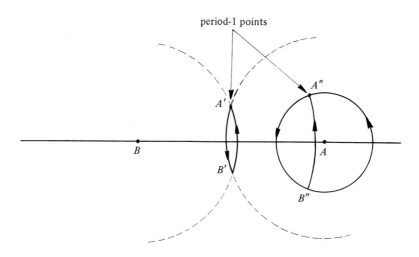

are the location of the fixed points (Figure 7.3.5(a)). The reader should verify that this is indeed the case.[12]

Figures 7.3.5(b)–(d) show the location of the period-1 periodic points for different values of the flow strength μ. The periodic points are labelled by integers n_1, n_2 which are the number of complete rotations of right and left vortices respectively. At low values of μ (Figure 7.3.5(a)) there are many periodic points, all of which lie above the x-axis. However, as μ is increased (Figure 7.3.5(b)) all but three intersections disappear, and finally, at high strengths (Figure 7.3.5(c)) new periodic points are created, some of which lie below the x-axis.

The question to ask next is which of the periodic points are hyperbolic, which are elliptic, etc. This entails the computation of the Jacobian of the mapping and an analysis of the eigenvalues at each periodic point. As with the TW mapping, we can write the trace of the Jacobian as

$$\mathrm{tr}[\mathrm{DF}(x^*, y^*)] = 2 + G.$$

In this case G is defined as

$$G = 4\sin(\theta_1 + \theta_2)[(\Delta\theta_1\Delta\theta_2 - 1)\sin(\theta_1 + \theta_2) + (\Delta\theta_1 + \Delta\theta_2)\cos(\theta_1 + \theta_2)]$$

Hence, the same analysis applies, as G is varied, but in this case the points have to be tested one by one. The analysis becomes cumbersome almost immediately (which limits the anlysis to period-1 points) and here we describe qualitatively just a few of the major results.[13] For example, consider the outermost periodic point located at the centerline, $x = 0$ and $y > 0$ (for additional details, see Khakhar, 1986). At low flow strengths, $\mu \to 0^+$, the point is hyperbolic. As μ is increased the point becomes parabolic at $\mu \approx 15$ and then elliptic (this kind of bifurcation is characterized by an exchange of stability and the birth of two fixed points of the same period; the bifurcation is called 'pitchfork', Guckenheimer and Holmes, 1983, p. 156). An analysis for the remaining period-1 points yields the following conclusions: If the graphs of the segments n_i and n_j are just tangent to one another, at the point of formation the periodic point is initially parabolic, and on decreasing μ, splits into two period-1 periodic points, the one closer to the origin is hyperbolic, the other elliptic (see Guckenheimer and Holmes, 1983, p. 146).

7.3.3. Horseshoe maps in the BV flow

As we have seen in Chapter 5 there is a relationship between homoclinic/ heteroclinic behavior and the existence of horseshoe maps. In a similar

Figure 7.3.5. (*a*) Schematic representation of conjugate lines; the intersection between the lines corresponds to period-1 points; the point P is mapped as shown by the broken lines, (*b*) case corresponding to $\mu = 0.5$, (*c*) case corresponding to $\mu = 3$, (*d*) case corresponding to $\mu = 10$. (Figures (*b*) and (*c*), reproduced with permission from Khakhar, Rising, and Ottino (1986).)

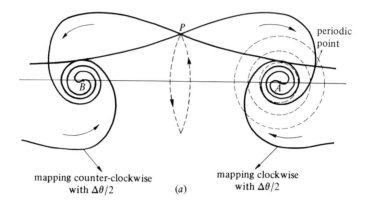

(*a*)

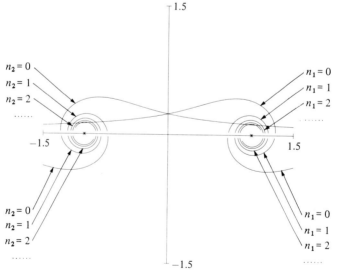

(*b*)

Figure 7.3.5 *continued*

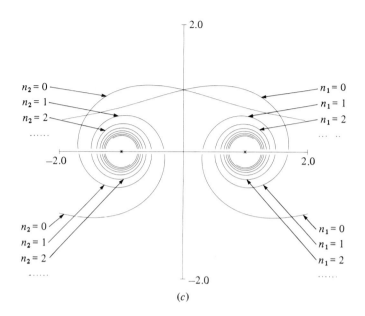

(c)

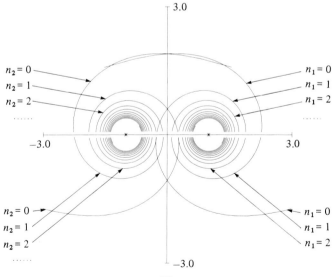

(d)

fashion as with the TW mapping it is possible to prove, by construction, the existence of period-1 horseshoe functions in the flow. The construction, which is shown in Figure 7.3.6 is based on the existence of two nearby period-1 periodic points. These points and the streamlines of the flow define a quadrilateral which is deformed as indicated in the figure forming horizontal striations by the forward mapping and vertical striations by the backward mapping. The superposition of the figures yields a period-1 horseshoe, and hence, the system is chaotic.[14] A similar idea, discussed in some detail in Section 7.5 can be used to examine experimental data.

7.3.4. Liapunov exponents, average efficiency, and 'irreversibility'

The Liapunov exponent calculated based on an infinitesimal segment initially placed at $X \equiv x$, and with orientation $M \equiv m$, is given by[15]

$$\sigma(\mathbf{x}, \mathbf{m}) = \lim_{t \to \infty} \left[\frac{1}{t} \ln |\mathbf{Df}(\mathbf{x}) \cdot \mathbf{m}| \right]$$

Computationally, we proceed in the following way: The length stretch of an infinitesimal segment with initial orientation $\mathbf{m}_i$ is calculated as (see Chapter 4)

$$\lambda_i = |\mathbf{Df}(\mathbf{x}_i) \cdot \mathbf{m}_i|$$

where $\mathbf{f}$ is the mapping

$$\mathbf{x}_i = \mathbf{f}(\mathbf{x}_{i-1})$$

and

$$\mathbf{m}_i = \mathbf{Df}(\mathbf{x}_{i-1}) \cdot \mathbf{m}_{i-1} / \lambda_{i-1}$$

where λ_{i-1} is the length stretch in the $i-1$ to i cycle. The total length stretch, $\lambda^{(n)}$, from $i = 1$ to $i = n$ is

$$\lambda^{(n)} = \lambda_0 \cdot \lambda_1 \cdot \ldots \cdot \lambda_{n-1}$$

and since the period is constant, the Liapunov exponent is calculated as the limit

$$\sigma(\mathbf{x}, \mathbf{m}) = \lim_{n \to \infty} \left(\frac{1}{nT} \sum_{i=0}^{n-1} \ln \lambda_i \right).$$

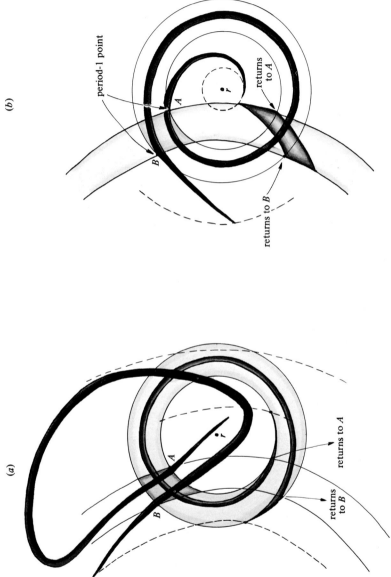

Figure 7.3.6. Horseshoe construction. The points A and B are periodic points. In (a) the right vortex (r) acts counter-clockwise for two turns, then the left vortex, not shown, acts counter-clockwise; in (b) the left vortex acts clockwise for two turns, then the right vortex (r) acts clockwise for two turns.

For the calculation of the efficiency we need the magnitude of $\mathbf{D}$ along the path of the particle. In this case it is constant for each half cycle and is given by $(\mathbf{D}:\mathbf{D})^{1/2} = 2^{1/2}\mu/r^2$.

Computational results for both the Liapunov exponent (the positive one) and the average efficiency are shown in Figure 7.3.7. Both quantities tend to a positive limit value for points initially located in the chaotic region, the values being independent of the initial location and orientation of the material element.[16] As we saw in Chapter 5, positive Liapunov exponents imply exponential rate of stretching of material elements, and hence, good mixing. The computations in Figure 7.3.7 were carried out using over 50,000 cycles of the flow. The points were located close to one of the vortices, so that prior to the transition to global chaos, the particle is restricted to the chaotic region around the vortex. The behavior of the Liapunov exponent is quite complex (Figure 7.3.7(a)); initially it decreases with increasing μ, and after the transition to global chaos ($\mu \approx 0.36$) increases with increasing flow strength. At higher values of μ there is another dip in the graph. Beyond $\mu \approx 5$ the calculations, up to a value of $\mu = 15$, show that the exponent increases slowly.

On the other hand, the average efficiency, shown in Figure 7.3.7(b), is better behaved and presents a single maximum, as might have been expected based on the results of Section 4.5, with a value of $\langle e \rangle_\infty = 0.16$ at $\mu = 0.8$. The average efficiency seems to level off at a value $\langle e \rangle_\infty \approx 0.07$ beyond $\mu \approx 3$ and the calculations indicate that it remains almost constant up to $\mu = 15$. It is important to place the previous results in perspective. From a practical viewpoint, 50,000 cycles is a very impractical way to mix. The values obtained should be considered as a bound for the performance of the flow.

Let us reconsider now the question posed in Chapter 2 about kinematical reversibility of flows. Figures 7.3.8(a)–(c) show a line composed of 10,000 points. The system is operated at $\mu = 1$ (the Liapunov exponent is approximately 0.62) and the line is stretched by the flow for different number of cycles forward and then the flow reversed to its original state.[17] The computations were done with a computational accuracy of 10^{-7}. Figures 7.3.8(a)–(c) show the results of forward and backward cycles ($n = 5$, 10, and 20, respectively). Each plot shows the initial line plus all the points that fail to return it. A point is considered to have returned to its initial location if it falls within a given radius of its initial location (in this case 0.01). These results are not surprising; note that the round-off error grows approximately as $\{10^{-7}\exp[0.62(2n)]\}$. Thus, $n = 5$, 10, and 20, give errors of the order of 5×10^{-5}, 2.4×10^{-2}, and 5.9×10^3, respectively.[18]

Figure 7.3.7. (*a*) Liapunov exponent as a function of flow strength μ, (*b*) flow efficiency as a function of flow strength μ. The different symbols correspond to differential initial conditions and precisions [$+$ $\mathbf{x}_0 = (0.9, 0)$, $*$ $\mathbf{x}_0 = (0.9, 0)$, ⊕ $\mathbf{x}_0 = (0.99, 0)$ D, $\times$ $\mathbf{x}_0 = (-0.99, 0)$ D, ⊡ $\mathbf{x}_0 = (-0.99, 0)$]; D denotes double precision. (Reproduced with permission from Khakhar, Rising, and Ottino (1987).)

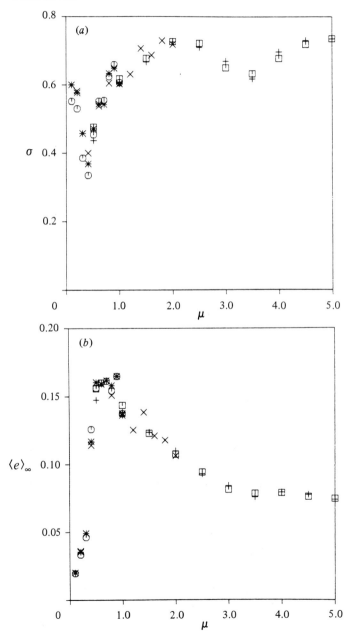

Figure 7.3.8. Stretching of a line by the blinking vortex operating at $\mu = 1$ for various number of iterations, (a) $n = 5$, (b) $n = 10$, (c) $n = 20$, using seven-digit precision.

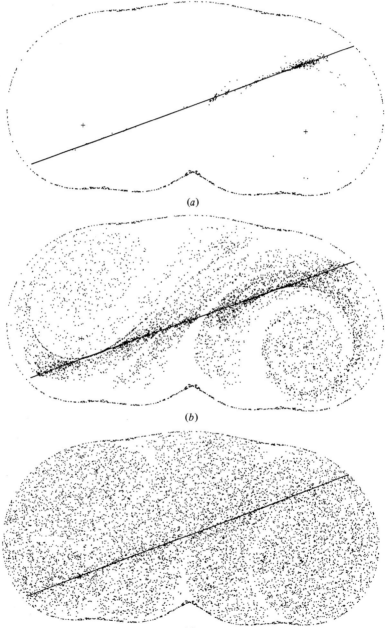

(a)

(b)

(c)

7.3.5. Macroscopic dispersion of tracer particles

Figure 7.3.9 shows the clear advantage of mixing in the globally chaotic region when compared with regular flows. First, a circular blob made up of a large number of points (9,500) is placed at a given location in the flow, which is operated at $\mu = 0.8$ for 30 cycles. In this case the flow is regular and the blob is stretched linearly. Second, we place the blob in the same location as the previous experiment but we operate the flow at $\mu = 1.0$ for 24 cycles. In this case the blob has been stretched in thin filaments which appear disconnected due to insufficient number of points, precision of the graphics output, etc. Note that the actual time of operation, μN, is the same in both cases. The drastic difference can be explained in terms of the Poincaré sections of Figure 7.3.3. In Figure 7.3.9(b) the blob lies outside the 'globally chaotic' region. When μ is increased the region grows, the blob finds itself inside the chaotic region and the mixing is effective.

However, placement of blobs in the chaotic region does not guarantee effective dispersion for short times and in fact, blob experiments provide some evidence for the degree of connectivity of manifolds belonging to different points. Consider the manifolds associated with the periodic points corresponding to the intersections of $n_1 = 0$ and $n_2 = 0$ (see Figure 7.3.5).[19] Figure 7.3.10 (see color plates) shows the stable and unstable manifolds for increasing values of μ (the manifolds of the periodic point at $x < 0$ are not drawn but they can be inferred by symmetry). For low flow strength, $\mu = 0.1$, the manifolds of the central periodic point seem to join smoothly, i.e., if there is homoclinic behavior it occurs below the resolution of the graphics device. However, the homoclinic behavior of the outer fixed point is apparent. At higher values of the flow strength, $\mu = 0.3$, the homoclinic behavior of the central periodic point becomes apparent while that of the outer periodic point increases in scale. At $\mu = 0.5$ there are heteroclinic intersections. More detailed computations indicate that the transition to global chaos occurs for $\mu \approx 0.36$ when the KAM surfaces separating the chaotic region around each vortex from the rest of the flow are destroyed and the outer manifolds intersect with the central ones forming heteroclinic points (see Figure 7.3.10(d)).

Figure 7.3.11 shows the obstruction to dispersion when the system operates at a value of $\mu = 0.5$, which is close to the bifurcation value (see manifolds in Figure 7.3.10). A blob is placed in two different locations and operated on for the same amount of time. Dispersion is very effective in the region near the vortices but only a few particles wander to the

Figure 7.3.9. Effect of regular and chaotic flows on a blob, (*a*) initial condition, (*b*) blob after mixing with $\mu = 0.8$ and 30 cycles, and (*c*) blob after mixing with $\mu = 1$ and 24 cycles. (Reproduced with permission from Khakhar, Rising, and Ottino (1986).)

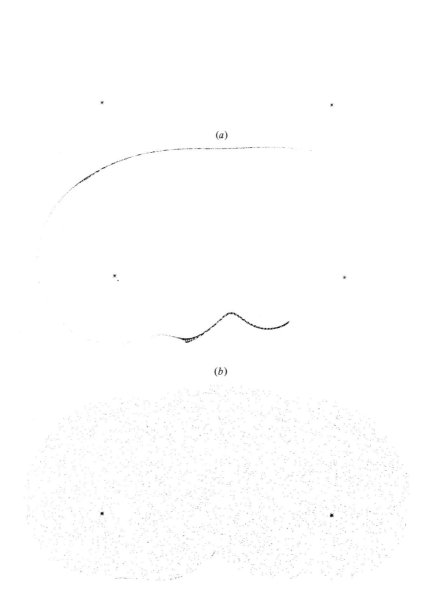

(*a*)

(*b*)

(*c*)

region occupied by the other vortex. The Poincaré section, Figure 7.3.2(*f*) shows that, eventually, particles will cover all the region.[20] However, the process is very slow. Usually, we cannot wait this long.[21]

Although, during mixing, there is always a well defined boundary separating the interior and exterior of the blob, it is hopeless to try to compute the exact location of the boundary even for a modest number of iterations ($O(10^1-10^2)$), and also, probably unnecessary. Having the length of the boundary, $L(t)$, allows the calculation of the intermaterial area as $a_v(t) \approx L(t)/$area and an average striation thickness, $s(t) \approx 1/a_v(t)$.[22] In many applications one might be interested in a map of the distribution of striation thicknesses in space, e.g., Example 9.2.1. A rough idea can be obtained in several ways. For example consider a blob as before, composed of many points. Consider also that we place a uniform grid on the flow region grid size δ. Counting the number of points in each pixel (or more conveniently color coding the results) gives an idea about the uniformity of the distribution (this was done by Khakhar, 1986). Note also that if the number of points in the blob tends to infinity and the pixel size tends to zero we should see a lamellar structure.[23]

Some of the above concepts can be quantified by the following parameter:

$$I = \frac{(\langle [C - \langle C \rangle]^2 \rangle)^{1/2}}{\langle C \rangle}$$

where C is the concentration of points in the pixel and the angular brackets represent a volume average. Since I is reminiscent of the intensity of segregation (Danckwerts, 1952) we will use this name. Figure 7.3.12 shows an example of such a computation where the value of the intensity of segregation is plotted as a function of μN. Even though the results of the computations are specific to the grid size used and the initial location of the blob, in most cases studied there is a rapid decrease of the intensity of segregation. It is possible however, for the particles to be 'demixed' at short times. The curve corresponding to $\mu = 0.5$, which shows an apparently anomalous behavior, can be explained in terms of the restrictions imposed by the degree of overlapping of manifolds. Recall that in this case during the first stage of mixing the particles are dispersed in the region near the vortices and I decays quickly. In the second stage, however, the dispersion of the particles is controlled by transport through the cantori, and I decreases slowly, as more and more particles leak through the gaps of the regular regions. At higher flow strengths ($\mu > 0.5$) the restriction is hardly

Figure 7.3.11. Obstruction to dispersion of blob due to poor communication of manifolds at $\mu = 0.5$ after 25 cycles of the flow. Placement as in (a) and (c) results in dispersion indicated in (b) and (d), respectively (see Figure 7.3.10 and compare with 7.3.2. (Reproduced with permission from Khakhar, Rising, and Ottino (1986).)

(a)

(b)

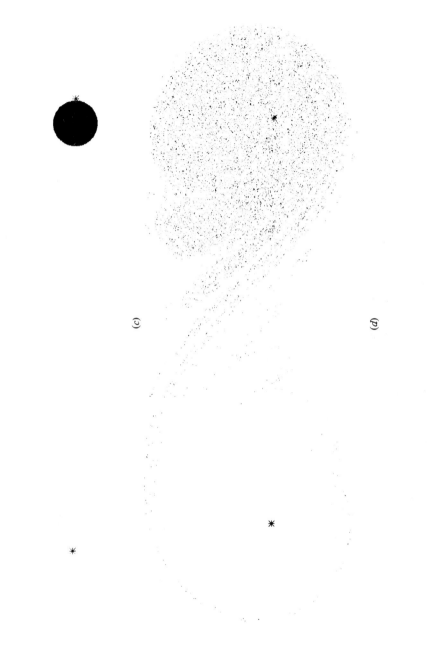

(c)

(d)

noticable. In the same spirit as standard experimental determinations we can define a mixing time as the time required for I to be within 5% of the asymptotic value.

Example 7.3.1
Rather than perturbing the integrable system of Figure 7.3.2(a) by turning vortices on and off in a discontinuous manner, consider the situation of Figure E7.3.1 where the system of two co-rotating vortices is perturbed by a linear flow

$$\mathbf{q} = \begin{bmatrix} 0 & G \\ KG & 0 \end{bmatrix} \begin{bmatrix} x \\ y \end{bmatrix}$$

where K and/or G are time-periodic. Let us consider a few aspects of this system using the Melnikov technique (Section 6.9). Consider two types of perturbations: (a) flow strength (G), and (b) flow type (K).

Figure 7.3.12. Decay of 'intensity of segregation'. (Reproduced with permission from Khakhar, Rising, and Ottino (1986).)

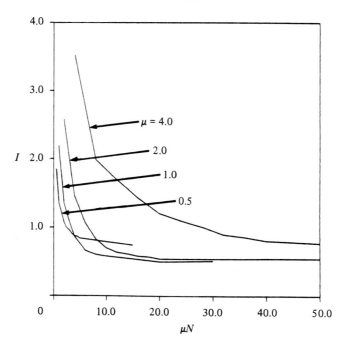

Figure 7.3.13. Schematic representation of Gibbs's argument regarding mixing irreversibiliy (see Endnote 23).

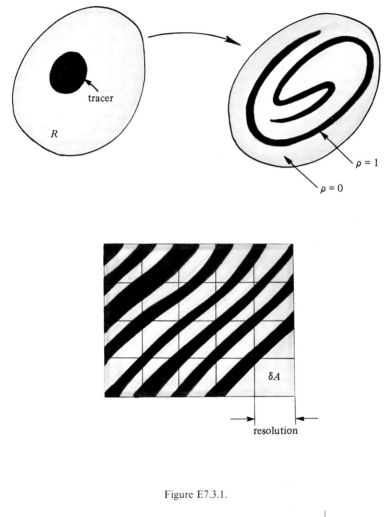

Figure E7.3.1.

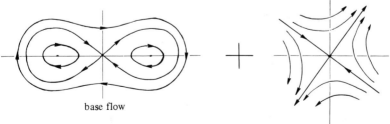

The streamfunction corresponding to the integrable case is

$$\psi = -\tfrac{1}{4}\{\ln[(x-1)^2 + y^2] + \ln[(x+1)^2 + y^2]\}$$

and the corresponding velocities are:

$$v_x = \frac{\partial \psi}{\partial y} = -\frac{y}{2}\left[\frac{1}{(x-1)^2 + y^2} + \frac{1}{(x+1)^2 + y^2}\right]$$

$$v_y = -\frac{\partial \psi}{\partial x} = \frac{1}{2}\left[\frac{x-1}{(x-1)^2 + y^2} + \frac{x+1}{(x+1)^2 + y^2}\right].$$

If we denote $(x-1)^2 + y^2 = D_1$, $(x+1)^2 + y^2 = D_2$, then $D_1 D_2 = 1$ on the homoclinic orbit ψ^0 associated with the central hyperbolic point. The evolution of a point (r, θ) belonging to the homoclinic orbit is governed by

$$dr/dt = -r\sin 2\theta, \qquad d\theta/dt = r^2 - \cos 2\theta,$$

which can be integrated to give

$$\theta = \tan^{-1}(\tanh t), \qquad r = (2/\cosh t)^{1/2},$$

or, in x, y co-ordinates

$$x = 2^{1/2}\cosh t/\cosh 2t, \qquad y = 2^{1/2}\sinh t/\cosh 2t.$$

These are the parametric equations of the homoclinic orbit ψ^0. In order to set-up the Melnikov function, we need to evaluate $\mathbf{f} = (f_1, f_2)$ on ψ^0 [$\mathbf{f}(\psi^0(t)]$ means the velocity of a point belonging to ψ^0). In this case this happens to be

$$f_1 = -2^{1/2}\frac{\sinh t(1 + 2\cosh^2 t)}{\cosh^2 2t}$$

$$f_2 = 2^{1/2}\frac{\cosh t(1 - 2\sinh^2 t)}{\cosh^2 2t}.$$

We need now to account for the effect of the perturbation $g_1 = Gx_2$, $g_2 = GKx_1$ on the homoclinic orbit, i.e., $(f_1 g_2 - f_2 g_1)$ evaluated on ψ^0.

$$f_1 g_2 = GKf_1 x = -2GK\frac{\sinh t \cosh t(1 + 2\cosh^2 t)}{\cosh^3 2t}$$

$$f_2 g_1 = Gf_2 y = 2G\frac{\sinh t \cosh t(1 - 2\sinh^2 t)}{\cosh^3 2t}.$$

Consider now perturbations of the type $G(t) = G_0(a + bh_1(t))$ where $h_1(t) = \sum c_n \exp(i\omega_n t)$. In this case the Melnikov's function is

$$M_1(t_0) = -G_0 b \exp(i\omega t_0)$$

$$\times \int_{-\infty}^{+\infty} \exp(i\omega t)\left[\frac{2(K+1)\sin 2t}{\cosh^3 2t} + \frac{(K-1)\sinh 2t}{\cosh^2 2t}\right]dt$$

whereas for a perturbation of the type $K(t) = K_0(a + dh_2(t))$ where $h_2(t) = \sum c_n \exp(i\omega_n t)$ the Melnikov's function is

$$M_2(t_0) = -GK_0 d \exp(i\omega t_0) \int_{-\infty}^{+\infty} \exp(i\omega t) \left[\frac{2 \sinh 2t}{\cosh^3 2t} + \frac{\sinh 2t}{\cosh^2 2t} \right] dt.$$

The integrations are cumbersome but it is possible to show that the system is capable of forming homoclinic points (Franjione, 1987).

Problem 7.3.1
Interpret physically the parameter μ.

Problem 7.3.2
Examine geometrically the flow in the neighborhood of A and B in Figure 7.3.6. Is the flow hyperbolic?

Problem 7.3.3*
Consider that a point x_0, y_0 has not returned to its original location after n forward and n backward iterations if $(x_n - x_0)^2 + (y_n - y_0)^2 \geq \varepsilon^2$. Study the fraction of points that do not return for different precisions and different values of ε.

Problem 7.3.4*
Study numerically the assumption of random reorientation used in the examples of Section 4.5. Introduce a blob of infinitesimal vectors and compute the orientation distribution function as a function of time.

Problem 7.3.5*
As a simple model for the decrease in striation thickness in the BV flow assume that $s = s_0 \exp(-\sigma(\mu)n)$ where $\sigma(\mu)$ is given by Figure 7.3.7(a) and s_0 is of the order of $2a$. Assume that the intensity of segregation levels off when $s = \delta/K$ where δ is the pixel size and K is a number of order 10^1–10^2. Calculate the values estimated in this way with the ones obtained in Figure 7.3.12. What is the best fit of K? Develop a more sophisticated model.

7.4. Mixing in a journal bearing flow

In this section we consider mixing in the flow region shown in Figure 7.4.1.[24] This case differs from the previous two sections in various respects. Since in this case we have an analytical solution for the streamfunction, and the system can be realized in the laboratory, emphasis will be placed on the comparison of numerical and experimental results. Also, we will concentrate on understanding only a few operating conditions rather than exploring a variety of flows in parameter space. The geometrical parameters of the system are given in Figure 7.4.1; there are two geometrical ratios:

R_{in}/R_{out} and δ/R_{out}; under creeping flow conditions the streamline portrait is determined by the ratio of angular velocities of the inner and outer cylinders, Ω_{in}/Ω_{out}. The numerical simulations are based on the solution corresponding to the creeping flow case obtained by Wannier[25] and the streamlines corresponding to various cases of interest are shown in Figure 7.4.2(a)–(d). Thus, according to the operating conditions, the flow might display one or two saddle points,[26] and time-periodic operation might give rise to homoclinic and heteroclinic trajectories. In what follows we will focus, exclusively, on the case in Figure 7.4.2(c).[27] The parameter values are $R_{in}/R_{out} = \frac{1}{3}$, $\delta/R_{out} = 0.3$, and $\Omega_{in}/\Omega_{out} = -2$.

The solution corresponding to the creeping flow problem is given by $\nabla^4 \psi = 0$. Since the problem is linear, the solution can be written as a linear combination of the forcings of the contributions corresponding to the inner and outer cylinders, i.e.,

$$\psi = \psi_{in}(x, y)\Omega_{in} + \psi_{out}(x, y)\Omega_{out}.$$

This suggests the adoption of a pseudo-steady state viewpoint for the case in which Ω_{in} and Ω_{out} are time-periodic:

$$\psi(x, y, t) = \psi_{in}(x, y)\Omega_{in}(t) + \psi_{out}(x, y)\Omega_{out}(t), \qquad (7.4.1)$$

i.e. $\psi(x, y, t)$ is determined by the instantaneous ratio of Ω_{in}/Ω_{out}. In order

Figure 7.4.1. Flow region in journal bearing flow and definition of parameters. *HP* denotes the approximate location of the period-1 hyperbolic point.

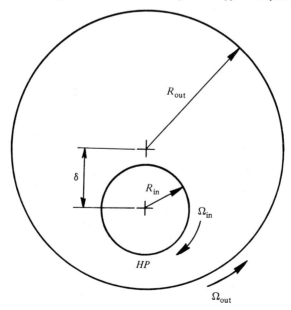

to achieve this condition in the laboratory the acceleration terms in the Navier–Stokes equation should be negligible. For a viscous dominated flow, i.e., a flow with characteristic pressure $= O(\mu V/L)$, the dimensionless version of the Navier–Stokes equation is

$$Re\left[Sr\frac{\partial \mathbf{v}}{\partial t} + \mathbf{v}\cdot\nabla\mathbf{v} \right] = -\nabla p + \nabla^2\mathbf{v},$$

where $Re = \rho VL/\mu$ is the Reynolds number, and $Sr = \omega L/V$ is the Strouhal number (V is characteristic velocity, $V = |V_{in}| + |V_{out}|$, with $V_{in} = \Omega_{in}R_{in}$ and $V_{out} = \Omega_{out}R_{out}$; L is a characteristic length, say $L = R_{in} - R_{out}$; and ω is a characteristic frequency of the motion of the boundaries). Obviously, if both $Sr \ll 1$ and $Re \ll 1$ the flow can be regarded as quasi-static, and time enters only as a parameter. From $Re \ll 1$ we get the condition $(V/L) \ll v/L^2$, and from $Sr \ll 1$ we get $\omega \ll V/L$. Combining the two

Figure 7.4.2. Streamlines corresponding to various cases of interest. The vaues of the parameters are: (a) $\Omega_{out} = 0$, $R_{in}/R_{out} = \frac{1}{3}$, $\delta/R_{out} = 0.5$; (b) $\Omega_{in} = 0$, $R_{in}/R_{out} = \frac{1}{3}$, $\delta/R_{out} = 0.5$; (c) $\Omega_{in}/\Omega_{out} = -2$, $R_{in}/R_{out} = \frac{1}{3}$, $\delta/R_{out} = 0.3$; (d) $\Omega_{in}/\Omega_{out} = 3.0$, $R_{in}/R_{out} = \frac{1}{3}$, $\delta/R_{out} = 0.5$.

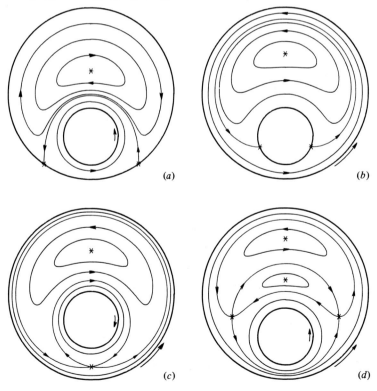

(a)   (b)

(c)   (d)

conditions requires that the time scale of the perturbation ($T = 1/\omega$) be much larger than the time-scale of the diffusion of momentum, i.e., $T \gg L^2/v$.[28] The computations are based on numerical solutions of

$$dx/dt = \partial\psi/\partial y, \qquad dy/dt = -\partial\psi/\partial x,$$

where $\Omega_{in}(t)$ and $\Omega_{out}(t)$ act as an input to the system via Equation (7.4.1).

The first observation to make is that the streamfunction is *independent of the actual speed* of the boundary. This implies that as long as the velocity histories do not overlap, i.e., $\Omega_{in}(t) = 0$ whenever $\Omega_{out}(t) \neq 0$ and *vice versa*, the results for different histories, $\Omega_{in}(t)$ and $\Omega_{out}(t)$, will be identical provided that the angular *displacements*

$$\theta_{in} = \int \Omega_{in}(t)\, dt \qquad \text{and} \qquad \theta_{out} = \int \Omega_{out}(t)\, dt$$

are kept the same. This point is indicated graphically in Figure 7.4.3; all discontinuous histories are equivalent to Figure 7.4.3(b) with suitable $\Omega_{in} = \text{const.}$ and $\Omega_{out} = \text{const.}$ Thus, for discontinuous histories the displacement θ_{in} or θ_{out} becomes an additional parameter (note that the ratio Ω_{in}/Ω_{out} determines θ_{in}/θ_{out}; in what follows we specify the value of θ_{out}).

The second observation is that it is convenient to start operation of the system by moving the first cylinder only one-half of its total angular displacement (see Figure 7.4.3). This results in symmetric Poincaré sections and savings of computational time. Figure 7.4.4 (see color plates) shows a comparison of a Poincaré section and an experiment produced by placing a blob near the location of the period-1 hyperbolic point (see Figure 7.4.1). In this case the agreement, after just 10 periods, is reasonably good and the blob invades nearly all the chaotic region with the exception of the regular island.

Figure 7.4.3. Angular velocity histories: $\Omega_{in}(t)$, $\Omega_{out}(t)$. As long as the motions of the inner and outer cylinders do not overlap and the displacements are kept the same, histories such as (a) and (b) are indistinguishable and produce identical results (e.g., identical Poincaré sections). Note also that the initial angular displacement is one-half of the total displacement. This results in symmetric Poincaré sections.[40]

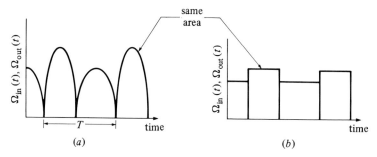

However, this kind of agreement is the exception rather than the rule. Generally, Poincaré sections do not convey a sense of the structure existent within the chaotic region and give no indication of the mixing rate. Figure 7.4.5 (see color plates) shows a similar comparison to that of Figure 7.4.4. In this case however, after the same number of periods, the blob invades only a small part of the chaotic region. A more careful analysis of the Poincaré sections reveals part of the reason. Initial conditions, identified by color, 'do not mix' due to poor communication of manifolds of low order hyperbolic points, even for a large number of periods (see Section 7.3). However, color-coding does not solve all the problems and Figure 7.4.6 (see color plates) shows an example where this is indeed the case. The figure shows two Poincaré sections, identical in all respects except that in Figure 7.4.6(a) the angular displacement, θ_{out}, is 165°, and in Figure 7.4.6(b) it is 166°. A naive interpretation of these results, especially after the discussion on tendrils and whorls of Section 6.11, might suggest that the mixing structure in both systems should be very different since (b) should produce whorls within whorls whereas (a) should produce just one large whorl (not taking into account the smallest islands invisible in the photographs). Furthermore, if this were the case the results would also indicate that extreme care should be exercised in experimental work since an error of 0.6% could determine the outcome of the result. However, a deeper analysis reveals that this is not so and these variations are unlikely to be of any consequence in the experiments. Indeed it can be shown that the rate of rotation within the large island is nearly a solid body rotation and 'whorls' are not noticeable even for a large number of iterations. The rate of rotation in the smaller islands is even slower and for all practical purposes they can be ignored in the stretching of lines placed in this region. In most cases the motion within the regular holes does not produce significant stretching (see experiments on reversibility in Section 7.5). It is apparent that a thorough understanding of this system requires a detailed analysis of *rate* processes. A first step towards this goal is the location of manifolds associated with the hyperbolic points of various orders. In order to accomplish this we first need to locate the periodic points of the system.

An important consequence of the symmetry of Poincaré sections (Figure 7.4.3) is that the search for periodic points is *one-dimensional* rather than two-dimensional. Figure 7.4.7 (see color plates) shows the location of periodic points up to order 8 for the case of a discontinuous velocity history, along with the corresponding Poincaré section.[29] The character of the points can be determined by examining the eigenvalues of the linearized flow; the crosses represent hyperbolic points whereas the circles represent elliptic points. Note that, as expected, the elliptic points fall within the islands, and there is a heavy concentration of points near the inner cylinder since the motion in the immediate neighborhood of the cylinders has (nearly) circular streamlines and therefore, points of all orders are possible.

The rate of spreading of a passive tracer is controlled by the unstable manifolds of the hyperbolic points and is roughly proportional to the value of the eigenvalues and inversely proportional to the period of the point. Figure 7.4.8(a) shows part of the manifolds associated with period-1 points after five periods of the flow. Figure 7.4.8(b) shows the manifolds associated with four period-4 hyperbolic points and sixteen periods. Note that the spreading is significantly less and that the period-1 manifolds act as a *template* for the structure of the system. A blob placed in the neighborhood of the point is 'captured' by the low-period manifolds. Indeed, Figure 7.4.5(b) (see color plates) can be regarded as an experimental manifestation of this effect.

All the previous results, including those of the examples of Sections 7.2 and 7.4, were for the case of a discontinuous velocity history of the boundaries. However, it might be argued that such a history is hard to achieve under laboratory conditions and that smoother histories might result in different phenomena; that is, the relevant question seems to be: How different are the results if the histories are different? Figure 7.4.9 shows several limit cases *producing the same wall displacement per period*. The main result insofar as Poincaré sections are concerned is that the results are remarkably similar and that the most important aspect of the flow prescription is the period rather than its shape[40] (see comparison in Figure 7.4.10, see color plates). Similar results hold for other systems such as the blinking vortex of Section 7.3.

The previous discussion suggests that most of the stretching, and therefore mixing, occurs in the neighborhood of the unstable manifolds of low period points. In order to investigate the regions of significant stretching we adopt the following procedure: We integrate the equation

$$D\mathbf{F}/Dt = (\nabla\mathbf{v})^{\mathrm{T}} \cdot \mathbf{F} \qquad \text{with } \mathbf{F}(\mathbf{X}, 0) = \mathbf{1}$$

for various initial conditions $\mathbf{X}$ and compute the stretching of a vector with initial orientation $\mathbf{M}$ as

$$\lambda = |\mathbf{F} \cdot \mathbf{M}|.$$

Figure 7.4.8. (a) Manifolds associated with the period-1 hyperbolic point, the stretching corresponds to four periods. (b) Manifolds associated with four period-4 points, the stretching corresponds to sixteen periods ($\theta_{\mathrm{out}} = 180°$).

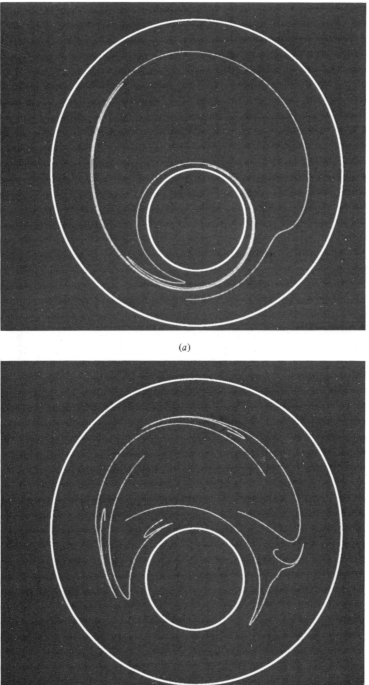

(a)

(b)

The *average stretching* is computed by averaging the stretchings produced by all orientations. Figure 7.4.11(a),(b) (see color plates) show the results of such computations for a small number of periods. The white region corresponds to large stretching (the cut-off value in (a) is 50, in (b) 5,000). The remarkable agreement of Figure 7.4.11(a) with the computations of manifolds shown in Figure 7.4.8, and the experimental results of Figure 7.4.5, is further evidence that the manifolds of the period-1 point provide a template for the stretching occurring in the flow (similarly, Figure 7.4.11(b) should be compared wth 7.4.4(a)). It is important to point out that for a few periods, say 5 or 10, it is possible to find initial conditions such that the stretching in the chaotic regions is less than or comparable to, that occurring in the regular regions. The results presented in this section give an idea of the possible comparisons between experiments and computations and should not be regarded as a complete analysis of this system.

Problem 7.4.1
Discuss the possibility of using the method of conjugate lines of Section

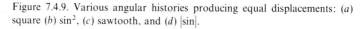

Figure 7.4.9. Various angular histories producing equal displacements: (a) square (b) $\sin^2$, (c) sawtooth, and (d) $|\sin|$.

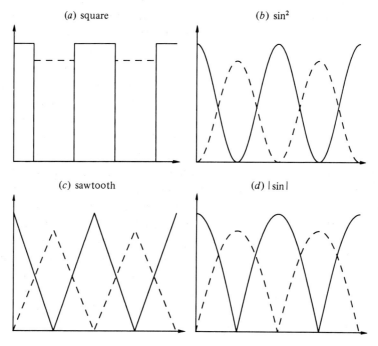

7.3 to locate period-1 periodic points in the alternating cavity flow and the eccentric journal bearing flow.

Problem 7.4.2*
Study the possibility and consequences of having periodic points with trace < -2 in the journal bearing flow (Swanson, 1987).

Problem 7.4.3
Consider that the angular velocities of the inner and outer cylinder are of the form

$$\Omega_{in}(t) = \Omega_{in}^0(1 + \varepsilon \cos(\omega t + \phi))$$
$$\Omega_{out}(t) = \Omega_{out}^0(1 + \varepsilon \cos \omega t)$$

where ε is small. Show that the Melnikov function takes the form

$$M(t_0) = F_a(\omega, \phi) \cos \omega t_0 + F_b(\omega, \phi) \sin \omega t_0.$$

Compute the functions F_a and F_b (Swanson, 1987).

7.5. Mixing in cavity flows

In this section we present experimental results of mixing in several classes of steady and time-periodic cavity flows. However, a number of general remarks will be made regarding the comparison of analysis, computations, and experiments. The experimental system consists of a computer controlled rectangular region capable of producing a two-dimensional velocity field in the x–y plane. The flow region of the experimental system is rectangular with width W and height H (see Figure E4.2.3); the experiments described here were conducted in an improved computer-controlled version of the system described by Chien, Rising, and Ottino (1986) (see Leong, 1990). The system consists of two sets of roller-pairs connected by timing belts driven independently by reversible motors, and two neoprene bands that act as moving walls. The entire system is immersed in a plexiglass tank which has a depth of approximately one foot and is entirely filled with glycerine.[31] By means of slots accompanying the rollers and suitable partition blocks made of acrylic, the flow regions (Figure 7.5.1) can be adjusted in size to a maximum area of 5 inches by 5.5 inches. (For details the reader is referred to Chien, 1986, and Leong, 1990.)

The tracer consists of a mixture of glycerine and a fluorescent dye and is injected 2–5 mm below the free surface of the fluid by means of a syringe (the diffusion coefficient of the tracer in glycerine is of the order of 10^{-8} cm^2/s). The Reynolds number used in the experiments is the highest compatible with two competing effects: creeping flow and two-dimensionality (i.e., absence of inertia and secondary flows) and minimum dye diffusion during the time of the experiment. An order of magnitude calculation based on diffusional effects gives a Reynolds number of order one.[32]

The cavity flow apparatus is used to study three classes of steady-state flows and two classes of time-periodic flows. The steady flows are the following: (i) the standard cavity flow (only one wall moving); (ii) a flow in which the walls are moved in the same direction; and (iii) a flow in which the walls are moved in opposite directions. The velocity of the top wall is denoted as v_{top} and the velocity of the lower wall is denoted as v_{bot}. Figures 7.5.2(b)–(d) show the result of mixing a line of tracer placed vertically in flows (i)–(iii) for equal amounts of time. In this case, the flows are steady and therefore, integrable. It is apparent that the line is trapped by the streamlines and, therefore, the mixing is poor. Indeed, this is a good technique for generating pictures of the streamlines of the steady flows as shown in Figure 7.5.3(a)–(d).

The two classes of periodic flows are: (iv) discontinuous operation of the boundaries in a co-rotating sense, i.e., jumping between Figures 7.5.3(a)–(c) with $v_{\text{top}} = -v_{\text{bot}} = U$; the top and bottom walls move for a time $(1/2)T$ [33]; and finally, (v), a periodic flow corresponding to wall motions of the form

$$v_{\text{top}} = U_{\text{top}} \sin^2(\pi t / T_{\text{top}} + \alpha) \tag{7.5.1a}$$

$$v_{\text{bot}} = -U_{\text{bot}} \sin^2(\pi t / T_{\text{bot}}) \tag{7.5.1b}$$

with equal amplitudes, $U_{\text{top}} = U_{\text{bot}} = U$, periods $T_{\text{top}} = T_{\text{bot}} = T$, and a phase angle, $\alpha = \pi/2$.[34] Thus the system evolves smoothly in the direction

Figure 7.5.1. Flow region. The dimensions of the cavity are $W = 10.3$ cm and $H = 6.2$ cm.

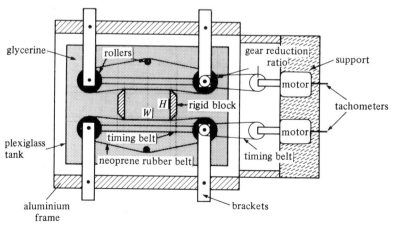

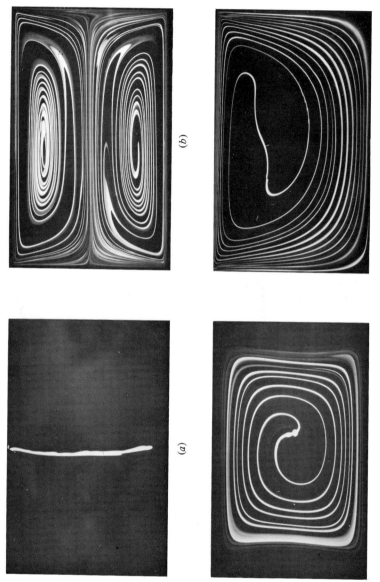

Figure 7.5.2. Stretching of a line placed vertically, as shown in (a), in various steady flows (b)–(d). The fluid is glycerine and the tracer is a fluorescent dye. The experimental conditions are the following: $Re = 1.0$ and the velocities of the top and bottom walls are $v_{top} = v_{bot} = 1.58$ cm/s. In (b) both walls move in the same direction, in (c) both walls move in opposite directions, and (d) only the top wall moves. The total time of the experiment is 5 min.

Figure 7.5.3. Instantaneous picture of the streamlines corresponding to the prescription (7.5.1(a,b)). The experimental conditions are: (a) $v_{\text{top}} = U$ and $v_{\text{bot}} = 0$, (b) $v_{\text{top}} = -v_{\text{bot}}$, (c) $v_{\text{top}} = 0$ and $v_{\text{bot}} = -U$, and (d) $-v_{\text{top}} = v_{\text{bot}}$. The photographs correspond to the case $U = 2.69$ cm/s. The dimensions of the cavity, constant in all the experiments, are $W = 10.3$ cm and $H = 6.2$ cm. The Reynolds number is 1.7.

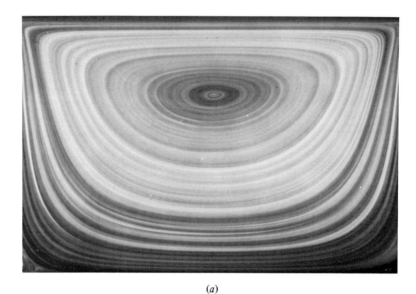

(*a*)

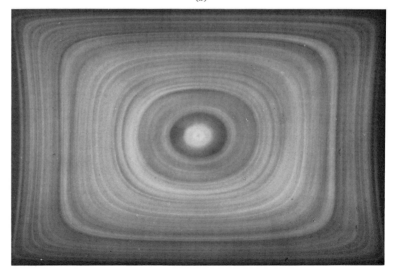

(*b*)

of Figures 7.5.3(*a*), (*b*), (*c*), (*d*) and returns to (*a*) at the end of the period. We regard the period *T* as the governing parameter.

Figures 7.5.4 show the result of placing a blob as initial condition when the system operates in a time-periodic mode. It is clear that the periodic operation improves the mixing (otherwise the blob would remain trapped

Figure 7.5.3 *continued*

(*c*)

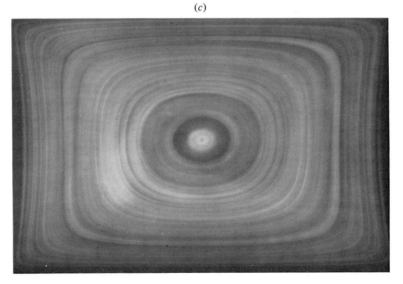

(*d*)

Figure 7.5.4. Evolution of a blob of initial conditions originally placed at the position $x = 2.2$ cm, $y = 3.1$ cm (diameter $= 0.5$ cm). Visualization is provided by a fluorescent dye dissolved in glycerine, excited by long-wave ultraviolet light of 365 nm. The top and bottom walls move according to the prescription of equation (7.5.1) with $U = 2.69$ cm/s ($Re = 1.7$ and $0.05 < Sr < 0.09$). The pictures are taken at the instant when the top wall and bottom wall are moving at maximum and minimum (zero) speed, respectively. The pictures correspond to time periods T: (*a*) 15 s (a second blob, near the center of the cavity was added in this case), (*b*) 20 s, (*c*) 25 s, and (*d*) 30 s; and to wall displacements (the distance travelled by the walls in one period). (*a*) 1612 cm, (*b*) 752 cm, (*c*) 951 cm, and (*d*) 1452 cm. Note the symmetry in Figure (*a*).

(*a*)

(*b*)

by confining streamlines). It appears that, as compared with the journal bearing flow, the cavity flow mixes better and fewer periods are necessary to achieve significant mixing. However, this system is a bit more susceptible to experimental error due to corners and the flexibility of the belts. Another significant difference is that there are no fixed hyperbolic points

Figure 7.5.4 *continued*

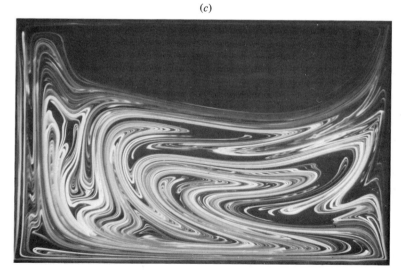

(c)

(d)

in the *velocity field* and that we cannot properly speak of small perturbations with respect to steady flow.

An important difference with the previous sections is that in this case there is no analytical solution for the velocity field. However, computations based on a discretization of the velocity field provide only a crude picture of the flow (e.g., exact location of periodic points, manifolds) and we are therefore left with experiments as the main investigative tool. Therefore, the experimental observations are limited to macroscopic structures, such as islands, produced by mixing, and the question is how do these structures behave under change of the governing parameter (in this case T). Also, any phenomena which are relatively long term (asymptotic) are not directly observable, unless they form structures which are more or less clearly delineated from the start. As we have seen in the previous section, the objects which can be detected are those with macroscopic spatial extent and low periods.

The comparison between analysis, computations, and actual experiments is not trivial and is often misunderstood. In theory there are always three systems; a *mathematical* system, a *computer simulated* system, and an *experimental* system. However, it is naive to expect that all the information extracted from experiments is easily obtainable from computations, and *vice versa*. For example, asymptotic phenomena and sets with measure zero are mathematical objects, and clearly a set with measure zero is not observable by means of computations. The requirements for computational *observability* are also different from those of experimental *observability*. In general an object will be observable in computer experiments if it is robust, i.e., relatively insensitive to round-off error, and if it occurs with high probability, i.e., if its presence can be ascertained regardless of the discretization choice (pixel size in the computer screen). On the other hand, an experimental system is fundamentally analog so that non-robust, low probability events can be seen, though perhaps not easily repeated.[35]

Consider now a few experiments comparing the discontinuous and $\sin^2$ histories (Equations 7.5.1(a,b)). The comparisons reinforce the need for careful experimental interpretation and point out some of the pitfalls of improper and incomplete comparison. Figure 7.5.4 shows the state of the system operating for various values of T under the prescription of Equation 7.5.1(a,b) (i.e. $\sin^2$). The initial condition in all the experiments is a small blob of tracer.[36] In this case the system evolves from no holes to a rather large hole for the highest value of T. Note that the large scale structures of all cases are similar. By analogy with the previous section we might conjecture that this is because the initial stretching and folding is dominated by low period hyperbolic points and their manifolds. The additional folds which occur as the number of periods is increased are indeed contained within the already striated region and very quickly the initial folding serves as a template for further stretching, and the structure of holes shown in the photographs is very similar to the one observed for long times. The holes (and also the large folds) form 'coherent regions' which can be followed in space and time; they are translated and deformed by the flow but conserve identity and do not disappear upon further mixing. In terms of qualitative appearance, there seems to be an optimum mixing at $T = 20$ s. Beyond that, a large regular region grows with increasing T.

The mixing structures produced by the discontinuous protocol are substantially different (Figure 7.5.5). A large hole is present at the lowest value of T used in the experiments and the results of experiments placing blobs inside and outside the island offer a nice demonstration of the speed of mixing within the islands and in chaotic regions. In Figure 7.5.5(a) the blob was placed inside the island; in (b) the blob was placed outside the island; it is clear that the interior of the island does not communicate with the rest of the flow. An increase in the value of T decreases the size of the island and the best mixing is obtained at approximately $T = 58$ s. It thus appears that the two prescriptions, the discontinuous protocol and the sin^2 protocol, give opposite results, and that the conclusions of the previous section, robustness of Poincaré sections (Figure 7.4.10) and so on, cannot possibly hold, since in one case we get hole-opening with increasing T whereas in the other we get hole-closing with increasing T. However, this is only partially correct. Indeed if the pictures are compared at *equal displacements per period*, there is a region (values of displacements in both flows) which produce similar results. Figure 7.5.6 shows a comparison at equal displacements for both prescriptions. Good agreement, in terms of similarity of large scale structures, is seen, especially at low values of T. However, both systems behave quite differently at large values of T.

The most important consequences of the coherence present in these pictures is that careful observations (video recording) allow the detection of periodic points, and that the evolution of the system through a series of changes in the parameter T gives information regarding the bifurcations occurring in the system. Figure 7.5.7 (see color plates) shows an example of a bifurcation which would be hard to capture in computer simulations. It shows the behavior of the island boundary at the time of island collapse. This result is a consequence of the breakup of the last KAM tori (rotation

Figure 7.5.5. Similar to Figure 7.5.4 ($Re = 1.2$, $0.07 < Sr < 0.11$) except that now the motion of the upper and lower walls is discontinuous; the top wall moves from left to right for half a period and then stops, then the bottom wall moves from right to left for half a period and then stops, and so on. There is a five second pause between each half period to reduce inertial effects. The pictures are taken at the end of the desired period. The periods T are: (a) 34 s, (b) 34 s, (c) 48 s, and (d) 58 s; the corresponding displacements are: (a) 1290 cm, (b) 650 cm, (c) 760 cm, and (d) 1100 cm. Note that the structure shown in (a) fits within the 'holes' left in (b) and that relatively little stretching is observed in the interior of the island.

(a)

(b)

number of the island equal to or multiple of golden mean). This type of smooth visualization is consistent with what is mathematically predicted, but it is something which might be missed by computer simulations.

Figure 7.5.8 (see color plates) shows part of the structure of periodic points corresponding to the system of Figure 7.5.4(*d*) and part of the bifurcations that took place up to this point. The entire system might be visualized as a sort of planetary system with the planets (hyperbolic points)

Figure 7.5.5 *continued*

(*c*)

(*d*)

Figure 7.5.6. Comparison of chaotic mixing between the discontinuous (Figure 7.5.5) and the sin² (Figure 7.5.4) prescriptions, at equal wall displacement per period. The initial condition is a blob of fluorescent dye located at $x = 2.2$ cm, $y = 3.1$ cm. The conditions for the sin² flow are the same as Figure 7.5.4 with periods T: ($b1$) 20 s, ($b2$) 35 s, ($b3$) 40 s; and corresponding wall displacements: ($b1$) 752 cm, ($b2$) 1129 cm, and ($b3$) 1075 cm. For comparison with the sin² flow, the discontinuous flow is started in the following way: the top wall moves for $(1/4)T$ and the bottom wall moves for $(1/2)T$, and then the top wall moves for $(1/2)T$, and so on. The other conditions are the same as in Figure 7.5.5 with T of: ($a1$) 29.2 s, ($a2$) 51.2 s, and ($a3$) 59.2 s; and wall displacements: ($a1$) 555 cm, ($a2$) 778 cm, and ($a3$) 1125 cm. In general, the results show some similarity of macroscopic structures at low T (up to $T = 25$ s). Compare ($a1$) and ($b1$).

(a1)   (b1)

(a2)   (b2)

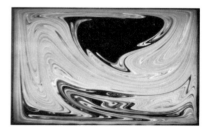

(a3)   (b3)

discontinuous   sin²

and their moons (elliptic points) returning to their initial locations after the periods characterizing the points.

We have seen that the Smale horseshoe map involves the stretching and folding of a blob onto itself. In the context of mixing in periodic two-dimensional flows such a map has a clear physical significance: A very fine subdivision of the blob in the region initially occupied by the blob. Obviously it is desirable that the flow forms many horseshoes, possibly interacting, in such a way that they influence a large region of the flow. From the point of view of mixing it is desirable that they be of low period (number of transformations needed to produce the horseshoe) since we want to achieve mixing as quickly as possible, and also since the domain of a high period horseshoe is usually small. The investigation of horseshoes involves the superposition of forward and backward transformations with the initial location of the blob. The construction is similar to the one given in Sections 7.2 and 7.3. There are several properties which must be verified in order to deduce the presence of a horseshoe (Moser, 1973). All the conditions must be carefully examined. For example, if the placement of the blob is not exact (and it actually never is) the striations will be *partially* filled with the tracer. In such a case the system might appear to violate some of the conditions (Chien, Rising, and Ottino, 1986).

Chaos magnifies errors and one might wonder to what extent it is possible to unscramble pictures such as those of Figures 7.5.4 and 7.5.5. We anticipate that the configurations in Figures 7.5.2(*a*)–(*c*), which are integrable, can be reversed within experimental error, and there have been experiments showing that this is indeed possible in creeping flows (see Bibliography, Chapter 2). However, we have already seen that this might be impossible, even numerically, if the flow is chaotic (Section 7.3.4). Figure 7.5.9 (see color plates) shows what happens to initial conditions placed in the interior of an island and in a chaotic region (see Figure 7.5.4(*d*)). Figure 7.5.9(*b*) shows the result of forward mappings and Figure 7.5.9(*c*) shows the results of backward mappings. It is apparent that the line placed in the hole undergoes relatively little stretching[37] and returns to its initial location. However, the line placed in the chaotic region (presumably near a horseshoe) does not return to its initial location; every time the line passes by the horseshoe it is stretched exponentially and the experimental errors are magnified. Apparently, two periods are enough to magnify significantly the errors of this particular experimental system.

We have emphasized that the degree of smoothness achieved in these pictures would be hard to achieve in straightforward numerical simulations (see Examples and 9.2.4). Consider now the experimental determination of a quantity that presents problems from a numerical viewpoint: the estimation of the interfacial area generation (see Example 9.2.3). At the same time, we will consider a feature of the experiments that we have avoided mentioning so far. Even though the flow is two-dimensional, since the camera is not at infinity, the area of the initial blob appears to increase in the photographs until it occupies the entire chaotic region.[38] Figure 7.5.10 shows the measurements of area and perimeter growth utilizing image analysis for three different cases: a steady flow with one moving wall, and the two time-periodic flows. Both time-periodic flows produce exponential growth of the area, A, and the perimeter, P. Both A and P follow approximately the same growth law, $\exp(\beta t)$, with similar values of β: for the discontinuous flow $\beta \approx 0.022 \text{ s}^{-1}$, for the $\sin^2$ flow $\beta \approx 0.019 \text{ s}^{-1}$. On the other hand, in steady flow the growth of both A and P is linear in time.

It is then clear that the actual fluid system provides an alternative visual tool, less controllable perhaps than a mapping, but with the advantage that much of the behavior of the system can be directly observed, particularly in the regions of hyperbolic sets. The smoothness and regularity shown in the experiments are hard to achieve in computer simulations where finite pixel size (resolution) and round-off errors govern the process and the regions often degenerate into pixel clouds by which the underlying structure is obscured.[39]

Problem 7.5.1
Using order of magnitude arguments obtain the optimal Re for the above experiments. Consider the lines stretch exponentially and use an efficiency of order 10^{-1}–10^{-2}. Assume that the striations are unrecognizable if the diffusion distances are of the same order of magnitude as the striation thickness itself.

Problem 7.5.2
Justify the exponential growth of dye area in the experiments.

Problem 7.5.3
Estimate the average mixing efficiency in the cavity flow.

Problem 7.5.4*
Rationalize the similarity of the results obtained in 7.5.6 a1 and b1.[40]

Bibliography

The first two sections are an abbreviated version of the presentation given in Khakhar, Rising, and Ottino (1986). The original paper treating the

Figure 7.5.10. Measurement of area growth and perimeter growth by image analysis. The resolution of the image is 512×512 pixels and 256 grey levels. The measurements are for three classes of flows: steady, with one wall moving; discontinuous (Figure 7.5.5(d)); and $\sin^2$ (Figure 7.5.4(c)) time-periodic flows. Both time-periodic flows produce an exponential growth ($\exp(\beta t)$) in area, A, and perimeter, P, with approximately the same exponent β. For the discontinuous flow $\beta \approx 0.022$ s^{-1}, whereas for the $\sin^2$ flow $\beta \approx 0.019$ s^{-1}. On the other hand, in the steady state flow the growth is linear in time.

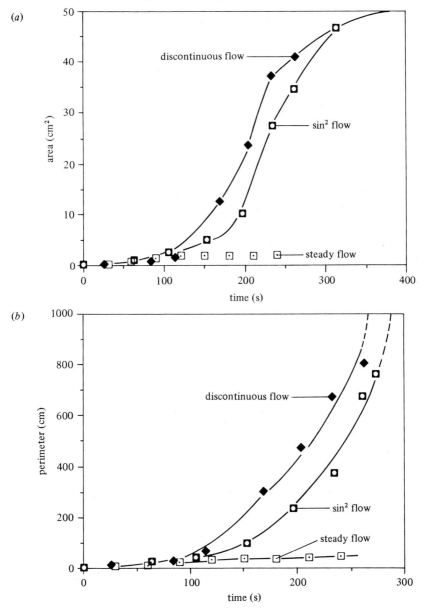

blinking vortex system is Aref (1984); the first treatment of the tendril–whorl map is given in the paper by Khakhar, Rising, and Ottino (1986). Preliminary results for both systems were given in various presentations: The horseshoe construction of the tendril–whorl system was given in Rising and Ottino (1985, abstract only); several results regarding efficiency of stretching were given in Khakhar and Ottino (1985, abstract only). The material of Section 7.4, journal bearing flow, is based on the presentation of Swanson and Ottino (1985, abstract only). A substantially more complete presentation is given in Swanson and Ottino (1990). The first study of the journal bearing flow is the computational study of Aref and Balachandar (1986). Several nice experimental results are presented by Chaiken *et al.* (1986), however, most studies involve a very large number of periods (order 10^2 and higher). Section 7.5, cavity flow, is based on experimental results by Leong (Ph.D. thesis, University of Massachusetts, Amherst); the original paper is by Chien, Rising, and Ottino (1986). The relationship between analysis, computations, and experiments is discussed by Ottino *et al.* (1988) in terms of the journal bearing flow and the cavity flow. A related topic is dispersion in Rayleigh–Bénard flows; even though there is a number of recent papers in this area, the treatments of mixing are rather scarce (e.g., Solomon and Gollub, 1988). A paper addressing mixing in the context of geophysics is given in Hoffman and McKenzie (1985).

Notes

1 Note that due to widespread usage the word 'flow' is used in two different ways throughout this work; the first one in the mathematical sense considered in Chapter 6, the second in the accepted fluid mechanical way, as for example, in the terms 'flow field' or 'fluid flow'.

2 For example, fractal dimensions in the chaotic regions, correlation functions, etc.

3 These terms were defined by Berry *et al.* (1979) in the context of quantum maps. In practice, whorls are hard to obtain in most chaotic flows (see Section 7.4).

4 In a saddle-node bifurcation for area preserving systems, as the parameter of the system increases the flow goes from no fixed points to two fixed points, one hyperbolic and one elliptic.

5 This kind of limitation is common to most of the computational studies described here.

6 Notice that there is efficient mixing in a region where the flow is approximately linear due to the interaction with the outside flow. This should be useful in understanding mixing at the microscale of turbulence, where the flows are, at some scale, linear and laminar.

7 Statements like this must always be qualified by saying 'within the resolution of the graphics output, etc.'.

8 Again, several possible generalizations are possible such as distribution of periods, etc.

9 However, an important difference is that, near the origin, the cavity flow behavior is $v_\theta \approx r$ whereas in the blinking vortex $v_\theta \approx 1/r$.

10 As we have seen in Section 4.6 this is an indication that this flow might have an optimal operating period for which the efficiency is maximized.

11 Another difference is that $\mathbf{D}:\mathbf{D}$ diverges at the center of the vortices and the integral of $\mathbf{D}:\mathbf{D}$ is unbounded. By contrast, the energy dissipation in the TW mapping is well behaved. Nevertheless, the efficiency in the BV is well behaved.

12 A similar construction can be used to find the periodic points in the TW map, and more generally, of generalized blinking flows (Rising and Ottino, 1985).

13 The reader might gain an appreciation of the complexity of chaotic systems by considering that a similar analysis applies to period-n points, where n can be a very large number!

14 This idea was used in Rising and Ottino (1985).

15 The reader should notice that in this case, since we do have an explicit mapping, the calculation is free of numerical errors which might arise due to renormalization of trajectories (see Lichtenberg and Lieberman, 1983, p. 280).

16 There are still some questions regarding the computation of long time averages due to the finite precision nature of the computations. However, when the motion is 'sufficiently chaotic' and numerical precision high (see Figure 7.3.10) the time averages are expected to be reliable.

17 Note that the accurate computation of length is not trivial; we can label points and introduce additional ones if they become too separated and join nearby points, say i and $i+1$, by short segments to approximate the length. However, as the number of cycles increases we need to add points at an exponential rate (see Example 9.2.3).

18 These numbers are only approximate. Regular regions and chaotic regions are intricately connected. If a point gets 'trapped' in a regular region it remains there for a long time.

19 Although the points are elliptic at the point of formation they are hyperbolic for the range of μ-values considered here.

20 This is one of the reasons why it is convenient to construct Poincaré sections using particles labelled with a color representing the initial condition.

21 This suggests the need for transport studies in Hamiltonian systems, possibly using renormalization methods, as indicated earlier, or in terms of the dynamics of manifolds (see Fig. 5.8.3.).

22 In the numerical experiments of Section 7.2.6 we can consider the line separating two different materials. Then if A is the area of the globally chaotic region and $L(0)$ the initial length of the line, we can write $A = L(0)s_0 = L(t)s(t)$, where s is a distance of the order of magnitude between the vortices. Thus, $s(t) \approx L(t)^{-1}$.

23 It is interesting to point out that J. W. Gibbs used a fluid mixing analogy, a regular flow actually, to explain the origin of irreversibility in statistical mechanics (Gibbs, 1948). His argument was re-interpreted by Welander (1955) and is closely related to the content of the previous sections. Although the argument is not entirely rigorous it brings up some important issues and it is convenient to repeat it here.

Consider some two-dimensional region with area R where we have placed a blob of a non-diffusive tracer with density $\rho(\equiv dm/da)$. We consider that $\rho(\mathbf{x}, t) = 1$ if $\mathbf{x}$ belongs to the tracer region and $\rho(\mathbf{x}, t) = 0$ anywhere else. The total mass of the tracer in R and the integral of the square density are both equal, i.e.,

$$\int_R \rho \, da = M, \qquad \int_R \rho^2 \, da = M$$

where M is less than R. Given these choices, no matter how effective the mixing, both these quantities remain equal (the particles of the tracer always carry the same density, $\rho = \rho(\mathbf{X}, t) \equiv 1$). However, what happens if the resolution of the integration is less than the striation thickness? (see Figure 7.3.13). At this scale the density is $\langle \rho \rangle \equiv \delta m/\delta A$, and the square density, $\langle \rho \rangle^2 = (\delta m/\delta A)^2$. Therefore the integral of the square density is

$$\int_R \langle\rho\rangle^2 \delta A = \int_R \left(\frac{\delta m}{\delta A}\right)^2 \delta A = \int_R \frac{\delta m}{\delta A} \delta m$$

but since $\delta m/\delta A$ cannot be equal to one everywhere (since $M < R$) we conclude that

$$\int_R \langle\rho\rangle^2 \delta A < 1.$$

In other words, the density distribution becomes more homogeneous due to poor resolution. In this context mixing can be viewed as the minimization of $\langle\rho\rangle^2$ in the domain R.

24 This system has been experimentally built by Chaiken *et al.* (1986) and Swanson and Leong, Ph.D. theses (in progress), University of Massachusetts, Amherst, 1986.

25 Wannier (1950). Other solutions were presented by Jeffery (1922) and Duffing (1924). Wannier's method is based on a complex variable technique and gives possibly, the most managable solution. A solution also is presented by Ballal and Rivlin (1977); this article considers also a first-order inertial perturbation and indicates that the streamlines are only distorted even for relatively large Reynolds numbers ($O(50)$). A related article focusing on the same problem is Kazakia and Rivlin (1978).

26 Note that in creeping flow, the streamline portrait is independent of the actual direction of rotation of the cylinders (thus, if the boundaries are moved in the opposite direction, all we need to do is reverse the direction of all the arrows). Note also that as long as the flow is creeping, only the ratio Ω_{in}/Ω_{out} is important, not the actual value of the velocities.

27 The experiments reported here have been obtained in a computer controlled apparatus (Swanson, 1988). The most important design feature of the apparatus is that it allows for an unobstructed view of the flow region since the outer cylinder is rotated from the 'outside' with a large bearing enclosing the outer cylinder. This is particularly important in the case under study since the period-1 hyperbolic point would not be visible in conventional designs (see Figure 7.4.1). Observations are made from the bottom of the apparatus, which is made of glass. The working fluid is glycerine floated on carbon tetrachloride.

28 This condition can be easily satisfied in laboratory experiments, see also comments in Section 7.5.

29 Note however, that we can not prove that *all* the points up to this order have been found; for example, it is possible that there might be more period-8 points than those shown in the figure.

30 The closest experimental studies are those of Ryu, Chang, and Lee (1986). They focused exclusively on steady flows. Other related studies were conducted by Pan and Acrivos (1967); and Bigg and Middleman (1974). Bigg and Middleman studied laminar mixing of two fluids with different viscosities in the standard cavity flow configuration, i.e., with only one wall moving. Pan and Acrivos focused on tall cavities, $H/W > 1$, and streamline visualization for high Reynolds number flows. Some of the results presented in this section, and a comparison with the system of Section 7.4, can be found in Ottino *et al.* (1988).

31 As a working fluid, glycerine presents some difficulties; its viscosity is not as high as it should be to accomplish comfortably the condition $Re \ll 1$. However, by using a transparent fluid we can check on the two-dimensionality of the flow.

32 Both the Reynolds number (Re) and the Strouhal number (Sr), which are a measure of the inertial effects in the flow, are kept low enough in the experiments as to be of no importance (in the computation of Re and Sr the characteristic velocity is taken as U and the characteristic length as H^2/W).

33 In the experiments it takes some time for the motion to set in, $O(H^2/\nu) = 5 \times 10^{-2}$ s. In order to minimize transient effects we wait for approximately 5 s between the motion of the upper and lower bands.

34 Obviously, there are infinitely many other possibilities for the motion of the walls, for example, even within the restricted class of motions, Figures (7.5.1) and (7.5.2), we can vary the relative amplitudes and direction of motion of the boundaries, the phase angle, time periods, etc.; only a few of these possibilities have been explored to date.

35 Thse points are discussed in Ottino *et al.* (1988). Note that there is a factor of two error in the periods T reported in this paper.

36 Note that if the flow were not two-dimensional the segments might cross in the photograph.

37 See comments in the previous section; it is indeed very hard to achieve a situation leading to the formation of visible whorls in just a few periods.

38 A calculation indicates that for our system the camera will be unable to detect striations when they are closer than 10 μm. Also, glowing effects and the (small) molecular diffusion present during the time of the experiments contribute to the area increase. To a first order approximation, these effects are proportional to the interfacial area between the tracer and the clear fluid.

39 A detailed study of breakup of islands in terms of area preserving maps, such as those presented by Ottino *et al.* (1988), is given in the Ph.D. thesis by Rising (1989).

40 Both flows have the same symmetries provided that they are examined at suitable times. For our purposes a map $\mathbf{M}$ with inverse $\mathbf{M}^{-1}$ is said to be *symmetric* if there exists a map $\mathbf{S}$, with $\mathbf{S}^2 = \mathbf{1}$, such that $\mathbf{M}$ and its inverse are related by $\mathbf{M} = \mathbf{SM}^{-1}\mathbf{S}$. The set of points $\{\mathbf{x}\}$ such that $\{\mathbf{x}\} = \mathbf{M}\{\mathbf{x}\}$ is called the *fixed line* of $\mathbf{M}$. The symmetries of time-periodic flows can be easily deduced using rules from map algebra. Consider for example the discontinuous history of Figure 7.4.9(a). Let the motion induced by moving the top wall during a time t be denoted $\mathbf{T}_t$ (*i.e.*, a set of particles $\{\mathbf{x}\}$ is mapped to $\mathbf{T}_t\{\mathbf{x}\}$ at time t). Similarly, denote by $\mathbf{B}_t$ the motion induced in the cavity by moving the bottom wall. Since the flow is a stokes flow and the streamlines are symmetric with respect to the y-axis it follows that $\mathbf{T}_t = \mathbf{S}_y \mathbf{T}_t^{-1} \mathbf{S}_y$ and $\mathbf{B}_{t'} = \mathbf{S}_y \mathbf{B}_{t'}^{-1} \mathbf{S}_y$, where $\mathbf{S}_y$ denotes the map $(x, y) \rightarrow (-x, y)$. In this case the fixed line of $\mathbf{S}_y$ is the y-axis itself. Furthermore if $t = t'$ then the top and bottom motions are related by $\mathbf{T}_t = \mathbf{S}_x \mathbf{B}_t^{-1} \mathbf{S}_x$ and $\mathbf{T}_t = \mathbf{R} \mathbf{B}_t \mathbf{R}$, where $\mathbf{R} \equiv \mathbf{S}_y \mathbf{S}_x = \mathbf{S}_x \mathbf{S}_y$ denotes at $180°$ rotation $(x, y) \rightarrow (-x, -y)$. The symmetries of composition of flows can be deduced by using these rules. For example, the flow $\mathbf{F}_{2t} \equiv \mathbf{B}_t \mathbf{T}_t$; *i.e.*, a series of top and bottom wall motions, has symmetry with respect to the x-axis, $\mathbf{F}_{2t} = \mathbf{S}_x \mathbf{F}_{2t}^{-1} \mathbf{S}_x$ whereas the flow $\mathbf{G}_{2t} \equiv \mathbf{B}_{t/2} \mathbf{T}_t \mathbf{B}_{t/2}$; *i.e.*, a series of top and bottom wall motions, but where observations are made half-way through the bottom wall displacement, has y-symmetry instead. This means that the islands in the flow are located in pairs across the y-axis or on the axis itself. Note that a flow can have more than one symmetry. For example, the $\mathbf{F}_{2t}$ flow can be written also as $\mathbf{F}_{2t} = \mathbf{S}^* \mathbf{F}_{2t}^{-1} \mathbf{S}^*$ where the symmetry $\mathbf{S}^*$ is given by $\mathbf{S}^* \equiv \mathbf{S}_y \mathbf{T}_t = \mathbf{T}_t^{-1} \mathbf{S}_y$. The fixed line of $\mathbf{S}^*$ is not obvious; it corresponds to the mapping of the y-axis with $\mathbf{T}_{t/2}^{-1}$; that is, the inverse of the top wall flow with half the displacement. In this case the symmetry line is a curve (these ideas are developed in the articles by J. G. Franjione, C. W. Leong, and J. M. Ottino, 'Symmetries within chaos: a route to effective mixing', *Phys. Fluids A*, **1**, 1772–83, 1989 and C. W. Leong, and J. M. Ottino, 'Experiments on mixing due to chaotic advection in a cavity', *J. Fluid Mech.*, **209**, 463–99, 1989).

Mixing and chaos in three-dimensional and open flows

In this chapter we study mixing in several three-dimensional fluid flows and open flows with varying degrees of complexity. We stress qualitative differences with respect to the time-periodic two-dimensional case and focus on the topology of the flows and in their mixing ability.

8.1. Introduction

So far we have studied mixing in two-dimensional time-periodic flows and have seen that considerable theoretical guidance exists for this case. With the exception of the tendril–whorl flow, in which fluid entered and left through conduits, the systems were closed. We now move to the case of three-dimensional and open flows. Several cases are discussed. The chapter starts by analyzing mixing under creeping flow conditions in two continuous mixing systems. In the first example – the partitioned-pipe mixer – the velocity field is periodic in space; in the second example – the eccentric helical annular mixer – the velocity field is time-periodic. These two flows are analyzed in some detail since they have many potential applications. As we shall see, intuition based on the two-dimensional case can be somewhat misleading in the understanding of three-dimensional flows in general. The next example, discussed in less detail, includes inertial effects and corresponds to mixing in a channel with sinusoidal walls. The treatment of this case, largely based on the work by Sobey (1985), is computational in nature and relatively little analytical work is possible. The next examples are of relevance to turbulent flows. The first one of this class is a perturbation of the Kelvin's cat eyes flow, which is important, as a first approximation, to mixing in shear layers. The next example considers the class of admissible flows obtained by constructing solutions of the Navier–Stokes equations by expanding the velocity field in a Taylor series expansion near solid walls. This example is of relevance to perturbed flows near walls and has obvious relevance to turbulence. Finally, the last example, originally due to Hénon (1966), is the analysis of mixing in an inviscid fluid by Dombre *et al.* (1986) and our presentation is largely based on their analysis.

8.2. Mixing in the partitioned-pipe mixer

The partitioned-pipe mixer consists of a pipe partitioned into a sequence of semi-circular ducts by means of rectangular plates placed orthogonally to each other (Figure 8.2.1). The fluid is forced through the pipe by means of an axial pressure gradient while the pipe is rotated about its axis relative to the assembly of plates, thus resulting in a cross-sectional flow in the (r, θ) plane in each semi-circular element. We study the mechanical mixing of a Newtonian fluid in the mixer when it operates under creeping flow conditions. The expectation that the flow in the partitioned-pipe mixer is chaotic is based on its similarity to the Kenics[R] static mixer which, as was mentioned in Chapter 5 (see Endnote 24), resembles the baker's transform in terms of its cross-sectional mixing (Middleman, 1977). Under ideal conditions, each stream is divided into two in each element (see Figure 8.2.2). The main difference between the two mixers is that the sense of rotation of the cross-sectional flow is the same in adjacent elements of the partitioned-pipe mixer while it is opposite in the static mixer.

8.2.1. Approximate velocity field

We consider an approximate solution to the fully developed Stokes flow of a Newtonian fluid in the semi-circular compartment of an element of the mixer under creeping flow. In this case the axial and cross-sectional components of the velocity field are independent of each other and can be considered separately. The axial flow is simply a pressure driven flow in a semi-circular duct and in cylindrical coordinates is given by

$$v_z = \langle v_z \rangle \sum_{k=1}^{\infty} \left\{ \left[\left(\frac{r}{R} \right)^{2k-1} - \left(\frac{r}{R} \right)^2 \right] \frac{\sin[(2k-1)\theta]}{(2k-1)[4-(2k-1)^2]} \right\}$$

where R is the radius of the pipe, and the average axial velocity $\langle v_z \rangle$ is

$$\langle v_z \rangle = \frac{8 - \pi^2}{4\pi^2 \mu} \frac{\partial p}{\partial z} R^2$$

where μ is the viscosity, and p the pressure.[1] The cross-sectional component is a two-dimensional flow in a semi-circular cavity, and is given as a solution of

$$\nabla^4 \psi = 0$$

where

$$\nabla^2 \equiv \frac{\partial}{\partial r} + \frac{1}{r} \frac{\partial}{\partial r} + \frac{1}{r^2} \frac{\partial}{\partial \theta^2}$$

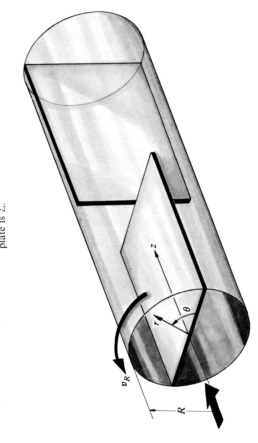

Figure 8.2.1. Schematic view of the partitioned-pipe mixer; the length of the plate is L.

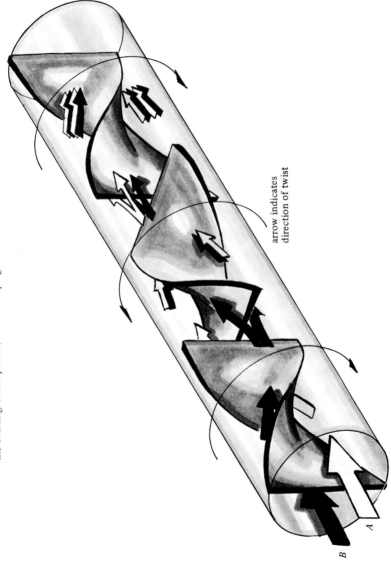

Figure 8.2.2. Kenics® static mixer, idealized view of the system. Two streams, *A* and *B*, enter segregated. After the first twist (clockwise looking from the entrance) each stream is halved and joined by half of the other stream in the second twist (counter-clockwise); the process is repeated periodically. The actual system is more complicated due to flows in the *r*, *θ* plane, produced by the twisting of the planes, and developing flows.

arrow indicates
direction of twist

B

A

with the velocity field given by

$$v_r = \frac{1}{r}\frac{\partial \psi}{\partial \theta}; \qquad v_\theta = -\frac{\partial \psi}{\partial r}.$$

The boundary conditions are

$$\psi = \frac{1}{r}\frac{\partial \psi}{\partial \theta} = 0; \qquad -\frac{\partial \psi}{\partial r} = v_R \qquad \text{for } r = R, \text{ and } \theta \in [0, \pi]$$

$$\psi = \frac{1}{r}\frac{\partial \psi}{\partial \theta} = \frac{\partial \psi}{\partial r} = 0 \qquad \text{for } \theta = 0, \pi, \text{ and } r \in [0, R]$$

where $v_R = v_\theta(R)$.

An approximate solution to the above problem can be obtained by the method of weighted residuals (Finlayson, 1972). The streamfunction is taken to be of the form

$$\psi = \sum_{m=1}^{N} \xi_m(r)\sigma_m(\theta)$$

where $\{\sigma_m\}$ is a set of known trial functions which satisfy the boundary conditions, and $\{\xi_m\}$ is a set of unknown functions to be found by solving the differential equations, obtained by minimizing the residual over the interior as in the Kantorovich–Galerkin method (Kantorovich and Krylov, 1964). The weighting functions in this case are the trial functions themselves, and are taken to be

$$\sigma_m(\theta) = \sin^2(m\theta).$$

The one-term solution for the stream function for the cross-sectional flow is

$$\psi = \frac{4v_R R}{3v}\left(\frac{r}{R}\right)^2\left[1 - \left(\frac{r}{R}\right)^v\right]\sin^2\theta$$

where $v = (11/3)^{1/2} - 1$.

The streamlines for the different values of the normalized stream function $\psi^* = \psi/[4v_R R/3v]$ are shown in Figure 8.2.3. The motion is obtained by integrating the following set of equations which represent the approximate velocity field

$$v_r = \frac{dr}{dt} = \beta r(1 - r^v)\sin 2\theta$$

$$\frac{v_\theta}{r} = \frac{d\theta}{dt} = -\beta[2 - (2 + v)r^v]\sin^2\theta$$

$$v_z = \frac{dz}{dt} = \frac{16\pi}{\pi^2 - 8}\sum_{k=1}^{3}\left\{(r^{2k-1} - r^2)\frac{\sin[(2k-1)\theta]}{(2k-1)[4 - (2k-1)^2]}\right\}.$$

The above equations are dimensionless, with radial distances made dimensionless with respect to R, axial distances with respect to the length of an element L, and time with respect to $L/\langle v_z \rangle$. The dimensionless parameter β, which we refer to as the *mixing strength*, is defined as

$$\beta = \frac{4v_R L}{3v\langle v_z \rangle R}$$

and is essentially a measure of the *cross-sectional* stretching per element (as opposed to *axial* stretching). The analogous parameter in the static mixer is related to the pitch and aspect ratio of each helical unit.

Neglecting developing flows, a fluid particle 'jumps' from streamsurface to streamsurface in the manner shown in Figure 8.2.4. The trajectory can be obtained by repeatedly carrying out the following steps: integration from the beginning to the end of an element, turning the co-ordinate system by 90°, carrying out the integration until the end of the next element, and returning the co-ordinate system to its original orientation.

Figure 8.2.3. Streamlines for the cross-sectional flow in the partitioned-pipe mixer for different values of the normalized stream function ψ^*. Starting from the innermost streamlines, the values of ψ^* are 0.009, 0.031, 0.060, 0.091, 0.117, and 0.134.

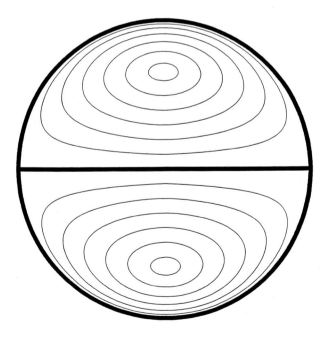

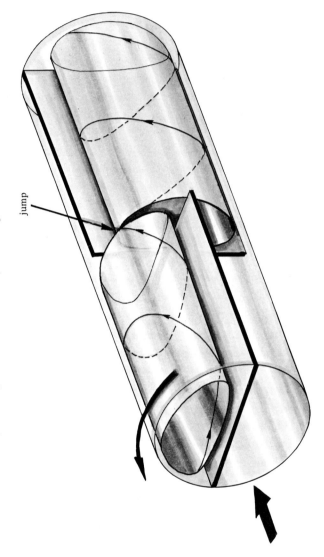

Figure 8.2.4. Qualitative picture of a particle trajectory in the flow field of the partitioned-pipe mixer; the particle jumps from streamsurface to streamsurface of adjacent elements. The streaklines and particle paths coincide; in particular they can not go backwards as in the system of Figure 8.3.3.

jump

It is clear from this procedure that though the particle path is continuous, its derivatives are not, and this results in infinite stresses. However, including the developing flows would significantly complicate the problem (a flow involving smooth trajectories is analyzed in Section 8.3).

8.2.2. *Poincaré sections and three-dimensional structure*

The flow in this case is periodic in axial distance rather than time (as for example, all cases of Chapter 7), so that the most convenient choice for the surfaces of section is the cross-sectional planes at the end of each periodic unit (consisting of two adjacent elements). Poincaré maps are then generated by recording every intersection of a trajectory with the surfaces of section in a very long (ideally, infinitely long) mixer, and projecting all the intersections onto a plane parallel to the surfaces. Every trajectory intersects with each surface of section, and the Poincaré map should capture some of the mixing in the cross-sectional flow.

Quite conceivably, any set of cross-sectional planes separated from each other by the length of a periodic unit could be used to generate a Poincaré section by this criterion. However, in all other cases it would not be possible to exploit the symmetries of the mapping (which reduces the computational effort by a factor of 8, Khakhar, Franjione, and Ottino, 1987).

Figure 8.2.5 shows the Poincaré sections for the partitioned-pipe mixer for various values of the mixing strength ($\beta = 2, 4, 8, 10$). The velocity field in the partitioned-pipe mixer is three-dimensional and it is important to study the relation of the Poincaré sections to the overall flow. We visualize the three-dimensional flow by plotting Poincaré sections at intermediate lengths, which enables us to follow the progress of the KAM curves through the mixer. In Figure 8.2.6 we have plotted these for $\beta = 2$, for increasing values of the axial distance, $z = 0.2, 0.8, 1.4, 2.0$ (notice the asymmetry of intermediate Poincaré sections). Each KAM curve then represents the intersecton of a *tube* with a surface of section, so that the tubes can be reconstructed by joining the KAM curves with their images in neighboring Poincaré sections by smooth surfaces (e.g., curve A in Figure 8.2.6). The last figure in the series (8.2.6(*d*)) gives the periodicity of various islands; apparently, the smaller islands have higher periods. Notice also in Figure 8.2.6 that the cross-sectional area of the tubes *is not* constant. This results from changes in the average axial velocity in the tube as it winds through the mixer; the total flow in the tube (average velocity in the tube multiplied by the cross-sectional area of the tube),

Figure 8.2.5. Poincaré sections for the partitioned-pipe mixer for various values of the mixing strength β: (a) $\beta = 2.0$, (b) $\beta = 4$, (c) $\beta = 8$, and (d) $\beta = 10$. Initial positions for the different trajectories were placed on an $(r-\theta)$ grid: eight r-values uniformly spaced between 0 and 1, and five θ-values uniformly spaced between $45°$ and $135°$. Each point was mapped for 300 iterations. (Reproduced with permission from Khakhar, Franjione, and Ottino (1987).)

(a) $\beta = 2.0$

Figure 8.2.5 *continued*

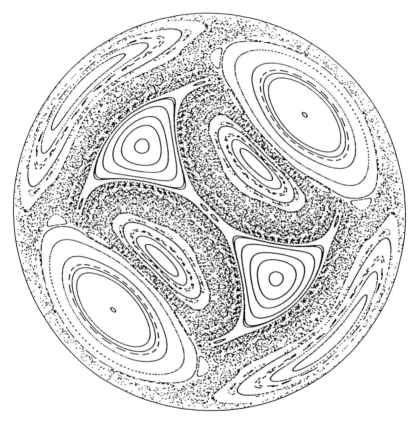

(*b*) $\beta = 4.0$

Figure 8.2.5 *continued*

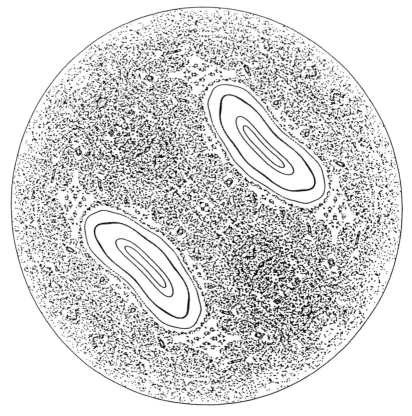

(c) $\beta = 8.0$

Figure 8.2.5 *continued*

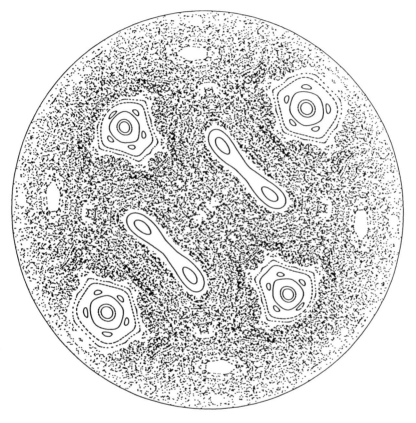

(*d*) $\beta = 10.0$

Figure 8.2.6. Poincaré sections for $\beta = 2.0$ at intermediate lengths along a periodic unit of length 2. Notice the asymmetry of the Poincaré sections. (a) $z = 0.2$, (b) $z = 0.8$, (c) $z = 1.4$, (d) $z = 2.0$. The letter A in (a) through (d) follows the same island down the length of the mixer (the reader might try to follow the evolution of some of the other islands). The numbers in (d) refer to the periodicity of the islands. (Reproduced with permission from Khakhar, Franjione, and Ottino (1987).)

(a)

Figure 8.2.6 *continued*

(*b*)

Figure 8.2.6 *continued*

(c)

Figure 8.2.6 *continued*

(*d*)

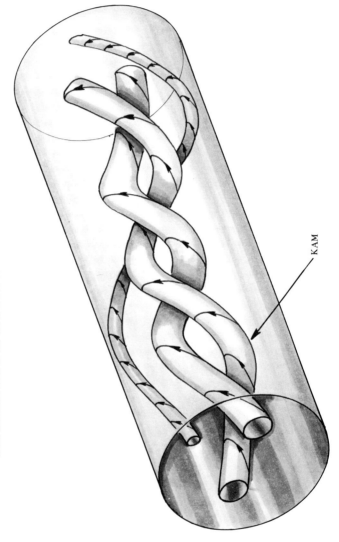

Figure 8.2.7. Schematic view of invariant tubes winding through the mixer. According to Figure 8.2.6(d) the smaller islands have higher periods (note that axial length scale is different from that of Figures 8.2.1 and 8.2.4). The cross sectional area of the tubes is not constant.

KAM

however, remains constant. A qualitative three-dimensional view of the tubes is shown in Figure 8.2.7.

Obviously the tubes corresponding to the KAM curves are invariant surfaces, and cannot be crossed by fluid particles. Consequently, the fluid flowing in a particular tube remains in the tube and cannot mix with the rest. Chaotic trajectories, on the other hand, wander in the regions left free by the tubes on two-dimensional homoclinic manifolds. Chaotic trajectories come close to $r = R$; such a device should be efficient in mass transfer operations between the bulk of the fluid and the wall of the tube.

We should note that the axial flow has a major effect on the Poincaré sections, and thus on the cross-sectional mixing. For example, the Poincaré section corresponding to plug axial flow (Figure 8.2.8(a)) and the same average axial velocity is quite different from the corresponding Poiseuille flow ($\beta = 2$, Figure 8.2.5). Another parameter that has a considerable effect on the Poincaré section, and thus the mixing, is the sense of rotation in the adjacent elements. The Poincaré section (Figure 8.2.8(b)) for the counter-rotating case, which corresponds to the configuration in the Kenics[R] mixer, indicates that the flow is chaotic over most of the cross-section, and seems to mix better than the co-rotating case ($\beta = 2$). It is not completely clear why this is so and a deeper study is required.

8.2.3. Exit time distributions

A pertinent question in open systems is whether or not the phenomena described in the previous section can be captured by various types of exit time distributions. In order to obtain an exit age distribution, we calculate the trajectories and residence times for a large number of particles, initially distributed uniformly (on a square grid) over the cross-section at the entrance of the mixer.[3] The particles then represent a pulse at the entrance of the mixer, and their exit age is found by integrating

$$\frac{\mathrm{d}t}{\mathrm{d}z} = \frac{1}{v_z}$$

along with the equations of motion.[2]

The exit age distributions for a mixer with 10 elements, and two different values of $\beta(2, 8)$ are shown in Figure 8.2.9, along with the corresponding contour plots of iso-residence times showing the residence times of particles based on their initial location in the cross-section. At low values of β (Figure 8.2.9(a)) when the regular islands are abundant and occupy much of the cross-section, we see that the exit age distribution has two peaks.

In some sense, the exit age distribution gives an indication of the inhomogeneity of the cross-sectional mixing: fluid streams in some tubes emerge faster than others, and there is no mixing between the streams in the different tubes. This is confirmed by a contour plot of iso-exit times and it is possible to identify the regions responsible for the two peaks. At higher values of the mixing strength, $\beta = 8$ (Figure 8.2.9(b)), the second peak disappears, in spite of the rather non-uniform cross-sectional mixing reflected in the Poincaré section and the contour plot of iso-exit times. In this case, some particles of the peak belong to the regular island, others to the 'chaotic' region outside. *In fact, the lowest residence time corresponds*

Figure 8.2.8. (a) Poincaré section for the partitioned-pipe mixer with plug flow. (b) Poincaré section for the partitioned-pipe mixer for counter-rotating cross-sectional flows in adjacent elements. The initial positions of the trajectories are the same as in the Poincaré sections of Figure 8.2.5. In both cases, $\beta = 2$. (Reproduced with permission from Khakhar, Franjione, and Ottino (1987).)

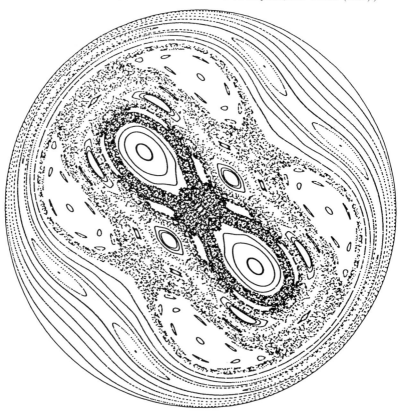

(a) $\beta = 2.0$

to the region outside the islands. In this case, the peak is more spread out
and is shifted towards higher residence times. The results here show
some of the limitations of exit age distributions for capturing the complex
structure of the mixing in continuous flows. It is clear that residence
time distributions alone are insufficient for an analysis of the mixing.

8.2.4. Local stretching of material lines

Let us examine now if the Poincaré sections are able to capture the details
of the stretching produced in the flow (only a few details of the analysis
are given here; for a more complete treatment the reader should consult
Khakhar, Franjione, and Ottino (1987)). The total length stretch (ln λ),
average specific rate of stretching (α), and average efficiency $\langle e \rangle$, were
calculated for a mixer consisting of 10 elements, for a number of material

Figure 8.2.8 *continued*

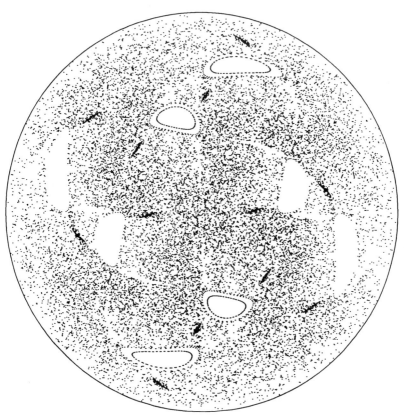

(*b*) $\beta = 2.0$

Figure 8.2.9. (a1, b1) Exit age distributions for the partitioned-pipe mixer, and (a2, b2) the corresponding contour maps showing iso-residence time curves based on the initial positions of the particles in the cross-section. Of 5,000 particles on a rectangular grid in the top half of the mixer, those which fell inside the cross-section were used in the calculations. In (a1, a2) $\beta = 2$. The initial location of the particles corresponding to each peak is shown in the contour plot by the numbers 1 and 2. In (b1, b2) $\beta = 8$ (reproduced with permission from Khakhar, Franjione, and Ottino, 1987).

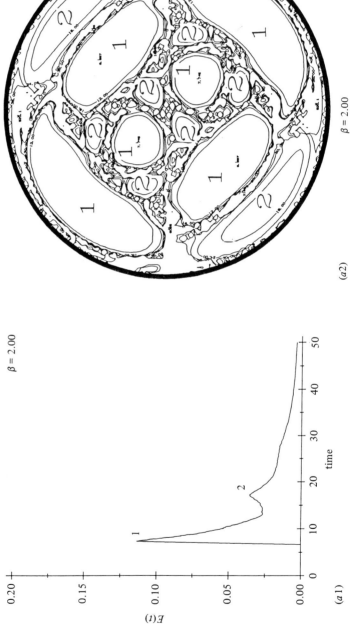

$\beta = 2.00$

(a1)

$\beta = 2.00$

(a2)

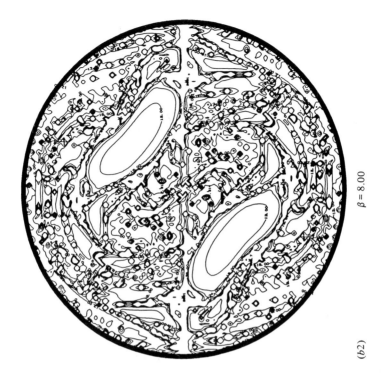

$\beta = 8.00$

(b2)

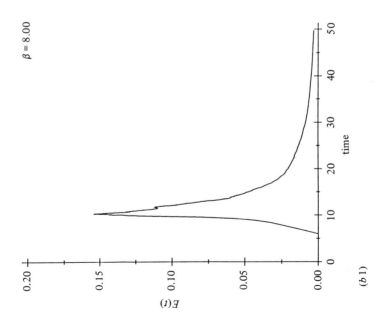

$\beta = 8.00$

(b1)

elements distributed on the same grid used for the residence time calculations. The results for two values of $\beta(2, 8)$ are shown in Figure 8.2.10, which indicates the initial positions of elements with the highest and lowest stretching. The patterns are reminiscent of the Poincaré sections. The material element is represented by a dot if the quantity being displayed is less than the specified value, and by a slash if it is greater than the specified value. Our expectation, mostly based on results of the

Figure 8.2.10. Initial location in the cross-section of material elements with the largest and smallest: length stretch $(a1, b1)$, and average efficiency $(a2, b2)$. The grid is the same as used for the residence time distribution, the initial orientation of all material elements is $\mathbf{m}_0 = (1, 0, 0)$ in cylindrical co-ordinates. Dot, quantity is smaller than the specified value; slash, quantity is larger than the specified value. In $(a1, a2)$ $\beta = 2$. $(a1)$ $\log \lambda$: dot, <3, slash, >4; $(a2)$ $\langle e \rangle$: $(b2) \langle e \rangle$: dot, <0.16, slash, >0.17 (reproduced with permission from Khakhar, Franjione, and Ottino, 1987).

$(a1)$ $\beta = 2.0$

two-dimensional time-periodic flows of Chapter 7, is that the fastest stretching takes place in the chaotic regions of the two-dimensional Poincaré sections, while the stretching in the regular regions is slow and inefficient. However, these calculations indicate that this is not true, in general, in three-dimensional flows. In Figure 8.2.10(*a*) the stretching is larger in the chaotic regions (as expected) but the efficiency is lower. Figure 8.2.10(*b*) is even more dramatic; both the stretching and the efficiency are larger in the regular regions. These seemingly contradictory results are easily rationalized: The jumping from streamsurface to streamsurface results in a continual reorientation of material elements in the regular regions (see Section 4.7). It thus appears that Poincaré sections alone cannot give a qualitative description of the entire mixing process, and must be augmented by other analyses.

Figure 8.2.10 *continued*

(*a*2) $\beta = 2.0$

Problem 8.2.1*

Consider a mixer of field dimensions (L, R). What is the optimal value of β to maximize the mixing efficiency? Attempt an analysis.

8.3. Mixing in the eccentric helical annular mixer

The eccentric helical annular mixer (EHAM) provides an illuminating counterpart for comparison of the results obtained in the previous section. The cross-section of this system corresponds to the journal bearing flow of Section 7.4 and the axial flow is a pressure driven Poiseuille flow (Figure 8.3.1).[4] In this case the system is time-periodic rather than spatially periodic and the solution for the flow field, under creeping flow, involves

Figure 8.2.10 *continued*

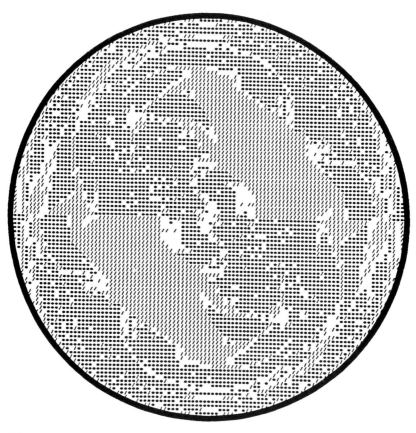

(*b*1) $\beta = 8.0$

no additional approximations. Furthermore, there is no discontinuous jumping of particles between streamsurface and streamsurface, and the particle paths have continuous derivatives. It is therefore of importance to investigate the analogies and differences that might exist with respect to the partitioned-pipe mixer.

In this particular example we will consider that the inner and outer cylinder rotate with a period T as

$$V_{\theta,\text{outer}} = U \cos^2(\pi/T)$$
$$V_{\theta,\text{inner}} = U \sin^2(\pi t/T)$$

and the relevant parameter is UT/R_{outer} (the only effect of the Poisseuille flow is to compress or stretch the particle paths in the axial direction).

Figure 8.2.10 *continued*

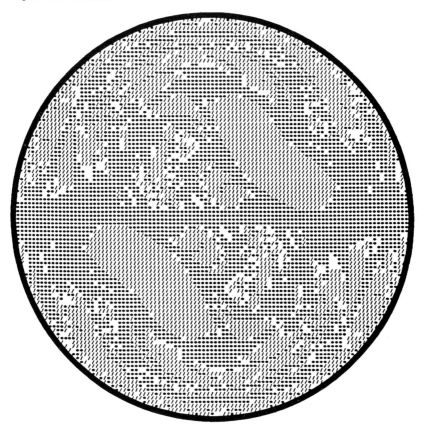

(b2) $\beta = 8.0$

In this particular case a literal 'surface of section', i.e., the marking of intersections of trajectories of initially designated initial conditions with periodically spaced planes perpendicular to the flow, results in an 'out-of-focus' picture (see Figure 8.3.2; in some cases 'islands' might disappear completely). A section in which intersections are recorded at designated time intervals, i.e., a Poincaré section, however, reveals no information about the axial structure of the flow since it is identical to those obtained for the journal bearing flow (Section 7.4). It is thus apparent that, in contrast with the case of the partitioned-pipe mixer, conventional Poincaré sections are not useful in investigating the structure of the flow and we have to resort to other diagnostic tools.

The most revealing visualization in this case is provided by streaklines (Figure 8.3.3). The calculations however, are considerably more difficult than those of the Poincaré sections, since the storage and computational requirements for smoothness increase exponentially with flow time. According to the location of injection the streaklines can undergo complex trajectories reminiscent of the Reynolds's experiment,[5] or 'shoot through' the mixer undergoing relatively little stretching. The dynamics of the streaklines have been relatively unexplored and more

Figure 8.3.1. Schematic of the eccentric helical annular mixer. The length of the mixer is L. In the case of the figure the cross flow corresponds to 7.4.2(c).

work is necessary. It is significant that even though the axial flow always moves forward, the streaklines can 'go backward' since they can wander into regions of low axial velocity, whereas other parts of the streaklines can bulge forward. Figure 8.3.3(*a*) shows an instantaneous snapshot of a lateral projection of a streakline injected in a chaotic region, as well as a view from the end of the mixer. It is significant to notice that in the 'end view' the streaklines crosses itself at various points, and that these streaklines would be the ones observed for a system with plug flow, or equivalently the journal bearing flow (see also Example 8.5.1). Figure 8.3.3(*b*) shows a similar computation but now the streakline is injected in a regular region. (The situation however is considerably more complex; the 'regular regions' are not static but move in a time periodic fashion;

Figure 8.3.2. The marking of intersections of trajectories of initially designated initial conditions with periodically spaced planes results in an 'out-of-focus' picture. Close observation reveals that the picture is not symmetric.

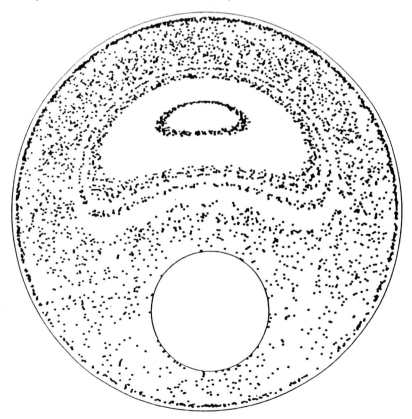

thus the dye injection apparatus can inject material in both regular and chaotic regions depending upon the injection time.)

Various other diagnostic tests used for the partitioned-pipe mixer in the previous section are valid also for the EHAM. In particular, it is possible to inject a dense grid of passive particles or vectors with various orientations and to compute contours of constant age, stretching, efficiency, and so forth. Multiple peaks in the exit distribution are possible in this mixer. Figure 8.3.4 shows an instantaneous picture of the axial concentration distribution created by the injection of a square array of particles at the entrance of the mixer (the concentration has been averaged with respect to the cross section). It thus appears that the peaks present in the partitioned-pipe mixer are not due to any artificiality present in the mixer and that similar behavior is indeed possible in other systems. Contours of constant stretch reveal that in this case, as opposed to the partitioned-pipe

Figure 8.3.3. Instantaneous picture of a streakline injected in: (*a*) chaotic region, and (*b*) regular region. In both cases the flow moves from left to right (the axial scale is different in both figures).

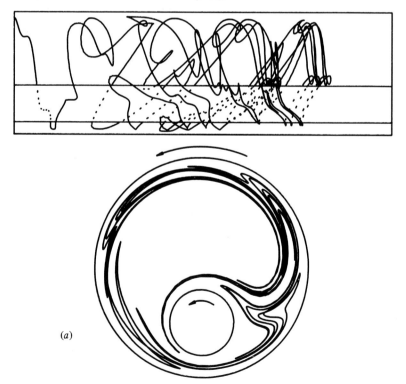

(*a*)

mixer, the stretching in the chaotic regions is exponential and linear in the regular regions.

8.4. Mixing and dispersion in a furrowed channel

The example described here was considered by Sobey (1985). The system consists of a channel with a minimum gap $2h$ and with boundaries given by

$$x_2 = 1 + f(x_1) \qquad \text{and} \qquad x_2 = -1 - g(x_1)$$

where $f(x_1)$ and $g(x_1)$ are sinusoidal functions given by

$$f(x_1) = (D/2)[1 - \cos(2\pi x_1/L)]$$
$$g(x_1) = (K/2)\{1 - \cos[(2\pi x_1/L) + \phi]\}.$$

Thus, the case $D = K$ and $\phi = 0$ is a symmetric channel; if $K = 0$ one of the walls is flat. Sobey solved the Navier–Stokes equations in this geometry and focused on the case of no net oscillatory flow with a frequency ω. The volumetric flow rate, $Q(t)$ is given by[6]

$$Q(t) = 2hU \sin(2\pi\omega t).$$

For a fixed geometry, the flow is described by two parameters: the Reynolds number, $Re = Uh/v$, and the Strouhal number, $Sr = \omega h/U$, which

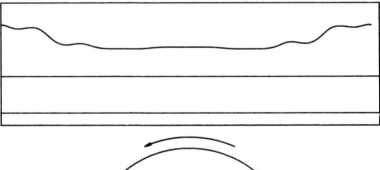

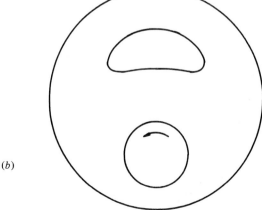

(b)

can be thought of as h/(average distance travelled axially by a fluid particle). Computations indicate that there is a region in the Sr–Re plane where flow separation occurs. At very small values of Sr the flow is quasi-steady and if separation is present (due to R) the vortices grow and decay in phase with $Q(t)$. At intermediate values of Sr the vortices do not decrease in size when $Q(t)$ decreases, but instead continue to grow and entrain further fluid. However, if Sr is very large, the flow is viscous dominated and vortices do not form. In this case the distance travelled by a typical particle, between cycle and cycle of the flow, tends to zero and there is no time for separation to occur.

The basic flow patterns obtained by Sobey by solving the Navier–Stokes equations, using a vorticity–streamfunction formulation (Gillani and Swanson, 1976), are given in Figure 8.4.1(a)–(f) which show the instantaneous streamlines for the flow at $Re = 75$ and $Sr = 0.01$ as a function of time. Figure 8.4.1(a) represents the beginning of the cycle, 8.4.1(b) the

Figure 8.3.4. Instantaneous picture of the axial concentration (averaged over the cross-section) created by the injection of a square array of particles at the entrance of the mixer ($UTR_{out} = 2$). In this case the particles in the regular regions move faster than the average speed.

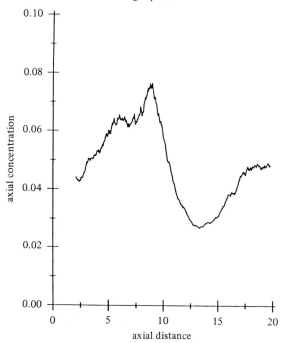

middle of it. The same flow was studied by Ralph (1986) for Reynolds numbers up to 300 and a range of Strouhal numbers of 10^{-4} to 1.

As we have seen in previous examples (notably, Section 7.4) the streamlines present only part of the picture insofar as mixing is concerned. Nevertheless, a superficial analysis indicates that the flow will be able to mix well since it consists of two interacting co-rotating vortices as the blinking vortex flow (Section 7.3) and the cavity flows (Section 7.5). In a rough sense we might interpret *Sr* as $1/\mu$, if the circulation of the vortex is taken to be *Uh*. Also in the symmetric case we expect the mixing to be rather poor since it is obvious that the centerline is an invariant surface of the flow, and prevents interaction between the regions $x_2 > 0$ and $x_2 < 0$.

As we have seen in previous examples, two-dimensional time-periodic flows are far richer in structure than what might be revealed by a cursory analysis and an in-depth analysis requires an entire arsenal of tools. It is apparent that this flow contains periodic points and is capable of homoclinic and heteroclinic behavior. However, an analysis on the level of that presented for the TW and BV flows is tremendously complicated and only a few aspects of the program of Table 7.1 can be completed. Here we present the results of Sobey for macroscopic dispersion of tracer

Figure 8.4.1. Instantaneous streamlines at various points in time for periodic flow at $Re = 75$ and $Sr = 0.01$, for various values of ωt: (*a*) 0.1, (*b*) 0.25, (*c*) 0.4, (*d*) 0.5, (*e*) 0.55, and (*f*) 0.75 (reproduced with permission from Sobey, 1985).

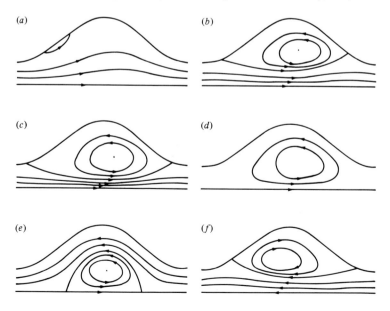

particles. The equations to be solved (numerically) are:

$$Sr \, dx_1/dt = v_1(x_1, x_2, t; Re)$$
$$Sr \, dx_2/dt = v_2(x_1, x_2, t; Re).$$

Note that, in general, v_1 and v_2 are periodic but the period need not be equal to ω^{-1}. Also, the identification of periodic points of the velocity field is far from trivial.

Figures 8.4.2–4 show the dispersion, and initial condition, of a plug of 3,000 passive particles for $Sr = 0.02$ and $Re = 10$, 25, and 100, *after one cycle of the flow*. Qualitatively, it appears that there is an optimum Re for a given Sr. Figures 8.4.5–6 show similar processes occurring in a channel with a phase angle $\phi = \pi$ and equal amplitudes and another case corresponding to one wall with amplitude $K = 0$. A measure of the axial dispersion is given by the variance of the axial particle positions $(x_1)_i$, $i = 1, \ldots, N$,

$$\sigma^2 = \frac{1}{N-1} \sum_{i=1}^{N} [(x_1)_i - \langle x_1 \rangle]^2$$

Figure 8.4.2. Evolution of an initial condition consisting of 3,000 passive particles, for $Sr = 0.02$ and $Re = 10$, after one cycle of the flow. The frames correspond to: (a) $\omega t = 0$, (b) $\omega t = 0.2$, (c) $\omega t = 0.5$, (d) $\omega t = 0.6$, (e) $\omega t = 0.75$, and (f) $\omega t = 1.0$. (Reproduced with permission from Sobey, 1985.)

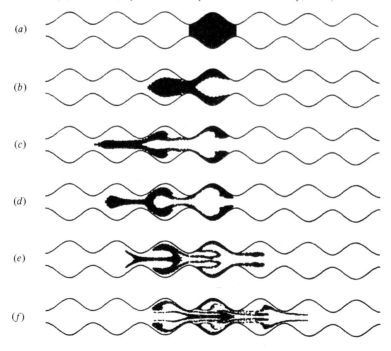

where $\langle x_1 \rangle$ is given by

$$\langle x_1 \rangle = \frac{1}{N} \sum_{i=1}^{N} (x_1)_i.$$

The value of σ^2 depends on the initial placement of the particles $(x_1)_i$ and the time at which they are placed in the periodic flow (release time, t_0). Computations indicate that there can be some contraction of the cloud of particles, i.e., the growth of σ^2 is not monotonic, but that if we integrate over t_0 the variance grows almost linearly in time.[7] It is unclear however, if this behavior is general or if other types of dispersion laws are possible in this system as well as the systems of Sections 8.2 and 8.3.

Problem 8.4.1

Consider that the particles are not material particles but rather actual tracer particles. Discuss the effect of the particle size on the process, i.e., account for Brownian diffusion and/or inertial effects.

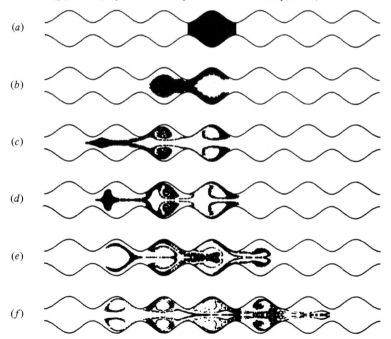

Figure 8.4.3. Evolution of an initial condition consisting of 3,000 passive particles for $Sr = 0.02$ and $Re = 25$, after one cycle of the flow. The frames correspond to (a) : $\omega t = 0$, (b) $\omega t = 0.25$, (c) $\omega t = 0.5$, (d) $\omega t = 0.6$, (e) $\omega t = 0.75$, and (f) = 1.0. (Reproduced with permission from Sobey, 1985.)

8.5. Mixing in the Kelvin cat eyes flow

Consider again the flow of Example 6.7.1. As we have seen, the stream-function with respect to a fixed frame is of the form

$$\psi(x_1, x_2, t) = ux_2 + \ln[\cosh x_2 + A \cos(x_1 - ut)] \qquad (8.5.1)$$

(here we have made the streamfunction dimensionless with respect to $\Delta U h/2$, the coordinates x_1 and x_2 with respect to h, the velocity U with respect to $\Delta U/2$, and the time with respect to $2h/\Delta U$). In frame x_1', x_2' moving with the mean flow the streamfunction $\psi(x_1, x_2, t)$ can be reduced to

$$\psi'(x_1', x_2') = \ln(\cosh x_2' + A \cos x_1') \qquad (8.5.2)$$

which has a streamline portrait known as the 'cat eyes' (Figure 8.5.1). The velocity of a fluid particle with respect to the moving frame is given by

$$f_1 = dx_1'/dt = \sinh x_2'/(\cosh x_2' + A \cos x_1')$$
$$f_2 = dx_2'/dt = A \sin x_1'/(\cosh x_2' + A \cos x_1').$$

Since the velocity field has heteroclinic trajectories we expect that

Figure 8.4.4. Evolution of an initial condition consisting of 3,000 passive particles, for $Sr = 0.02$ and $Re = 10$, after one cycle of the flow. The frames correspond to: (a) $\omega t = 0$, (b) $\omega t = 0.25$, (c) $\omega t = 0.5$, (d) $\omega t = 0.6$, (e) $\omega t = 0.75$, and (f) $\omega t = 1.0$. (Reproduced with permission from Sobey, 1985.)

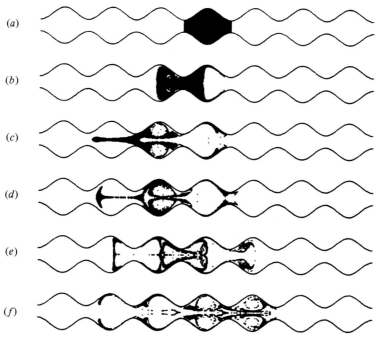

perturbations would lead to transversal intersections of stable and unstable manifolds. As we have seen, a useful technique to determine the existence of transversal intersections is the Melnikov method (Section 6.9). As a simple area preserving perturbation[8] we take

$$g_1 = \varepsilon \sin(\omega t)$$
$$g_2 = 0.$$

which gives a Melnikov integral of the form

$$M(t_0) = [-\varepsilon A \cos(\omega t_0)/(1 + A)]F(\omega),$$

where $F(\omega)$ is given by

$$F(\omega) = \int_{-\infty}^{\infty} \sin(x_1)|_{\text{manifold}} \sin(\omega t)\, dt.$$

The function $F(\omega)$ is plotted in Figure 8.5.2. Since $M(t_0)$ represents the distance between the perturbed manifolds, we expect that an extreme in $F(\omega)$ should maximize the 'extent of chaos'. In this particular case, the optimal frequency, ω, appears to be in the neighborhood of 0.3. It is important to check whether or not such a prediction is substantiated by

Figure 8.4.5. Similar initial conditions as Figures 8.4.2–4, but in a channel with a phase angle $\phi = \pi$ and equal amplitudes. Dispersion after one cycle of the flow. The frames correspond to: (a) $\omega t = 0$, (b) $\omega t = 0.25$, (c) $\omega t = 0.5$, (d) $\omega t = 0.6$, (e) $\omega t = 0.75$, and (f) $\omega t = 1.0$. (Reproduced with permission from Sobey, 1985.)

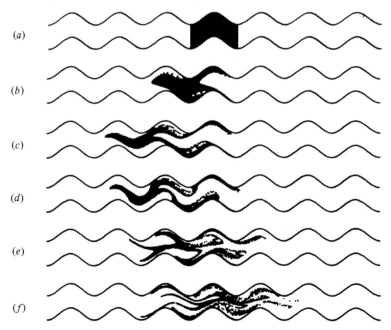

(a)

(b)

(c)

(d)

(e)

(f)

other techniques, such as Poincaré sections. A Poincaré section for this problem consists of recording intersections of initially designated fluid particles with the x_1'–x_2' plane every 2π time units. Also, taking advantage of the spatial periodicity of the flow in the x_1'-direction, we record the intersections making x_1' mod 2π. Visual inspection of the results shown in Figure 8.5.3 indicates that, indeed, chaos is maximized for $\omega \approx 0.3$ (substantially higher values of the frequency, $\omega \approx 4$, indicate a reduction in the degree of chaos and mixing).

Since streaklines would be the tool of choice in experimental studies, it is important to examine how they look in the perturbed case. Experimentally,

Figure 8.4.6. Similar initial conditions as Figures 8.4.2–4, but in a channel with a phase angle $\phi = \pi$ and one wall with $K = 0$. Dispersion after one cycle of the flow. The frames correspond to: (a) $\omega t = 0$, (b) $\omega t = 0.25$, (c) $\omega t = 0.5$, (d) $\omega t = 0.6$, (e) $\omega t = 0.75$, and $(f) = 1.0$. (Reproduced with permission from Sobey, 1985.)

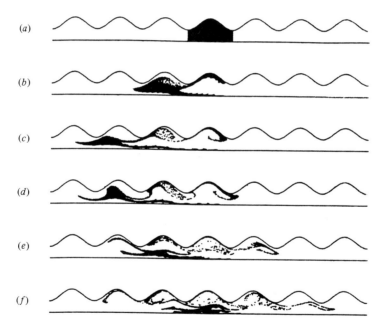

Figure 8.5.1. Streamline portrait corresponding to the streamfunction of Equation (8.5.2) for $A = 0.8$ (the spacing between vortices is 2π); $\Delta\psi$ is constant between streamlines.

the streaklines would be injected with respect to the fixed frame, but computationally it is possible also to investigate injection in a frame moving with the cat eyes (Figure 8.5.4). A comparison between the streaklines and the corresponding Poincaré section reveals that the streakline injected within the chaotic region undergoes significant stretching and folding whereas the streaklines injected within the 'cat eyes' and outside 'cat eyes' undergo little stretching. Figure 8.5.5 shows the stable and unstable manifolds associated with the hyperbolic points.

The streakline injected with respect to the fixed frame reveals also significant stretching and folding (Figure 8.5.6). However, in this case, as opposed to the Hama flow of Example 2.5.1, the streaklines serve as a good indicator of the position of the vortices (there are none in the Hama flow). It is important to notice that given the time-periodicity of the flow, it is possible to simplify substantially the problem of streakline tracking

Figure 8.5.2. Function $F(\omega)$ in the Melnikov integral.

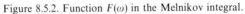
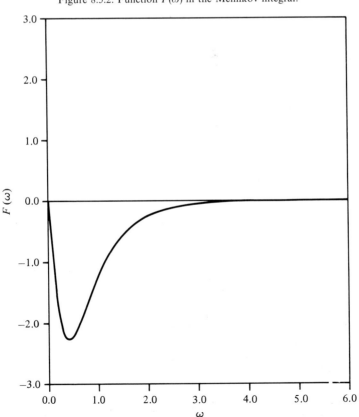

(obviously, a similar simplification is possible in the Hama flow); the points defining the streakline injected during the second period undergo the same stretching history as the line formed during injection in the first period, and so on.

A more realistic simulation of this flow might involve a sequence of vortices interacting in the fashion described in Example 6.7.2. In this case, however, the attack has to be necessarily computational and the utility of the Melnikov method is somewhat lost.

Example 8.5.1

A careful examination of Figure 8.5.4(b) shows the crossing of two streaklines injected with the chaotic region. Is this only possible in a chaotic region or could it happen in an integrable system? Examine the system

$$dx_1/dt = -x_2 + \sin \omega t, \qquad dx_2/dt = x_1,$$

Figure 8.5.3. Poincaré sections for the perturbed cat eyes flow ($A = 0.8, \varepsilon = 0.1$). The values of ω are as follows: (a) 0.1, (b) 0.2, (c) 0.3, (d) 0.4, (e) 0.5, and (f) 0.6. The maps consist of eight initial conditions and 100 iterations.

(a) (b)

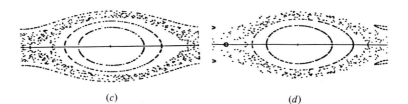

(c) (d)

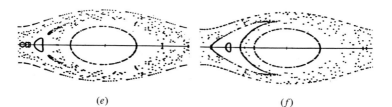

(e) (f)

with solution

$$x_1 = (x_1)_0 \cos t - (x_2)_0 \sin t + [\omega/(1 - \omega^2)](\cos \omega t - \cos t)$$
$$x_2 = (x_1)_0 \sin t + (x_2)_0 \cos t + [1/(1 - \omega^2)](\sin \omega t - \omega \sin t),$$

which is of the form

$$\mathbf{x} = \mathbf{Q}(t) \cdot \mathbf{x}_0 + \mathbf{f}(t).$$

The streakline passing through the point $\mathbf{x}'$ at time t, parametrized by t', is given by

$$\mathbf{x} = \mathbf{Q}(t - t') \cdot \mathbf{x}_0 - \mathbf{Q}(t - t') \cdot \mathbf{f}(t') + \mathbf{f}(t).$$

It is possible to show that this flow possesses streaklines which cross and return to their points of injection. For example, for $\omega = 3.1$ a streaklines injected at $t = 0$ at $(0.5, 0.5)$ returns to its initial location after

Figure 8.5.4. Snapshots of streaklines injected in a moving frame in the cat eyes flow: (a) outside the eyes, (b) within the chaotic region, and (c) within the eyes. The initial conditions (x_1', x_2') are: (a) (0.0, 2.0), (0.0, 1.8), (b) (0.0, 1.0), (0.0, 0.5), (0.5, 0.0), and (c) (2.0, 0.0). For comparison, a Poincaré section is given in (d). The parameters are: $A = 0.8$, $\varepsilon = 0.1$, $\omega = 0.1$. The total time of injection is four periods of the perturbation, $8\pi/\omega$.

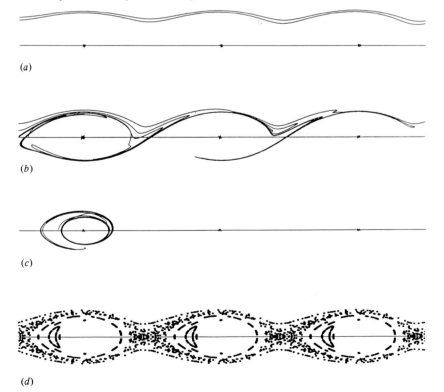

(a)

(b)

(c)

(d)

$t = 20\pi$. Compute some typical streaklines (Franjione and Danielson, 1987). Propose other examples.

8.6. Flows near walls

In this section we consider a few examples of structurally unstable flow fields near walls. The importance of this class of flows lies in the conjecture

Figure 8.5.5. Stable and unstable manifolds corresponding to the case $A = 0.8$, $\varepsilon = 0.1$, $\omega = 0.3$, for various injection times (the 'most chaotic system', see Figure 8.5.3(c)).

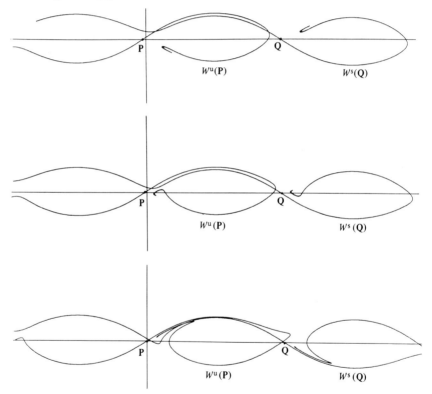

Figure 8.5.6. Snapshot of streaklines injected in a fixed frame in the cat eyes flow. The time of injection corresponds to two periods (spanning two cat 'eyes'). The position of the vortices is indicated by crosses. The parameters are: $A = 0.8$, $\varepsilon = 0.1$, $\omega = 0.1$

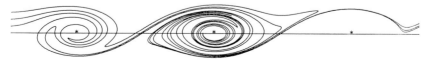

that chaotic particle trajectories might be a precursor to Eulerian turbulence. Also, since many of the flows presented here involve flow separation, they serve to highlight the role of separation and recirculation bubbles in creating favorable conditions for chaos. Even though the development of this section is not as complete as some of the preceding sections and Chapter 7, it should be clear that the possiblities are many and that some of the ideas presented here have a clear connection with the concepts of Section 5.10. The flows considered are asymptotically exact solutions of the Navier–Stokes and continuity equations in two- and three-dimensions. The presentation is based on recent work of Perry *et al.*[9] The starting point is to expand the Eulerian velocity field in a Taylor series around a point $\mathbf{p}$, i.e.,

$$\mathbf{u}(\mathbf{x}) = \mathbf{u}(\mathbf{p}) + (\mathbf{x} - \mathbf{p}) \cdot \nabla \mathbf{u}(\mathbf{x})\big|_{\mathbf{x}=\mathbf{p}} + \tfrac{1}{2}(\mathbf{x} - \mathbf{p})(\mathbf{x} - \mathbf{p}) : \nabla \nabla \mathbf{u}(\mathbf{x})\big|_{\mathbf{x}=\mathbf{p}} + \cdots$$

Each term $\nabla^{(n)}\mathbf{u}(\mathbf{x})\big|_{\mathbf{x}=\mathbf{p}}$ in the expansion represents a tensor of order $n + 1$. Denoting

$$A_{ijk\ldots} = (1/n!)\, \partial^n u_i / \partial x_j \partial x_k \ldots,$$

the expansion can be written as

$$u_i = A_i + A_{ij}x_j + A_{ijk}x_j x_k + A_{ijkl}x_j x_k x_l + A_{ijklm}x_j x_k x_l x_m + \cdots$$

where the variables associated with the $(n + 1)$ order tensor $A_{ijk\ldots}$ are of the form $(x_1)^a(x_2)^b(x_3)^c$ with $n = a + b + c$, where a, b, c are integers or zero ($c = 0$ for a two-dimensional flow).[10] It is clear that the tensors $A_{ijk\ldots}$ are symmetric in all indices but the first one due to the fact that the order of differentiation is immaterial. These tensors constitute the unknowns and are to be found by forcing the series to satisfy the continuity and Navier–Stokes equations as well as the boundary conditions of the problem in question. The coefficients may have time dependence or not, depending on whether the flow is steady or time varying (due to some external perturbation).

The number of independent coefficients, N_c, generated by an Nth order expansion of a three-dimensional flow field is given by

$$N_c = 3 \sum_{K=0}^{N} \sum_{J=0}^{K+1} J.$$

Substituting the velocity expansion into the continuity and Navier–Stokes equations and equating coefficients of equal power generates a series of *independent* relationships between the coefficients $A_{ijk\ldots}$. For example, the continuity equation generates just one relationship for $n = 1$, three for $n = 2$, and six for $n = 3$, and so on. For $n = 4$ the relationship between cofficients is given by

$$A_{11ijk} + A_{22ijk} + A_{33ijk} = 0,$$

Table 8.6.1

N	Two-dimensional			Three-dimensional		
	N_c	E_c	E_{NS}	N_c	E_c	E_{NS}
2	12	3	0	30	4	0
3	20	6	1	60	10	3
4	30	10	3	105	20	11
5	42	15	6	168	35	26
. . .	. . .	. . .	. . .	. . .	. . .	. . .
15	272	120	91	2448	680	1001

and the number of independent relationships is 20. The corresponding expression for a two-dimensional flow is

$$A_{11ijk} + A_{22ijk} = 0.$$

The general rule for the number of independent relations between coefficients generated by the continuity equation, E_c, is

$$E_c = \sum_{n=0}^{N} \sum_{J=0}^{n} J.$$

Similarly the number of independent relations generated by the Navier–Stokes equations, E_{NS}, is given by

$$E_{NS} = \sum_{n=3}^{N} \sum_{J=2}^{n-1} (2J - 1).$$

The corresponding relations for two-dimensional flow are:

$$N_c = 2 \sum_{J=1}^{N+1} J$$

$$E_c = \sum_{J=0}^{N} J$$

$$E_{NS} = \sum_{J=2}^{N-1} (J - 1).$$

A few tabulated values, taken from the work of Perry and Chong (1986) are given in Table 8.6.1. An examination of Table 8.6.1 shows that, in general, several boundary conditions should be specified in order to obtain the numerical value of the remaining coefficients. For example, for $n = 5$ we need a number of conditions to obtain the unspecified $42 - (15 + 6) = 21$ coefficients. Boundary conditions such as no-slip at the wall and a specification of the surface vorticity are commonly used and generate a series of non-linear equations in terms of the unknown tensor components. For time varying velocity fields the equations become a series of non-linear ordinary differential equations.

As an example, consider the flows displayed in Figure 8.6.1. In this case the velocity field was expanded to 5th order in $\mathbf{x}$, the boundary conditions are no-slip at the wall, and the vorticity at $x_2 = 0$ is specified to be

$$\omega_3 = K(x_1^2 - x_s^2)$$

where K is a constant which determines the strength of the vorticity, and x_s determines the points of flow separation and reattachment of the flow.

Figure 8.6.1. Two-dimensional flow generated by a fifth order perturbation with $K = 0.5$, $\theta_1 = 45°$, $x_s = 1$, $Re = 50$. In (a) the position of the elliptic point, (x_1^c, x_2^c) is $(0, 0.5)$ and a homoclinic orbit is produced; in (b) the point is closer to the wall $(0, 0.25)$ and a heteroclinic orbit appears, connecting the stagnation flows at the walls.

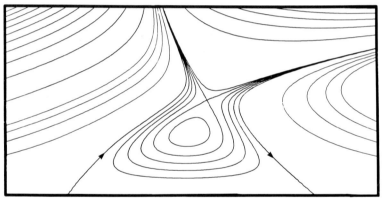

(a)

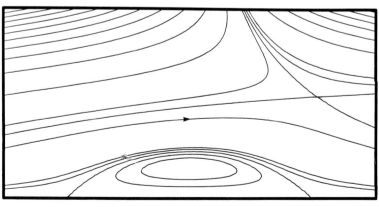

(b)

Since the vorticity changes sign at $\pm x_s$ there are two separation points. We also specify the angle of separation and reattachment of the streamlines θ. The relationship between the angle at the point of separation, measured in the clockwise direction from the wall, and the derivatives of pressure and vorticity evaluated at $x_2 = 0 \, x_1 = \pm 1$ (separation point), is given by

$$\tan \theta = 3\mu(\partial \omega_3/\partial x_1)/(\partial p/\partial x_1),$$

which specifies a relationship between the angle, the separation points, and the gradient of pressure. However, these conditions are not sufficient to specify all the coefficients. In order to close the equations the location of a critical point (for example an elliptic point) also must be specified. In this case we can specify the location of an elliptic point; the second critical point (i.e., the hyperbolic point in Figure 8.6.1(a)) arises as a result of solving the problem with the boundary conditions as stated above. Thus, in this problem, a fifth order expansion, the total number of unknowns is 42, and according to Table 8.6.1, the number of independent relations between coefficients provided by the Navier–Stokes equation is 6, whereas the number of relations provided by the continuity equation is 15. Therefore the number of coefficients to be determined is 21. It can be shown that for a two-dimensional flow the no-slip condition provides $2(N + 1)$ relations, the specification of the surface vorticity N relations, and the angle relations and the location of the elliptic point (x_2^e, x_2^e), two conditions each. For $N = 5$, this provides an additional 21 relations, from which all coefficients can be determined.[11]

In this particular example the length scale L is selected as x_s, and the time scale T as $1/[K(x_s)^2]$. The resultant velocity field is given by

$$v_1 = A_{12}x_2 + A_{122}x_2^2 + 3A_{1112}x_1^2x_2 + 3A_{1122}x_1x_2^2 + A_{1222}x_2^3$$
$$+ 6A_{11122}x_1^2x_2^2 + A_{12222}x_2^4 + 10A_{111122}x_1^3x_2^2$$
$$+ 5A_{112222}x_1x_2^4 + A_{122222}x_2^5$$

$$v_2 = -3A_{1112}x_1x_2^2 - A_{1122}x_2^3 - 4A_{11122}x_1x_2^3 - 10A_{111122}x_1^2x_2^3 - A_{112222}x_2^5$$

where the coefficients, and relationships between coefficients, is given by

$$A_{12} = -1$$
$$A_{1112} = 1/3$$
$$A_{1222} = -2/3$$
$$A_{11122} = (3 - A_{1222})/6$$
$$A_{12222} = 1/(x_2^e)^3 + 2/(3x_2^e) + (1/10)RA_{1122}x_2^e - A_{122}/(x_2^e)^2$$
$$A_{111122} = -(3/10)A_{1122}$$
$$A_{112222} = -A_{1122}/(x_2^e)^2$$
$$A_{122222} = -(1/10)RA_{1122}$$

and where the values of A_{1122} and A_{122} are given by the solution of the system

$$\frac{1}{3} - 20A_{1122}\left(\frac{6}{10} + \frac{1}{(x_2^e)^2}\right) = 0$$

$$-A_{122}\left(\frac{1}{3} + \frac{1}{(x_2^e)^2}\right) + 1 + \frac{1}{(x_2^e)^3} + \frac{2}{3x_2^e} + \frac{A_{1122}}{10}Rx_2^e = 0$$

where R is the Reynolds number of the flow $R = (L^2/T)/v = K(x_s)^4/v$, which is taken equal to one in the computational results shown in Figure 8.6.1(a). The flow constructed has a homoclinic orbit. It appears that the flow is structurally unstable in the class of *all* perturbations and that a simple time-periodic perturbation will most likely give rise to Lagrangian turbulence.

Another example is shown in Figure 8.6.1(b). In this case the flow is subject to the same boundary conditions as Figure 8.6.1(a), with the exception that the arbitrarily designated elliptic point has been moved closer to the wall. The topology of the flow is radically altered. A streamline separates from the wall and reattaches again after some distance. In this case the connection is heteroclinic and the most likely place for chaos is in the region near the end of the unstable manifold (as it 'reattaches' to the wall).

The range of applicability of these truncated velocity field equations can be ascertained by means of various checks. For example, the pressure gradient field can be generated in two different ways. One possibility is to expand ∇p in terms of a Taylor series; the coefficients of the expansion can be obtained in terms of the known $A_{ijk...}$s. Another possibility is to write ∇p in terms of the Navier–Stokes equation and then to replace the truncated velocity field into $\mathbf{v} \cdot \nabla \mathbf{v}$ and $\nabla^2 \mathbf{v}$. This generates a second expansion for ∇p containing more terms. The percent difference at an arbitrary point between the two ∇p fields is taken to be a measure of the accuracy, or more properly, consistency, of the solution at that point. Another way to check the accuracy of a solution is to expand about a different point in the field and compare the percent difference in the magnitude of the velocity from both points. In fact, if great accuracy were desired in any particular point in the fluid, an expansion could be done at that point.[12] For instance, it may be advantageous to expand about a homoclinic or heteroclinic point in the flow if time dependent perturbations were being carried out. Figure 8.6.2 shows an example generated by this method in the case of a three-dimensional flow.

Time dependent perturbations can produce chaotic trajectories in two-dimensional flows with homoclinic and heteroclinic trajectories. Figure 8.6.3 shows an example generated by perturbing the location of the elliptic point according to

$$x_1^e = 0, \qquad x_2^e = d + \varepsilon \sin(2\pi\omega t).$$

The time evolution of the coefficients A_{122} and A_{1222} is given by

$$\frac{\mathrm{d}A_{1122}}{\mathrm{d}t} = \frac{1}{3} - 20A_{1122}\left(\frac{6}{10} + \frac{1}{(x_2^e)^2}\right)$$

$$\frac{\mathrm{d}A_{122}}{\mathrm{d}t} = 12\left[-A_{122}\left(\frac{1}{3} + \frac{1}{(x_2^e)^2}\right) + 1 + \frac{1}{(x_2^e)^3} + \frac{2}{3x_2^e} + \frac{A_{1122}}{10}x_2^e\right].$$

The system of equations can be investigated with an array of techniques suitable for ordinary differential equations. (See, for example, Guckenheimer and Holmes, 1983; Arnold, 1985, 1983.) This is one of the few instances within the work presented here that the systems can be volume

Figure 8.6.2. Three-dimensional flow field generated by a fifth order perturbation: (*a*) surface flow pattern, (*b*) three-dimensional view. (Reproduced with permission from Perry and Chong (1986).)

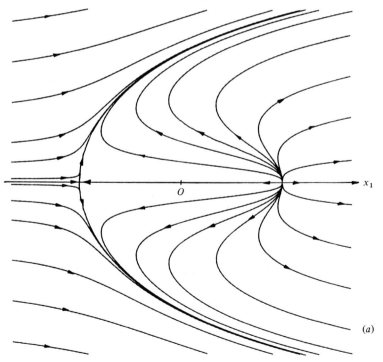

(*a*)

Figure 8.6.2 *continued*

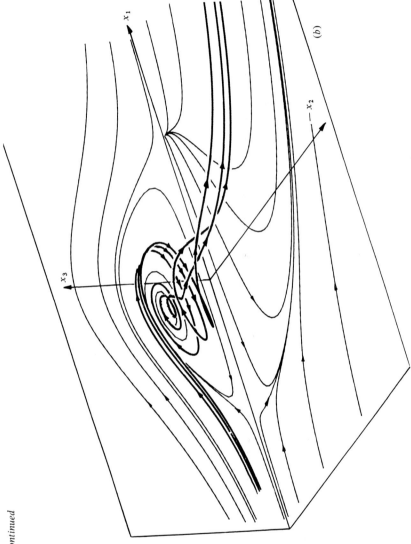

(b)

contracting and strange attractors appear to be possible (of course this depends upon the number of dimensions and the exact details of the systems). This brings up the possibility of complex Eulerian signals.

The computations correspond to the following values of the parameters: $d = 0.2$, $\varepsilon = 0.05$, $\omega = 8/2\pi$, $R = 1$. Figure 8.6.3(a) shows a Poincaré section corresponding to the integrable system ($\varepsilon = 0$), whereas Figure 8.6.3(b) shows a Poincaré section corresponding to similar conditions but for the time-dependent perturbation of the elliptic point. Note that the points leak as the unstable manifold 'reattaches' to the wall which is shown in Figure 8.6.3(c)

Problem 8.6.1

Study the restrictions imposed by non-slip on a wall with normal x_2. Prove that in a third-order expansion of the velocity field, the no-slip condition implies that the following coefficients are zero: A_1, A_2, A_{11}, A_{21}, A_{111}, A_{211}, A_{1111}, A_{2111}.

Problem 8.6.2

Show that the restriction imposed by the Navier–Stokes equation on a third order expansion of a two-dimensional velocity field takes the form

$$2(A_1 A_{112} + A_2 A_{122}) + (A_{12} A_{11} + A_{22} A_{12}) - 6\nu(A_{1112} + A_{1222}) + dA_{12}/dt =$$
$$2(A_1 A_{211} + A_2 A_{212}) + (A_{11} A_{21} + A_{21} A_{22}) - 6\nu(A_{2111} + A_{2122}) + dA_{21}/dt.$$

Figure 8.6.3. Behavior of the bubble under a time-periodic perturbation of the elliptic point, of the form $x_1^e = 0$, $x_2^e = d + \varepsilon \sin(2\pi\omega t)$ with $d = 0.2$, $\varepsilon = 0.05$, $\omega = 8/2\pi$, and $Re = 1$. Figure 8.6.3(a) shows a Poincaré section corresponding to the integrable system ($\varepsilon = 0$), whereas (b) shows a Poincaré section corresponding to similar conditions but for the time-dependent perturbation of the elliptic point. Note that the points leak as the unstable manifold 'reattaches' to the wall, which is shown in (c).

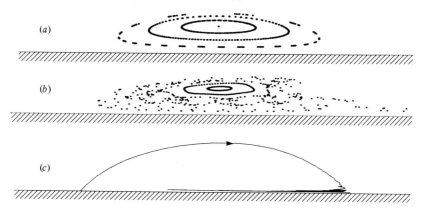

Problem 8.6.3

Show that the equations describing the time evolution of the coefficients A_{122} and A_{1222} are volume contracting.

Problem 8.6.4

Show that the angle of separation and attachment depends only on the viscous terms, and that the non-linear and time dependent terms are identically zero for all expansion orders.

8.7. Streamlines in an inviscid flow

As seen in Chapter 3, a steady isochoric flow of an inviscid fluid is governed by

$$\mathbf{v} \times \boldsymbol{\omega} = \nabla(\psi + p/\rho + \tfrac{1}{2}q^2)$$

and therefore $\psi + p/\rho + \tfrac{1}{2}q^2$ is constant over streamlines (and also pathlines and streaklines). The simplest case corresponds to $\psi + p/\rho + \tfrac{1}{2}q^2$ uniform over some region of space and this implies that the vorticity and velocity are co-linear at every possible $\mathbf{x}$, i.e.,

$$\boldsymbol{\omega} = \beta(\mathbf{x})\mathbf{v}$$

Furthermore, since $\nabla \cdot \boldsymbol{\omega} = 0$, we have

$$\mathbf{v} \cdot \nabla \beta = 0,$$

and the streamlines are constrained to belong to surfaces $\beta(\mathbf{x}) = $ constant, which act as a constant of the motion.[13] However, if β is uniform, i.e., $\nabla \beta = 0$, this requirement disappears and $\mathbf{v}(\mathbf{x})$ is not constrained to belong to any surface. Arnold conjectured that such flows might have a complex topology and Hénon, in a short note (Hénon, 1966), took on his suggestion and examined the case

$$dx_1/dt = A \sin x_3 + C \cos x_2$$
$$dx_2/dt = B \sin x_1 + A \cos x_3$$
$$dx_3/dt = C \sin x_2 + B \cos x_1$$

for $0 < x_i < 2\pi$, $i = 1, 2, 3$, which corresponds to the case $\lambda = +1$. Note that this is a steady flow and that the portrait of iso-vorticity (surfaces $|\omega| = $ constant), iso-dissipation (($\mathbf{D}:\mathbf{D})^{1/2} = $ constant), helicity ($\boldsymbol{\omega} \cdot \mathbf{v} = $ constant), etc. are all very simple looking and can be computed easily. The structure of *fixed* points of the *flow*, however, and their associated manifolds can be extremely complicated and a complete analysis is not attempted here. Hénon integrated the equations numerically and recorded the intersections with the plane $(0 < x < 2\pi, 0 < y < 2\pi, z = 0$, Poincaré section).

Figure 8.7.1 shows Hénon's results for the case $A = 3^{1/2}$, $B = 2^{1/2}$, $C = 1$. The points joined by curves correspond to the same pathline (i.e., they all originate from the same initial condition); all the isolated points originate also from one initial condition (this behavior was called 'semi-ergodic' by Hénon). The behavior of $A = B = C = 1$ is similar except that the 'semi-ergodic' region occupies more space.

The same flow was analyzed, in a more complete study, by Dombre *et al.* (1986). It is easy to check that the velocity field has eight fixed points corresponding to

$$-C \sin x_2 = B \cos x_1 = \pm [(B^2 + C^2 - A^2)/2]^{1/2}$$
$$-B \sin x_1 = A \cos x_3 = \pm [(A^2 + B^2 - C^2)/2]^{1/2}$$
$$-A \sin x_3 = C \cos x_2 = \pm [(C^2 + A^2 - B^2)/2]^{1/2}.$$

The condition that x_1, x_2, and x_3, be real is that a triangle be formed with sides A^2, B^2, and C^2. The condition $d\mathbf{x}/dt = 0$ implies $\boldsymbol{\omega} = 0$ and therefore at the fixed point $\nabla \mathbf{v}$ is symmetric (since $\Omega = 0$). It follows that the eigenvalues of $\nabla \mathbf{v}$ are real and the flow is hyperbolic (the sum of the eigenvalues is zero since $\nabla \cdot \mathbf{v} = 0$). Furthermore they are all different except

Figure 8.7.1. Poincaré section in the ABC flow for the case $A = 3^{1/2}$, $B = 2^{1/2}$, $C = 1$.

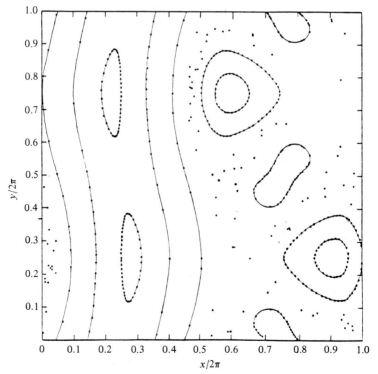

in the case $A = B = C = 0$. (For $C = 0$ the equations are integrable and the streamlines belong to the surfaces $B \sin x_1 + A \cos x_3 = \text{const.}$).

The stable and unstable manifolds of the fixed points (either two-dimensional surfaces or lines) have either to join smoothly or intersect transversally and complex possibilities of the type depicted in Section 5.10 are possible. Such a scenario is indicated by Dombre *et al.*

Extensive numerical simulations indicate that the system displays chaos.[14] If the flow were a truly effective mixing flow every filament of length λ_0 placed in any orientation whatever in the chaotic regions will increase its length exponentially. However, even though this is suggested by the numerical calculations, it is not true in general. Given the character of the flow we know that if the filament of length λ_0 is placed initially coinciding with a vortex line with vorticity ω_0 its length, when it moves to a region of vorticity ω is given by

$$\lambda = (|\omega|/|\omega_0|)\lambda_0.$$

However, ω is bounded (in this case its value is exactly equal to **v**) and we have to conclude that the stretching λ *cannot increase exponentially* as is expected in the chaotic region of a two-dimensional time-periodic flow no matter what the initial orientation. Note however, that the stretching history can be extremely complicated, and that filaments placed in any position, most likely, will be stretched exponentially.

Bibliography

Most of the work reported in this chapter is currently in progress and the presentation should be regarded as a sketch of the current state of the field and as an outline of possibilities. Some of the original material on the partitioned-pipe mixer appeared in the Ph.D. thesis by Khakhar (1986). A more thorough presentation appears in Khakhar, Franjione, and Ottino (1987). The work is continued and considerably extended in the Ph.D. thesis of Franjione. The treatment of Section 8.4 follows closely the presentation by Franjione and Ottino (1987, abstract only). The material in Section 8.4 follows entirely the article by Sobey (1985). Additional material can be found in Ralph (1986). The treatment of the Kelvin cat eyes flows (Section 8.5) is based on the Ph.D. thesis of Danielson. The treatment of flows near walls is based on the work by Perry *et al.* (see references in Section 8.6) and recent work by Danielson. The material in Section 8.7 is based on the short article by Hénon (1966) and Dombre *et al.* (1986) with some interjections of our own. A relevant paper for the analysis of three-dimensional flows is Crawford and Omohundro (1984) since it emphasizes dynamical aspects which are not visible in a Poincaré map.

Notes

1 In our calculations, we use a 3-term approximation of the series $v_z(r, \theta)$; such approximation was found to give a reasonably accurate description of the flow as compared to a 100 term approximation, see Khakhar, 1986.

2 Two other possibilities come to mind with regard to the calculation. The first is 'solving' the partial differential equations for the tracer concentration, a task which would require considerable computational effort. The second is adding stochasticity at each step of the mapping to mimic molecular diffusion.

3 There are several other standard possibilities, for example injecting particles with a speed proportional to the axial speed. However, this is hard to do experimentally.

4 An analytical solution is given by Snyder and Goldstein (1965). The cross-sectional flow is based on the solution by Wannier (see Section 7.4).

5 See photograph 103 in van Dyke (1982). The EHAM has been analyzed experimentally by Kusch (1988).

6 Note that the equation defines the speed U.

7 A related flow, with the same geometry which should produce similar results corresponds to the case in which one of the *walls* is moved in a time-periodic manner. Another example of a spatially periodic system producing efficient mixing due to secondary vortices is a succession of twisted pipe sections (elbows). See Jones, Thomas, and Aref (1990).

8 Obviously, a realistic perturbation and a full analysis of the problem would give rise to three-dimensional effects which are not considered here.

9 For a description of the method see Perry and Chong (1986), a short introduction is given in Perry and Fairlie (1974); a general review is given in Perry and Chong (1987).

10 The spatial coordinates are made dimensionless with respect to L and the time with respect to T, where L and T are suitably selected length and time scales, which depend on the boundary conditions of the problem.

11 Of course since the system is non-linear this need not be true in general.

12 However, a point worth mentioning is that we cannot prove that the solution generated by this method is unique since the equations are non-linear.

13 These surfaces can be very complicated but are, at least, differentiable.

14 Recent works examining this system are Galloway and Frisch (1986, 1987), which examine the linear stability in the presence of dissipation; and Feingold, Kadanoff and Piro (1987), which examines a discretized version of the flow of the form

$$x_1' = x_1 + A \sin x_3 + C \cos x_2 \ (\text{mod } 2\pi)$$
$$x_2' = x_1 + B \sin x_1 + A \cos x_3 \ (\text{mod } 2\pi)$$
$$x_3' = x_3 + C \sin x_2 + B \cos x_1 \ (\text{mod } 2\pi).$$

In this case the emphasis is on invariant structures and 'diffusion'.

9

Epilogue: diffusion and reaction in lamellar structures and microstructures in chaotic flows

The chaotic flows described in Chapters 7 and 8 provide a 'fabric' on which it is possible to superimpose several processes of interest such as interdiffusion of fluids and stretching and breakup of microstructures. A brief sketch of possibilities is discussed here. If the fluids are purely *passive*, the connectedness of material surfaces suggests a way of incorporating the effects of stretching on diffusion and reaction processes. In the first section of this chapter we study diffusion–reaction processes occurring at small scales in lamellar structures; the presentation is general and independent of specific flow fields. Another prototypical situation, discussed in the second part of the chapter, corresponds to *active* particles; in this case the fluid particles are endowed with some 'structure' in such a way that they can mimic the behavior of small droplets, an interfacial region, or macromolecules (which we shall refer to as microstructures). The microstructure is governed by some evolution equation, for example, a vector equation based on suitable physics, which coupled to the underlying flow, governs the processes of stretching, change of orientation, breakup, etc. The nature of the presentation is speculative and is intended primarily as a catalog of possibilities and as an outline of future problems.

9.1. Transport at striation thickness scales

The simplest problem of stretching of microstructures corresponds to the case of *passive* material elements. For example, in the case of passive mixing, which can be regarded as the case of two fluids of similar viscosity and without interfacial tension, the boundary between the two fluids acts as a marker of the flow; the motion is topological and the interface can be regarded as *passive*. In the case of *active interfaces*[1] the interfaces interact with the flow and modify it. Obviously, when describing mixing, it is too complicated to try to attack the problem in full. It is convenient to describe mixing in terms of passive interfaces and then add, possibly at small scales, the effect of active interfaces.[2] Let us consider first the purely passive case, the active case is considered in Section 9.3

   Within the framework of the motions described so far, $\mathbf{x} = \mathbf{\Phi}_t(\mathbf{X})$, and in particular all the flows of Chapters 7 and 8, initially designated material volumes remain connected, i.e., topological features are conserved and cuts of partially mixed materials reveal a lamellar structure. The striation thickness, s, serves as a measure of the mechanical mixing[3] (see Figure 1.1.1). We wish to preserve a similar structure for the case of diffusing fluids in such a way that a properly defined striation thickness can be thought to act as the underlying fabric on which the transport processes and reactions occur. We define a tracer as a hypothetical non-reactive material m that moves everywhere with the mean mass velocity $\mathbf{v} \equiv \sum (\rho_s/\rho)\mathbf{v}_s$ (see Section 2.3). Thus, if ω_m represents the mass fraction of the tracer ($\omega_m = \rho_m/\rho$) we have

$$D\omega_m/Dt = 0.$$

Under these conditions, material surfaces of the tracer remain connected and map the average mass motion of the flow.

   The patterns produced by the tracer are similar to those shown in the examples of Chapters 1, 7, and 8: tendrils, whorls, and primarily, stretched and folded structures. Diffusive species traverse the hypothetical surfaces of the tracer, concurrent with stretching and folding of the tracer surfaces, which is governed by the equations of Chapter 3. The entire process is governed by the convective–diffusion equation

$$\rho \frac{D\omega_s}{Dt} = -\nabla \cdot (\rho_s \mathbf{u}_s) + r_s \tag{9.1.1}$$

where r_s represents the rate of reaction of species-s (an important simplification, which we will adopt here, is that the transport processes and the chemical reaction do not affect the fluid mechanics). Since in general it is impossible to predict the stretching of surfaces in complete detail,[4] it is clear also that the diffusion and reaction problem cannot be solved in a completely general way, and that we have to search for simple representations that contain the essential physics.

   We take the structures of Figure 7.2.1 as the *morphological building blocks* of arbitrary flows. These material regions (labelled by the material particle $\mathbf{X}_m$) stretch and deform, in a Lagrangian sense, as indicated in Figure 9.1.1, aiding the diffusion and overall reaction processes. We assume that the gradient of concentration has (locally) only one non-vanishing component, i.e.,

$$\nabla \omega_s = (\partial \omega_s/\partial x, 0, 0) \tag{9.1.2}$$

where x is a co-ordinate direction which is normal to the planes in the lamellar structure (Figure 9.1.1).[5] The simplest case corresponds to a periodic structure with striation thicknesses s_A and s_B: in such a case we can study a single pair of striations thereby avoiding the complexities arising from distributions of striation thicknesses.[6] Consider now the application of (9.1.1) in a Lagrangian sense, i.e., by focusing our attention on a (small) identifiable material region of fluid labelled by the material particle $\mathbf{X}_m$ and denoted $S_{\mathbf{x}_m}$.

Figure 9.1.1. Deformations of structures in flows: (a) tendrils, (b) whorl, and (c) a pair of striations, with corresponding concentration fields along a cut. Note that the instantaneous location of the maximum direction of contraction is not necessarily normal to the striations.

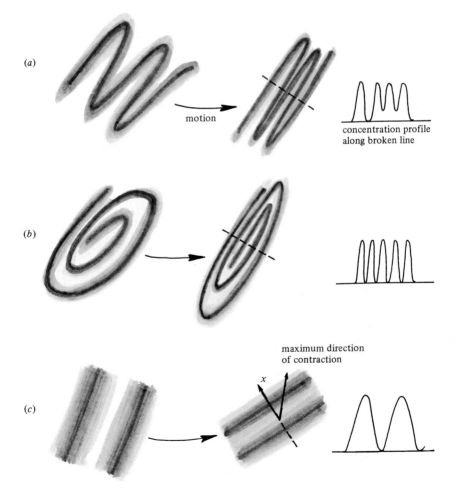

The simplest theories of diffusion are based on Fick's law, i.e., the relative diffusion velocities are given by

$$\mathbf{u}_s = -\omega_s^{-1} D_s \nabla \omega_s. \tag{9.1.3}$$

Since $\mathbf{u}_s$ transforms as $\mathbf{u}'_s = \mathbf{Q} \cdot \mathbf{u}_s$ and ∇ transforms as $\nabla' = \mathbf{Q} \cdot \nabla$ we conclude that the diffusion equation has the same form in any moving frame F' provided that the velocity field is referred to the frame F'. An identical argument applies to the energy equation.[7] In particular we can apply Equation (9.1.1) to a frame attached to a material particle $\mathbf{X}_m$ such that, at small distances from $\mathbf{X}_m$, in frame F', we have

$$\frac{\partial c_s}{\partial t} + \mathbf{v}_{rel} \cdot \nabla' c_s = D_s \nabla'^2 c_s + r_s \tag{9.1.4}$$

where c_s is the concentration of species-s and $\mathbf{v}_{rel}$ is the relative velocity field of Section 2.8.

According to the lamellar assumption (9.1.2) we have

$$\frac{\partial c_s}{\partial t} + (\mathbf{D}:\mathbf{nn})x \frac{\partial c_s}{\partial x} = D_s \frac{\partial^2 c_s}{\partial x^2} + r_s \tag{9.1.5}$$

where $\mathbf{n}$ is a normal vector in the direction of ∇c_s.[8] In general, $\mathbf{n}$ does not coincide with the instantaneous (local) maximum direction of contraction (see Figure 9.1.1(c)). Another way of writing Equation (9.1.5) is in terms of the surface area stretch (for $\nabla \cdot \mathbf{v} = 0$)

$$\frac{\partial c_s}{\partial t} - \left(\frac{D \ln \eta}{Dt}\right) x \frac{\partial c_s}{\partial x} = D_s \frac{\partial^2 c_s}{\partial x^2} + r_s.$$

The stretching function $\alpha(\mathbf{X}, t)$ $(\equiv D \ln \eta / Dt)$ depends on the fluid mechanics and provides the tie up with the previous chapters.[9] In the simplest case the problem is represented by one equation of the type (9.1.5) and the elements of Figure 9.1.1. According to the circumstances the complexity of the problem can be augmented by incorporating additional effects (for example a different stretching history for each microflow $S_{\mathbf{X}_m}$, see Section 9.2). However, even in the simplest cases, i.e., simple chemical reactions, the multiplicity of parameters makes the analysis quite complicated and it is hard to obtain results of general validity.

Example 9.1.1

The equation

$$\frac{\partial c_s}{\partial t} + (\mathbf{x} \cdot \mathbf{L}) \cdot \nabla c_s = D_s \nabla^2 c_s + r_s$$

where $\mathbf{L}$ is $\nabla \mathbf{v}$, which is a generalization of (9.1.4), and where c_s can be either concentration or temperature, has been extensively analyzed in the context of various applications. For example, Townsend (1951) considered the case

$$\frac{\partial T}{\partial t} + \sum_{i=1}^{3} D_{ii} x_i \frac{\partial T}{\partial x_i} = \alpha_T \nabla^2 T$$

where the D_{ii}s are the diagonal components of $\mathbf{D}$, which in general are functions of time, but are considered constant here. A heat spot is released at the origin, and according to the value of the determinant of $\mathbf{D}$, it is converted into a filament (det $\mathbf{D} > 0$) or into a flat disk (det $\mathbf{D} < 0$). We take tr $\mathbf{D} = 0$ and consider $D_{11} \geqslant D_{22} \geqslant D_{33}$; i.e., $D_{11} > 0$ and $D_{33} < 0$. The temperature evolves as

$$T = T_m(t) \exp\left[-\frac{1}{2}\left(\frac{x_1^2}{a_1^2} + \frac{x_2^2}{a_2^2} + \frac{x_3^2}{a_3^2} \right) \right]$$

where the a_is are given by

$$\frac{da_i^2}{dt} - 2D_{ii}a_i^2 = 2\alpha_T$$

and

$$T_m(t) = A/a_1 a_2 a_3,$$

where A is a constant determined by the initial conditions. Thus, the isotherms are ellipsoids, and for long times ($D_{11}t \gg 1$), $T_m(t)$ decays exponentially in time. Batchelor (1959) considered a similar problem, but with the initial condition

$$T = A_0 \sin(\mathbf{l} \cdot \mathbf{x})$$

where $\mathbf{l} = (l_1, l_2, l_3)$ is a constant vector and $\mathbf{x} = (x_1, x_2, x_3)$. In this case the solution is of the form

$$T(\mathbf{x}, t) = A(t) \sin(\mathbf{m}(t) \cdot \mathbf{x})$$

where $\mathbf{m}(t)$ is given by

$$d\mathbf{m}/dt = -\mathbf{D} \cdot \mathbf{m},$$

and therefore (assuming $l_3 \neq 0$) we obtain

$$\mathbf{m}^2 = m_1^2 + m_2^2 + m_3^2 \to l_3^2 \exp(-2D_{33}t), \qquad \text{as } t \to \infty$$

and

$$T \to A_0(\alpha_T \mathbf{m}^2/2D_{33}) \sin(\mathbf{m}(t) \cdot \mathbf{x}), \qquad \text{as } t \to \infty$$

which shows that the normal to surfaces of constant T (i.e., with normal ∇T) approaches the direction of the greatest rate of contraction. Note however, that this local result is valid only for linear velocity fields.[10]

Example 9.1.2
On many occasions it is sufficient to characterize the mixing in average or structured continuum terms, e.g., profile of striation thickness across a

mixing layer, average striation thickness in the cross-section of an extruder, etc. Consider a thin sheet of a tracer which at time $t = 0$ coincides with the boundary between materials. As mixing proceeds, and interdiffusion occurs, the thin layer S of the tracer moves with the mean mass velocity, and if the mixing is effective, it is finely distributed throughout a material volume V. We define $\rho^*(\mathbf{X}, t)$ as the mass of tracer per unit volume ($\rho^* = 0$ for points $\mathbf{x}$ not belonging to the thin layer). It follows that

$$\frac{D}{Dt} \int_V \rho^* \, dv = 0$$

and

$$\frac{D\rho^*}{Dt} + \rho^*(\nabla \cdot \mathbf{v}) = 0.$$

Similarly, we define $\rho''(\mathbf{X}, t)$ as the mass of tracer per unit area of the surface at the location $\mathbf{X}$ and with orientation $\mathbf{n}$. We have

$$\frac{D}{Dt} \int_S \rho'' \, ds = 0$$

and

$$\frac{D\rho''}{Dt} + \rho''(\nabla \cdot \mathbf{v} - \mathbf{D} : \mathbf{nn}) = 0.$$

We define $a_v(\mathbf{X}, t)$ (and similarly, $a_v(\mathbf{x}, t)$) as $a_v = \rho^*/\rho''$. The quantity a_v can be interpreted as the area of tracer/volume.[11] Furthermore a_v is proportional to the area stretch,

$$a_v = a_v^0 \eta / \det \mathbf{F}$$

and obeys (see Ottino, 1982)

$$\frac{Da_v}{Dt} + a_v \mathbf{D} : \mathbf{nn} = 0.$$

Problem 9.1.1
Show that material planes remain flat in the flow field $\mathbf{v}_{rel} = \mathbf{x} \cdot (\nabla \mathbf{v})^T$.

Problem 9.1.2
Confirm the invariance of the convective–diffusion equation.

Problem 9.1.3
Show that the distance $\delta(t)$ between two small, nearby, parallel planes with normal $\mathbf{n}$ varies as a function of time as (Ottino, Ranz, and Macosko, 1979)

$$\frac{1}{\delta(t)} \frac{D\delta(t)}{Dt} = \mathbf{D} : \mathbf{nn}.$$

Therefore, $1/\delta(t)$ is proportional to the intermaterial area per unit volume, a_v (Example 9.1.2).

Problem 9.1.4

Build a simple model to interpret the determination of mixing times in terms of the temperature decay of hot spots.[12]

9.1.1. *Parameters and variables characterizing transport at small scales*

It is important to recognize that the simplest problem of stretching with diffusion and reaction at small scales is characterized by, at least, three time scales and many parameters. Such multiplicity of parameters makes a general analysis difficult. There are three time-scales associated with *local* processes, the time-scale of diffusion, which is related to diffusional distances, $O(s^2(t)/D_s)$, the time scale of the reaction(s) (in the case of multiple reactions there can be several times scales), and the time-scale of the thinning of the striation thicknesses, $O(1/\alpha)$ where α is the stretching function (see Section 2.9). There is also at least one time-scale associated with the macroscopic processes of mixing, which can be taken as the exit age of the element, t_{res} (see Figure 9.1.2). Let us consider now the independent variables and parameters characterizing the system.

In order to cast the equations in dimensionless form it is convenient to define the following variables:

$$\Gamma = t/t_C, \qquad \text{reduced time}$$

$$\tau = \frac{1}{t_C} \int_0^t [\eta(t')]^2 \, dt', \qquad \text{'warped time'}$$

Figure 9.1.2. Un-premixed reactor. The reactants A and B enter the system and are mixed by complex motions. The microflow element $S_{\mathbf{x}_m}$ exits the system after a time t_{res}.

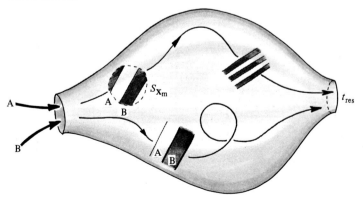

where η is the area stretch associated with the microflow element[13] and t_C is a characteristic time. The usual choices are $t_C = t_{D_0}$ (diffusional time) and $t_C = t_R$ (reaction time). A dimensionless distance is defined as

$\zeta = x/s(t)$,   *space scale based on the instantaneous value of the striation thickness.*

We also define the following fluid mechanical parameters:

$$\alpha(t) = D \ln \eta/Dt, \qquad \textit{stretching function}$$

$$\eta(t) = \exp\left[\int_0^t \alpha(t')\, dt'\right], \quad \textit{area stretch at time } t$$

$$\langle\alpha(t)\rangle = \frac{1}{t}\int_0^t \alpha(t')\, dt', \qquad \textit{time averaged value of the stretching function}$$

Based on these definitions we define the following characteristic times:

$$t_F(t) = 1/\langle\alpha(t)\rangle, \qquad \textit{characteristic fluid mechanical time (stretching)}$$

$$t_D = s(t)^2/D_K, \qquad \textit{instantaneous diffusional time}$$

and

$t_{D_0} = s_0^2/D_K$, *characteristic diffusion time* (K denotes a reference species). The characteristic time of reaction is denoted t_R.

The variables Γ, τ, and the fluid mechanical parameters, $s(t)$, $\alpha(t)$, and $\eta(t)$, are associated with the microflow element S_{X_m}, the label X_m being omitted without fear of confusion.

In terms of these variables the convective diffusion Equation (9.1.6) can be transformed into the following two forms (see Chella and Ottino, 1984):

$$\frac{\partial C_i}{\partial \Gamma} = \eta^2(\Gamma)\Delta_i\left(\frac{t_C}{t_{D_0}}\right)\frac{\partial^2 C_i}{\partial \zeta^2} + \beta_i\left(\frac{t_C}{t_R}\right)R_i \qquad (9.1.6.a)$$

$$\frac{\partial C_i}{\partial \tau} = \Delta_i\left(\frac{t_C}{t_{D_0}}\right)\frac{\partial^2 C_i}{\partial \zeta^2} + \eta^{-2}(\tau)\beta_i\left(\frac{t_C}{t_R}\right)R_i \qquad (9.1.6b)$$

where C_i represents the dimensionless concentration of species-i, Δ_i is a diffusion coefficient ratio, and β_i is the initial concentration ratio of species-i. R_i represents a reaction term (production) of species-i.

Note that the contribution of stretching is double; first, an increase in the area available for diffusion accelerates the interdiffusion process; second, the continual stretching does not allow the mass fluxes to decay. Equations (9.1.6a, and b) provide two different viewpoints of the influence of local stretching on diffusion and chemical reaction: Equation (9.1.6a) shows that in the time scale Γ and striation thickness based scale ζ, the apparent diffusion coefficient is augmented by a factor $\eta^2(\Gamma)$ due to the local stretching; Equation (9.1.6b) shows that in the warped time scale τ

and striation thickness based scale ξ, the apparent reaction rate is multiplied by a factor $\eta^{-2}(\tau)$ due to the local stretching.[14]

The usual initial conditions correspond to complete segregation of the streams; the physical situation might correspond to streams being fed to the reactor (in this case the interface is a streaksurface) or two segregated volumes initially placed in a batch reactor. As initial condition we establish that the streams are unmixed, i.e.,

$$\tau \text{ or } \Gamma \geqslant 0, \qquad \partial C_i/\partial \xi = 0, \qquad \text{for } \xi = 0, 1.$$

where $H_i(\xi)$ is a unit step function discontinuous at $\xi = \phi$, where ϕ is the volumetric fraction of the one-stream (one of the streams is arbitrarily designated as 'one').[15] The boundary conditions, in the periodic case, are obtained from symmetry considerations:

$$\tau \text{ or } \Gamma \geqslant 0, \qquad \partial C_i/\partial \xi = 0, \qquad \text{for } \xi = 0, 1.$$

Consider now the value of the initial striation thickness, s_0. In establishing this value we face two competing effects: We might argue that if s_0 is too large, and curvature effects are important, the lamellar structure assumption is probably not quite valid and consequently the model a poor approximation. On the other hand, if we wait longer and select a smaller s_0, we might miss some of the diffusion–reaction process and not be able to estimate the initial conditions. However, if s_0 is large, as for example during the first stages of mixing in a shear flow (see for example, the Hama flow of Example 2.5.1), the interfaces are essentially non-interacting and therefore it really does not matter that the lamellar structure has not been established.

Problem 9.1.5
Eliminate the convective term $(\mathbf{v} = \mathbf{x} \cdot \mathbf{D})$ in the equation

$$\frac{\partial \mathbf{c}}{\partial t} + \mathbf{v} \cdot \nabla \mathbf{c} = \nabla \cdot [\mathbf{K} \cdot \nabla \mathbf{c}] + \mathbf{R}$$

where $\mathbf{c}$ is a vector of concentrations and $\mathbf{K}$ is diagonal, by transforming the time t as $\Lambda = \Lambda(t)$ and $\mathbf{x}$ as $\xi = \mathbf{Q} \cdot \mathbf{x}$, where $\mathbf{Q}$ is diagonal. Show that the convective term disappears if we choose

$$\frac{D\xi}{Dt} + \mathbf{Q} \cdot [\mathbf{D} \cdot \mathbf{x}] = 0.$$

This leaves several choices for the selection of $\Lambda(t)$ (Chella, 1984).

9.1.2. Regimes

Equations (9.1.6a and b) form a system of coupled, non-linear partial differential equations, which can be solved, in general, only by numerical

Table 9.1

		Average striation thickness $\longrightarrow$ Stretching at small scales	
		Reactor size	
Description of diffusion and reaction at small scales	Age of microflow element	Entrance $\longrightarrow$	Exit
Kinetic control Slow reactions $t_D \ll t_R$	$t < t_D \ll t_R$ No appreciable reaction until reaching scales at which reaction behaves as homogeneous	$t \approx t_R$ System of ODEs with initial conditions (classical reactor analysis)	
Some reaction kinetics controlled, others diffusion controlled t_R, t_D	$t \ll t_R, t_D$ No appreciable reaction	Combination of ODEs and PDEs (some reactions might be in equilibrium whereas others might be diffusion controlled)	$t = O(t_R, t_D)$ System of coupled ODEs
Diffusion control Fast reactions $t_R \ll t_D$	$t \ll t_D$ No appreciable reaction	$t < t_D$ System of PDEs with matching flux conditions and initial conditions	$t_R \approx t_D$ Reaction practically complete

means. The limit cases, 'fast' or 'slow' reactions, are more tractable than the general case. The different regimes of reaction are shown, schematically, in Table 9.1.1.

It is evident that the presence of at least three major time scales produces two dimensionless ratios (Damköhler numbers) which characterize the processes occurring at small scales:

$$Da_1^{(L)} \equiv t_R/t_F \qquad 16$$

$$Da_{II} \equiv t_D/t_R.$$

Note that the characterization of a reaction as fast or slow depends, in general, on t_R, t_D, and t_F (see Figure 9.1.3).

Slow reactions

A reaction is called 'slow' if $t_D/t_R \ll 1$ during the course of the reaction, except possibly during the first stages of mixing (this is a conservative criterion since t_{D_0} is defined with respect to the initial segregation – the loosely defined s_0). Actually, the diffusion time-scale varies in time, since the striation thickness is decreasing in time. Consider a simple example corresponding to exponential thinning. Initially, the diffusional time-scale, $t_D = s(t)^2/D$, might be larger than the reaction time-scale, i.e., $s(0)^2/D > t_R$, but after some time, we might achieve $t_D \approx t_R$, since $s(t)$ decays as $s(0) \exp(-\alpha t)$, and quite rapidly afterwards we might be in a regime such that $t_D \ll t_R$. From this time onwards, intrinsic kinetics controls and local mixing has no influence on the course of the reaction. Under these conditions the mass balance for species-s reduces to:

$$dC_s/dt = \beta_s R_s. \qquad (9.1.7)$$

This limit corresponds to residence time theory (see Chapter 1).

Figure 9.1.3. Schematic representation of regimes corresponding to slow and fast reactions.

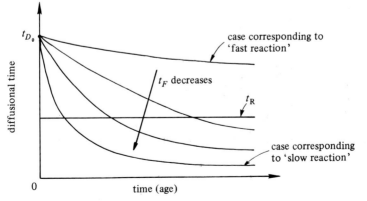

Very fast reactions

A reaction is called 'very fast' if $t_D/t_R \gg 1$ during the course of the reaction. In such a case the reacting species cannot co-exist and the reacting zone reduces to a plane. The rate of diffusion to this plane controls the overall reaction rate, and we lose the influence of the intrinsic kinetics. The controlling factor is the rate of stretching and the location of the reaction zone, which in the simplest case does not move and acts as a material surface.[17] In addition to the initial and boundary conditions, the flux of reactants to the plane must satisfy the overall reaction stoichiometry. Here the variable τ, with t_C selected as t_{D_0}, contains all the effects of the local convective mixing (through the stretching function $\alpha(\mathbf{X}, t)$) and the problem reduces to

$$\frac{\partial C_i}{\partial \tau} = \Delta_i \frac{\partial^2 C_i}{\partial \xi^2}. \tag{9.1.8}$$

Thus, the solution of (9.1.8) for a particular reaction scheme allows the superposition of any stretching history (choice of $\alpha(\mathbf{X}, t)$). This is the basic idea behind the use of fast reactions as tracers of the fluid motion.[18]

Example 9.1.3

Figure E9.1.3 represents the effect of the local flow on a series-parallel reaction $(A + B \rightarrow P)$, with a specific rate of reaction k_2, $R + B \rightarrow S$, with a specific rate of reaction k_3). The system is governed by the equations

$$\frac{\partial C_A}{\partial \Gamma} = \eta^2 \Delta_A \left(\frac{t_C}{t_{D_0}}\right) \frac{\partial^2 C_A}{\partial \xi^2} - \left(\frac{t_C}{t_R}\right) C_A C_B$$

$$\frac{\partial C_B}{\partial \Gamma} = \eta^2 \Delta_B \left(\frac{t_C}{t_{D_0}}\right) \frac{\partial^2 C_B}{\partial \xi^2} - \left(\frac{t_C}{t_R}\right) \beta_B C_B (C_A + \varepsilon C_R)$$

$$\frac{\partial C_R}{\partial \Gamma} = \eta^2 \left(\frac{t_C}{t_{D_0}}\right) \frac{\partial^2 C_R}{\partial \xi^2} + \left(\frac{t_C}{t_R}\right) C_B (c + \varepsilon C_R).$$

The parameters are: $\varepsilon = 0.1$, $\Delta_A = \Delta_B = 1$, $\beta_B = 0.5$, $\phi = 0.5$, initial diffusion time, $t_{D_0} = 2.25 \times 10^5$ s, final diffusion time, $t_{D_f} = 2.25 \times 10^{-1}$ s (i.e., the striation thickness reduction is 10^3), the characteristic reaction time is $t_R = 2.25 \times 10^{-1}$ s. Two types of flows are considered: a hyperbolic flow, $v_1 = \dot{\varepsilon} x_1$, $v_2 = -\dot{\varepsilon} x_2$, and a shear flow $v_1 = \dot{\gamma} x_2$, $v_2 = 0$; the initial orientation in both cases is $\mathbf{N} = (0, 1)$. The values of $\dot{\varepsilon}$ and $\dot{\gamma}$ are selected such that they produce the same striation thickness reduction in the total time, 10 s. The value of R in the ordinate in Figures E9.1.3 represents the average concentration of species-R in the $S_{\mathbf{X}_m}$ element. The two flows are compared with that of a premixed reactor, where the initial condition corresponds to A and B mixed molecularly. In this example, the two unpremixed systems have the same overall values of $Da_I^{(L)}$ when the mechanical time t_F is defined as $t_{res}/\ln(s_0/s_f)$. Extensive computations indicate that, for simple reactions, this definition of t_F gives a reasonable measure of convective mixing; for example, even though the fluid mechanical histories $\alpha(t)$ are substantially different, the final results do not seem to be very sensitive to the form of $\alpha(t)$ for values of $Da_I^{(L)}$ as high as 0.5. On the other hand, when $Da_I^{(L)} < 10^{-3}$, convective effects are not significant (Chella and Ottino, 1984).

Example 9.1.4
The previous example showed a case where the functional form of $\alpha(t)$ does not seem to be very important and flows with equal overall values of $Da_I^{(L)}$ give similar selectivities. In more complex reactions, however, the fluid mechanical path can produce drastic differences in the final product. Figure E9.1.4 shows the result of a more elaborate simulation (see Fields and Ottino, 1987a,b) using similar stretching paths to those used in Example 9.1.3. The physical situation corresponds to a co-polymerization between two species, A_1 and A_2, with a third one, B, leading to the

Figure E9.1.3. Effect of local kinematics on a series-parallel reaction. The figure shows the evolution of concentration of the intermediate reactant R.

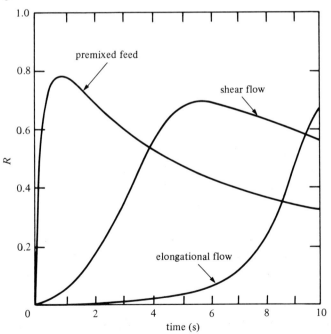

formation of (A_1-B) units (soft segments, denoted P_1) and (A_2-B) units (hard segments, denoted P_2). The reaction is exothermic and the diffusion coefficients are a function of the concentration. High molecular weight polymer is preferentially formed near the interface. Average properties (gross concentration profiles, average molecular weight, etc.) are not path dependent whereas other properties such as the maximum molecular weight across the striation are strongly path dependent.

Problem 9.1.6

Calculate the penetration thickness of a product P, δ_P, produced by a very fast reaction $A + B \rightarrow P$. The plane of reaction undergoes area stretch given by $\eta \approx \exp(\int \alpha \, dt)$. Show that for α constant δ_P reaches a steady state proportional to $(D/\alpha)^{1/2}$. Does this happen for any other functional form of $\alpha(t)$?

Figure E9.1.4. Effect of stretching path on a copolymerization. Molecular weight across the interface of reaction for (a) shear and (b) elongational histories corresponding to a dimensionless time of 250, (c) evolution of molecular weight for shear and elongational histories over a longer time scale; the initial and final values of the striation thicknesses in the shear and elongational prescriptions are the same. (Reproduced with permission from Fields and Ottino (1987b).)

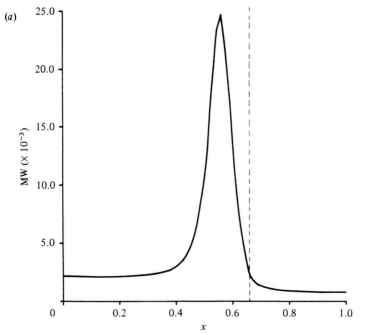

(*a*)

Figure E9.1.4 *continued*

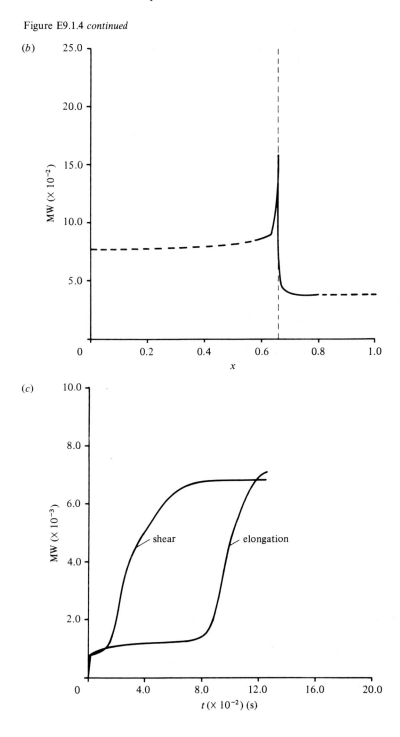

Problem 9.1.7

As a crude indicator, assume that reactions are diffusion controlled if $t_D/t_R > 10^4$ and kinetically controlled if $t_D/t_R < 10^{-1}$. Calculate the length scales necessary for these two regimes for the following second order reactions: $HCl + HONa$, $k_2 \approx 10^{11}$, $CO_2 + HONa$, $k_2 \approx 1.3 \times 10^4$, $HCOOCH_3 + HONa$, $k_2 \approx 4.7 \times 10^1$ (k_2 in $1/mol \cdot s$ and at 30°C).

9.2. Complications and illustrations

9.2.1. *Distortions of lamellar structures and distribution effects*

Here we consider briefly some of the additional physics necessary to complete the simplified picture described in Section 9.1. It will become apparent that while many effects can be considered we might get diminishing returns by their introduction. In any case we want to obtain results of general validity, and then, if necessary, according to the physical situation, complicate the picture.

Striation thickness distribution

In general, the striation thicknesses are distributed rather than spatially periodic, and the distributions can give rise to large scale diffusion effects and 'isolation of reactants' (Ottino, 1982). In some cases, such as combustion, this effect is referred to as flame shortening (Marble and Broadwell, 1977), in other problems such as polymerizations this effect produces permanent results due to variations in the diffusion coefficient and local 'quenching'. Numerical experiments (direct simulations) provide a glimpse into this effect.[19,34]

Changes in topology

By definition, the material surfaces of the tracer do not change the topology and remain connected throughout time (however, they might appear disconnected in two-dimensional cuts as in Figure 1.1.2). However, the iso-concenration surfaces of *diffusing* scalars move at a different velocity than the average mass velocity and might change topology (initially connected regions can break and form islands). (See Gibson, 1968.) This effect distorts the lamellar appearance.

Note also that Equations (9.1.6a and b) apply to only one microflow element; in practice we have the following complications:

● A distribution of fluid mechanical histories, $\alpha(\mathbf{X}, t)$: Each element might have a different stretching history – identified by its label $\mathbf{X}$ – although a given flow might be well represented by a 'typical type'. For example, all steady curvilinear flows (Section 4.4) have $\eta \approx t$, in a globally chaotic flow 'almost all **X**s' evolve as $\eta \approx \exp(t)$.

- A distribution of initial striation thicknesses, s_0s: This distribution might depend on the feeding or initial placement of the reactants. As an example consider the inlet condition in a multi-jet reactor. A large s_0 can be used to signify that the reacting surfaces do not interact.

- A distribution of ages: The microflow elements identifed by labels **X** will exit the reactor (or reaction zone) after a time t_{res} according to an age distribution function (Figure 9.1.2).[20]

In spite of these effects, it should be emphasized that the main idea is that the overall rate of reaction might be controlled *by the smallest scales*. Therefore, most of the physics is controlled by striation thickness scales and is modified only quantitatively, but not qualitatively, by distribution effects. If the reactions *are not* diffusionally controlled, then the problem becomes simpler and falls within the scope of classical reaction engineering.

Example 9.2.1

Consider a system, a simplification of the co-polymerization of Example 9.1.4 (Fields and Ottino, 1987c). The reaction is of the form $A_1 + B \rightarrow P_1$ and $A_1 + B \rightarrow P_2$. For simplicity the striation thickness is constant in time. Figure E9.2.1 shows two systems with the same mean value of the striation

Figure E9.2.1. Instantaneous concentration of reactant in a system with: (a) uniform and (b) distributed striation thicknesses. (Reproduced with permission from Fields and Ottino (1987b).)

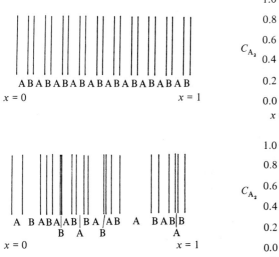
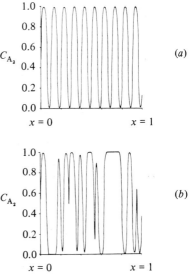

thickness, one of the systems has a uniform striation thickness whereas the other has a distribution of values. Figure E9.2.1 shows two snapshots of the reacting systems at the same real time; it is clear that thick striations remain unreacted for relatively long times (this effect has been called 'isolation of reactants').

Problem 9.2.1

The rate of variation of the concentration of a scalar c when moving at speed $\mathbf{v}$ in a fluid is given by

$$\frac{Dc}{Dt} = \frac{\partial c}{\partial t} + \mathbf{v} \cdot \nabla c.$$

Consequently, the velocity, $\mathbf{c}_{c=\text{const.}}$, corresponding to the speed of a surface $c(\mathbf{x}, t) = \text{const.}$ is given by

$$\frac{\partial c}{\partial t} = -\mathbf{v}_{c=\text{const.}} \cdot \nabla c.$$

Use this result to prove that, at the point $\mathbf{x}$, and at any instant t,

$$\mathbf{v}_{c=\text{const.}}(\mathbf{x}, t) = \mathbf{v}(\mathbf{x}, t) - D \frac{\nabla^2 c \nabla c}{|\nabla c|^2}.$$

Compute a similar expression for the velocity of regions of zero gradient, i.e., $\nabla c = 0$ (Gibson, 1968).

9.2.2. Illustrations

The following illustrations serve to reinforce some of the previous concepts and exemplify the limits of current computational efforts and the ability to resolve striation thickness scales.

Example 9.2.2

Figure E9.2.2 shows what happens during the wrapping of an interface due to the first stages of instability in a shear flow in the direction x, for times such that the layer is constrained to be two-dimensional (Corcos and Sherman, 1984). The initial velocity and concentration profiles are

$$u_x = U \, \text{erf}[\pi^{1/2} y/(2\delta)], \qquad \rho = -0.5\Delta\rho[(\pi Pr)^{1/2} y/(2\delta)]$$

i.e., the streamwise velocity changes from $+U$ to $-U$ and the scalar from $+\Delta\rho$ to $-\Delta\rho$. The value of δ is such that $\delta = U[(\partial u_x/\partial y)_{\text{max}}]^{-1}$; the Reynolds number is defined as $Re = U\delta/v$ and the Schmidt number is v/D. The time-scale is made dimensionless with respect to d/U where d is the length scale of the initial disturbance.

Figure E9.2.2(a) shows the wrapping of the interface along with the instantaneous picture of the streamlines, and iso-concentration curves of

Figure E9.2.2(*a*). Wrapping of the interface between two fluids due to the first stages of instability in a shear flow: (*a*1) location of the interface at dimensionless times 0.5, 1.0, 1.5, and 2.0, with $\alpha = (2\pi\delta)/d = 0.43$, $Re = 100$; (*a*2) iso-concentration curves of a passive scalar with $Sc = 1$, $\alpha = 0.4$, $Re = 100$, at times 1.5 and 2.0. (Reproduced with permission from Corcos and Sherman (1984).)

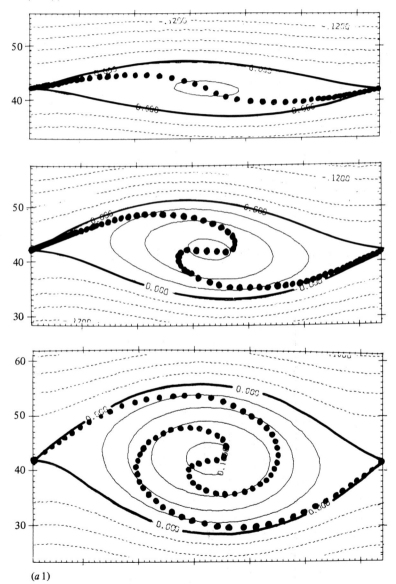

(*a*1)

the diffusive passive scalar. Note that the stretch of the interface is fairly small ($O(10)$) and that the lamellar structure assumption is a reasonable one except near the center of the 'cat eye'[21] (something more complicated occurs during vortex pairing). The details of the stretching history of the elements are highly complicated. With the exception of elements near the hyperbolic points, all other points experience a quasi-periodic stretching history where $\alpha(\mathbf{X}, t)$ becomes negative. Corcos and Sherman (1984) noted, as might have been expected, that the stretching history is highly sensitive to the initial conditions near the hyperbolic point.

Example 9.2.3
As we have seen in Section 9.1.3, the accurate computation of material interfaces is important in problems involving fast reactions producing thin reaction zones. The problem also has relevance to computations such as those described above and the direct computations given below. Order of magnitude calculations indicate formidable problems due to storage requirements. Computationally, the material interface is composed of a large number of points connected by small linear segments. However, due to the flow the interface stretches and folds, and if the distance between points becomes too large, the linear segments connecting adjacent points could cross, which is physically inadmissible. In order to maintain consistent resolution, the number of particles should be increased in such a way as to be able to resolve the striation thickness accurately. Obviously, such a calculation is a precondition for the resolution of the concentration

Figure 9.2.2(*a*) *continued*

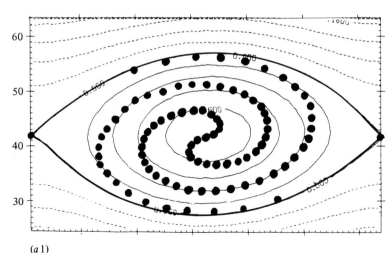

(*a* 1)

field within the striations and the computation of concentration gradients without loss of information due to averaging. The storage requirements can be estimated based on the idea of Figure 9.2.2(b). Consider a striation thickness reduction of 10^{-r}, maintaining throughout the process a ratio of particle separation (δ) to striation thickness (s) of 10^{-p}, where p is of

Figure 9.2.2(a) *continued*

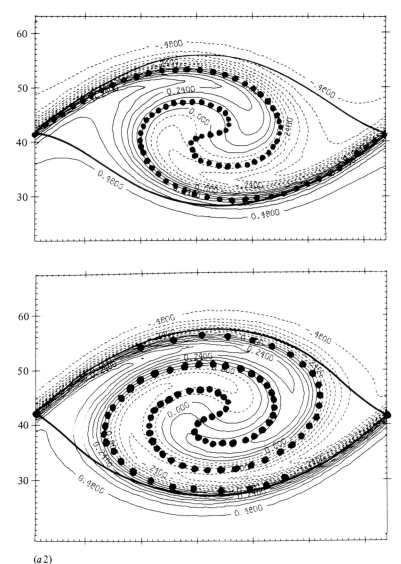

($a2$)

order one. The simplest possible case, which is a lower bound for more complicated situations, corresponds to a two-dimensional (deterministic, time-periodic) mapping in a 'globally chaotic' regime where the length of a line increases, on the average, as $L_0 \exp(\sigma k)$, where k refers to the number of mappings and σ is a suitable Liapunov exponent (in the case of integration of the Eulerian velocity field, we have estimated that the number of operations for each dimension, F, is of the order 10 to 100, and that each period requires 10^τ time steps with τ of order 3). An estimate of the number of floating point operations per mapping, taking into account the loss of precision due to growth of errors, for an accuracy of 0.01%, yields the results shown in Table 9.2 (Franjione and Ottino, 1987b). It should be noted that the bulk of the computational effort is not due to growth of errors. For example for the case of a striation reduction of 10^{-4}, relaxing the accuracy to 1% and 5%, reduces the storage requirements by only 1.5 and 2 Gigabytes respectively.

Example 9.2.4
There have been a few direct simulations incorporating chemical reactions.

Figure E9.2.2(b). Tracking of the interface between two fluids being mixed. The distance between material points is δ, the striation thickness is s. Material points are used to represent the blob itself or the interface between the blob and the surrounding fluid.

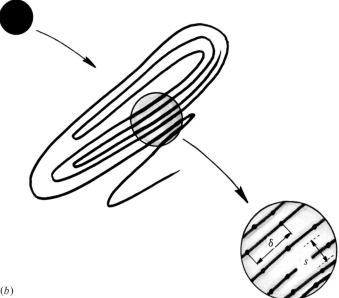

(b)

Table 9.2. *Computational estimates for material line tracking in a chaotic flow on a computer capable of 10^6 floating point operations per second. The calculations assume $\sigma = 1$, $p = 1$, $F = 20$, and $\tau = 3$*

Striation thickness reduction (r)	Computational time		Number of required digits of precision	Storage requirements (bytes)
	Explicit mapping	Integration of Eulerian velocity		
3	4 *minutes*	1.5 *weeks*	9	5.2×10^6
4	6.5 *hours*	36 *months*	11	600×10^9
5	1 *month*	300 *years*	13	70×10^{12}

The article by Riley *et al.* (1986) presents one of the most complete. They considered a binary chemical reaction with no heat release (i.e., the reaction does not alter the fluid mechanics) in a *temporally* growing mixing layer, which is simpler to analyze, and contains some of the same physics as a *spatially* developing mixing layer (i.e., their geometry is similar to that in Figure E9.2.2(*a*)). The initial condition was also qualitatively similar to that of Corcos and Sherman (1984). Riley, Metcalfe, and Orzag (1986) used a pseudo spectral method using 32^3 and 64^3 grids; typical runs involved between 500 and 1000 time steps and consumed one to two hours of Cray 1 computer time. The Schmidt number was kept of order one so that the resolution requirements for both velocity and concentration were roughly the same, the Reynolds number based on the Taylor microscale was approximately 50, and the diffusivities of all species equal. Figure E9.2.2(*c*) shows snapshots of iso-vorticity plots and iso-concentration contours, the iso-concentration contours being simpler and more lamellar-like (compare with the cat eyes flow). Note that the limit of resolution is set by the computational grid. Striations created below this scale will be averaged out and the lamellar structure, if present, will be lost. If the reactions are slow the loss of resolution is not serious. However, depending on time-scales, the reaction might be diffusion controlled at these length scales and in this case the resolution problems become serious (see previous example). Similar treatments have included compressibility, for low Mach numbers, and also heat release (McMurtry *et al.*, 1986).

Problem 9.2.2
A possible way of connecting the results for microflow elements with Eulerian measurements is by computing the spectrum of concentration of a product or a passive scalar. Lundgren (1985) used an axisymmetric microflow element to obtain the product spectrum corresponding to small

length scales (large wave numbers). To what extent is the spectrum dependent upon the characteristics of the microflow element?

Problem 9.2.3*

Study the computational problems associated with the determination of the striation thickness evolution by means of Eulerian sensors. Use the BV flow (Section 7.3) and consider a probe such that it marks 'one' if it is contained within a striation and 'zero' if it is not. Discuss the possibility of using this signal to determine a 'mixing time'.

Problem 9.2.4*

Consider the analysis of Kerr's direct simulation data (Kerr, 1985) done by Ashurst *et al.* (1986). Kerr conducted a three-dimensional

Figure E9.2.2(c). Direct simulation of a mixing layer with a bimolecular reaction: (c1) iso-vorticity; and (c2) iso-concentration, showing a single roll-up. The initial condition is similar to that of 9.2.2(a) and is taken as $u(z) = (U/2) \tanh(0.55z/z_m)$; the time is made dimensionless with respect to $(du/dz)^{-1}$ at $z = 0$, whereas distances are made dimensionless with respect to UT; the Reynolds number is defined as Uz_m/v and is equal to 50; the second Damköhler number based on a diffusional time defined with respect to the scale z_m is 12.5. (Reproduced with permission from Riley, Metcalfe, and Orzag (1986).) The frames show the state of the system at various times.

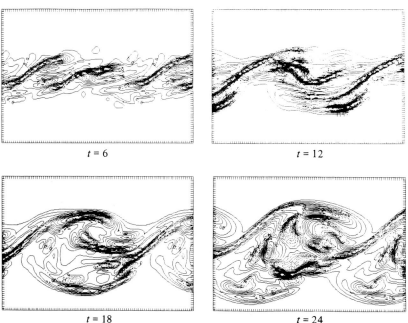

$t = 6$ $t = 12$

$t = 18$ $t = 24$

(c1)

simulation in a cubic box using up to 128^3 points to compute velocity and passive scalar (concentration, c) fields with periodic boundary conditions in all three directions (the equations were forced to maintain steady-state turbulence by driving the wave numbers less than a specific cut-off value). Ashurst *et al.* took 'snapshoots' of the flow and at various points **x** calculated the eigenvalues of the symmetric part of $\nabla \mathbf{v}$, denoted α, β, and γ ($\alpha \geqslant \beta \geqslant \gamma$) and the direction of ∇c (they analyzed 8 planes each of which consisted of 128^3 points and in each point they recorded the value of v_i, $\partial v_i / \partial x_j$, and ∇c). They found that 23% of the points analyzed were such that $\beta < 0$ and the eigenvalues were ordered approximately as $|\beta| \approx |\gamma|/4$ and $\alpha \approx 1.2|\gamma|$ (a sphere becomes a filament). In the remaining 77% of the cases $\beta > 0$ and $\beta \approx |\gamma|/3$ and $\alpha \approx (3/4)|\gamma|$ (a sphere becomes a disk). Moreover, in this case ∇c was found to be nearly parallel to the maximum direction of compression of the flow (γ). Attempt a rationalization of these findings in terms of the two-dimensional results of Sections 7.2 (tendril–whorl) and 7.3 (blinking vortex). Note that if the efficiency decreases, the flow becomes less extensional. Compare the results with the classical findings of Batchelor and Townsend.[22] Investigate also

Figure E9.2.2(*c*) *continued*

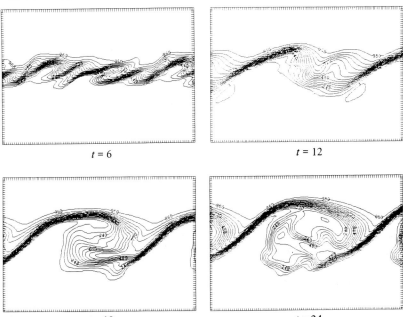

$t = 6$	$t = 12$
$t = 18$	$t = 24$

(*c2*)

the roll of loss of precision in similar computations in light of the Example 9.2.3. (For additional details see Ashurst *et al.*, 1987.)

9.3. Passive and active microstructures

The simplest problem of deformation and rotation of microstructures corresponds to the case of passive material elements. By definition, a *passive material filament* of length $d\mathbf{x}$ and orientation $\mathbf{m} = d\mathbf{x}/|d\mathbf{x}|$ evolves according to the equations of Chapter 3, i.e.,

$$D(\ln \lambda)/Dt = (\nabla \mathbf{v}):\mathbf{mm} \qquad (9.3.1a)$$

$$D\mathbf{m}/Dt = (\nabla \mathbf{v})^{\mathrm{T}} \cdot \mathbf{m} - (\mathbf{D}:\mathbf{mm})\mathbf{m}. \qquad (9.3.1b)$$

A similar viewpoint can be adopted in the case of *active microstructures* (both infinitesimal filaments and planes). For example, in the case of filaments, an active element described by a vector equation obeys a somewhat more complicated expression than (9.3.1a,b); if the element has some internal resistance, such as surface tension or viscosity in the case of a droplet or an 'entropic force' in the case of a coil, it stretches and rotates in a different way than a passive element.

In the context of mixing the most interesting case corresponds to the case of droplets. As we have seen, for miscible fluids, molecular diffusion becomes the controlling mechanism when the striations are reduced to small enough length scales. In the case of immiscible fluids at the beginning of the process, the blobs or striations fed into the flow are large and the stretching and breakup are dominated by inertial and viscous forces. However, as the length scales of the blobs decrease, surface tension forces might play a dominant role in preventing further stretching and breakup.

The deformation of the blobs is related in a complicated way to the velocity field. In practical mixing problems, the velocity field, even in the case of a single fluid, is not known, and in any case it would be perturbed by the presence of a second fluid. If the fluids are immiscible, coalescence and other hydrodynamic interactions further complicate the analysis, as does accounting for blobs of several length scales produced by the mixing process. In practice the fluids are more often than not, rheologically complex. In the same style as Section 9.1, it is desirable to start the analysis from a clearly defined point.

To a first order approximation, the velocity field with respect to a frame fixed on the drop's center of mass, denoted $\mathbf{X}$, and far away from it, denoted by the superscript ∞, can be approximated by

$$\mathbf{v}^{\infty} = \mathbf{x} \cdot \mathbf{L} + \textit{higher order terms}, \qquad (9.3.2)$$

where the tensor $\mathbf{L} = \mathbf{L}(\mathbf{X}, t) = \mathbf{D} + \mathbf{\Omega}$ is a function of the fluid mechanical path.[23] The central point is to investigate the role of $\mathbf{L}$ in the stretching and breakup of the drop.

Denote the viscosity of the drop as μ_i and the viscosity of the surrounding fluid as μ_e (the subindices 'i' and 'e' represent the *interior* of the drop and the *exterior* fluid, respectively). Neglecting any body forces that arise due to the rotation and translation of the moving frame and assuming that the Reynolds numbers $\rho_e S a^2 / \mu_e$ and $\rho_i S a^2 / \mu_i$ are vanishingly small ('a' is a characteristic length scale, e.g., the radius of the undisturbed drop, and S is characteristic shear rate, $|\mathbf{L}|$), the problem is governed by the creeping flow equations

$$\mu_m \nabla^2 \mathbf{v}_m = \nabla p_m, \qquad \nabla \cdot \mathbf{v}_m = 0, \qquad m = i, e \qquad (9.3.3)$$

with the boundary conditions

$$\mathbf{v} \to \mathbf{v}^\infty = \mathbf{x} \cdot \mathbf{L} \qquad \text{as } |\mathbf{x}| \to \infty$$

$\mathbf{v}_e = \mathbf{v}_i$ and $\mathbf{n} \cdot (\mathbf{T}_e - \mathbf{T}_i) = 2H\sigma\mathbf{n}$, at the surface of the drop (denoted S), where σ is the interfacial tension, and $2H$ is the mean curvature at the point with exterior unit normal $\mathbf{n}$. As formulated above the main parameters of the system are[24]:

(i) the dimensionless strain rate or *capillary number* $Ca = aS\mu_e/\sigma$, which is the ratio of the viscous to surface tension forces,
(ii) the viscosity ratio $p = \mu_i / \mu_e$,
(iii) the imposed flow, $\mathbf{L} = \mathbf{L}(\mathbf{X}, t)$, and
(iv) the initial shape of the drop S_0.

Note that the equations contain no explicit time dependence. This fact has an important consequence: the velocity field is instantaneously determined by the drop shape and the imposed flow, or equivalently, the drop shape at time t is uniquely determined by the initial condition S_0 and the value of $\mathbf{L}(\mathbf{X}, t)$ (Rallison, 1984).

There are almost no experimental data studying the deformation of droplets or microstructures with a prescribed deformation history (i.e., a specification of $\mathbf{L}(\mathbf{X}, t)$). Also, a general theoretical analysis is plainly impossible and simplifications are necessary. Therefore, the bulk of analysis and experimentation has centered on prototypical two-dimensional (steady) flows such as simple shear and plane hyperbolic flow. Theoretically it is possible to study also the case of axisymmetric extensional flow (Acrivos, 1983; Rallison, 1984). However, only recently have studies been conducted with linear two-dimensional flows spanning the range between pure shear and pure straining (Bentley and Leal, 1986b). (See Problem 2.5.3.)

9.3.1. *Experimental studies*

The focus of both theoretical and experimental investigations is to find the deformation produced by a given flow and how 'strong' it must be in order to break the droplet. Consider first some of the classical observations corresponding to the limits of *simple shear* and *planar extensional flow*. Experiments[25] indicate that when Ca is small, the deformation of an initially spherical droplet is small, and if the flow is steady the drop attains a steady shape (the time response to viscous stresses being of order $p\mu_c a/\sigma$). When the deviation from sphericity is small the usual measure of deformation is $D = (l-b)/(l+b)$; when the drops are significantly elongated (slender drops), the preferred measure is l/a (see Figure 9.3.1).

The most important experimental observations are the following[26]:

(i) At low Ca ($Ca < 0.1$ or so) and in both flows the deformation D increases linearly with Ca.

(ii) In both flows, for *low* values of the viscosity ratio p, and large values of Ca, drops are slender and have pointed ends.

(iii) In most cases there is a critical capillary number, Ca_c, such that if the strain rate is increased the extent of deformation increases until surface tension forces are surpassed by the viscous forces and the drop continues to elongate in the steady flow. The critical capillary number corresponds to the value of Ca (the shear rate being increased quasi-statically) such that

$$dl/dt > 0 \text{ for } Ca > Ca_c \qquad \text{and} \qquad dl/dt = 0 \text{ for } Ca < Ca_c.$$

However, in simple shear there is a value of $p_c \approx 4$ such that if $p > p_c$ drops would not break no matter how high the value of Ca.

There have been many experimental studies for the limit cases of simple shear and planar extensional flows. However, experimental studies in the class of linear flows studied in Example 4.2.3 (see also Figure P2.5.3) which span the limit cases of $\Omega = 0$ (pure stretching) and $\mathbf{D} = 0$ (pure rotation) as K is varied from 1 to -1, are relatively more recent. Simple shear corresponds to $K = 0$ (such flows can be generated in a four roller apparatus by varying the rotational speed of diagonally opposed pairs of cylinders, see Problem 2.5.3). In the first thorough study, Bentley (1985) considered the effect of such a flow by carrying out experiments under quasi-static conditions (dS/dt small) and for K in the range 0.2 to 1.[27] As might be expected, the curves of $Ca_c = S\mu_c a/\sigma$ versus p for flows with more vorticity than hyperbolic two-dimensional flow ($K < 1$) were found to lie between those of simple shear and hyperbolic two-dimensional flow (Figure 9.3.2). For $K = 1$, Ca_c was almost independent of p for $p > 1$. For

$K < 1$, however, Ca_c, increased with p, for $p > 1$. For the flow with the most vorticity considered, there was a limiting value ($p \approx 14.7$, based on theory), beyond which breakup was not possible. For high and intermediate viscosity ratios ($p > 0.05$), the data agreed well with the small deformation theory of Barthès-Biesel and Acrivos (1973), and for low viscosity ratios ($p < 0.01$) with an empirically modified version of the result of Hinch and Acrivos (1979) for hyperbolic extensional flow. In a related study, Stone, Bentley, and Leal (1986) focused on the interfacial driven motion which

Figure 9.3.1. (*a*) Drop being deformed in a flow; the flow in the neighborhood of the drop is $\mathbf{L} \cdot \mathbf{x}$ (the center of coordinates corresponds to the center of mass of the drop). (*b*) Deformation measures for nearly spherical drops, $D = (l - b)/(l + b)$, and elongated drops, l/a.

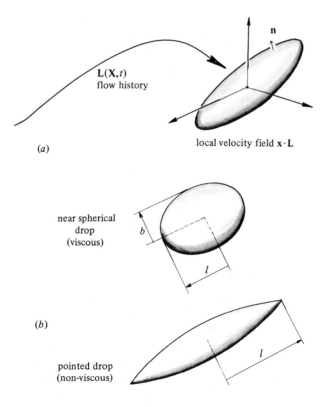

n

$\mathbf{L}(\mathbf{X}, t)$
flow history

(*a*)   local velocity field $\mathbf{x} \cdot \mathbf{L}$

near spherical
drop
(viscous)

b

l

(*b*)

pointed drop
(non-viscous)

l

occurs after the flow is stopped abruptly, for various flows with $K = 1.0$, 0.8, 0.6, 0.4, and 0.2 and viscosity ratios in the range $0.01 < p < 12$. They found that for modest extensions drop breakup does not occur with the flow but may occur as the result of a new mechanism, which they called 'end pinching'. In this mechanism the ends of the elongated drop become spherical, forming a dumbbell-like shape; the ends then proceed to pinch off leaving a cylindrical thread of fluid which can again repeat the process

Figure 9.3.2. Critical capillary number as a function of viscosity ratio $p = \mu_i/\mu_e$, with flow type, K, as a parameter. Comparison between experimental data and small ($O(\varepsilon^2)$) deformation theory (full line) and large deformation theory (broken line). (Reproduced with permission from Bentley and Leal (1986a).) The values of K increase from top to bottom (circles, $K = 1.0$; triangles; $K = 0.8$, diamonds, $K = 0.6$; inverted triangles, $K = 0.4$; crosses, $K = 0.2$).

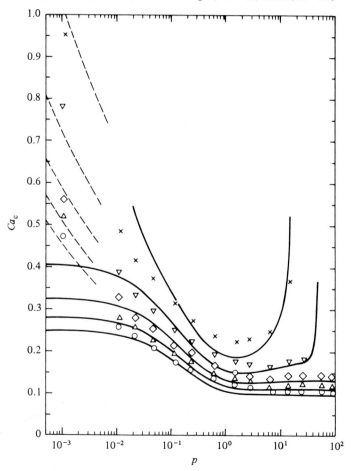

or relax back to a spherical shape. This mechanism was observed only in moderately elongated drops.[28] Figure 9.3.3 shows the critical elongation necessary to produce breakup after cessation of the flow. One of the conclusions of this study is that both *very viscous* and *inviscid* drops are hard to break; if the drops are very viscous they do not break in any flow, however, if the drops are inviscid, a very large elongation under quasi steady state conditions is needed before breakup can be achieved.

9.3.2. *Theoretical studies*

There is a strong correspondence between experimental and theoretical/ anaytical studies. Analytical studies can be divided into those applicable to high viscosity droplets, where deviation from sphericity is small (small deformation analyses), and those applicable to low viscosity droplets – highly elongated, only slightly bent and nearly axisymmetric – and where it is possible to use results obtained in the context of slender-body theory [for reviews through 1983 see Acrivos (1983) and Rallison (1984)]. In the

Figure 9.3.3. Necessary elongation to produce breakup following an abrupt halt of the flow, as a function of viscosity ratio $p = \mu_i/\mu_e$. The triangles denote the smallest elongation for which a drop was observed to break; the squares denote the largest elongation for which the droplet relaxed back to a sphere. (Reproduced with permission from Stone, Bentley, and Leal, (1986).)

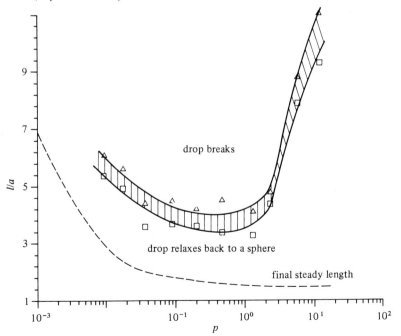

context of theoretical studies, breakup can be interpreted in various ways; according to the theory it might be unbounded growth, absence of a steady-state shape, or reaching a prescribed amplitude in the growth of capillary waves.

The basis of *small deformation analyses* is the exact general solutions of the creeping flow equations for spherical geometries, due to Lamb (1932). When the drop is only slightly deformed from a spherical shape, the flow field inside and outside the droplet can be obtained as a regular perturbation series, with the deformation from sphericity as the small parameter, and the evolution of the drop surface can be predicted. This method was first used by Taylor (1934) to study a drop in shear flow, assuming the drop to be deformed into an ellipsoid. Several years later Cox (1969) generalized Taylor's first order solution to the case of all linear flows. The shape of a slightly deformed drop is given by

$$(\mathbf{r} \cdot \mathbf{r})^{1/2} = 1 + \varepsilon \mathbf{A} : \mathbf{rr} + O(\varepsilon^2) \tag{9.3.4}$$

where ε is a small parameter. The shape of the drop is characterized by the time evolution of the tensor $\mathbf{A}$. Since the evolution equation is linear in $\mathbf{A}$, a steady shape exists for all Ca and therefore the model is not capable of predicting breakup. However, if some of the higher order terms are taken into account, the equation is non-linear and beyond a particular Ca no steady solution is possible. Using computer algebra to facilitate the anaysis, Barthès-Biesel and Acrivos (1973) obtained the second order solution to the problem. Even though the expansions should not be carried out to the same order of approximation, a point which was noted by the authors, the model was able to predict the breakup of drops. Their results were verified by the data of Bentley (1985).

The *large deformation analyses* theories pertain to the case of low viscosity drops in flows with sufficiently large Ca so that the drops are slender with pointed ends. The main component of the analysis is 'slender body theory' in which the outer flow is constructed approximately by replacing the drop with a distribution of singularities (e.g., sources, sinks, stresslets). The original paper in this area is by Taylor (1964), thirty years after observing the pointed drop shapes in his 1934 experiments. In line with previous work, he considered the simplest case corresponding to axisymmetric extensional flow; even though few details were given Taylor managed to obtain the correct expression for the critical strain rate ($Ca_c = 0.148p^{-1/6}$) as was shown later by Acrivos and Lo (1978). Buckmaster (1972, 1973) studied the same problem, within a somewhat more rigorous framework, and apparently without knowledge of Taylor's work, and found that there could be several possible steady-state solutions for the drop shape, but based on physical grounds he selected the one in which the drop was stretched the least as the only realistic solution (this was also the solution found by Taylor (1964)). Several years later, Acrivos and Lo (1978) obtained the same conclusion by means of a stability analysis. Hinch and Acrivos (1980) studied the case of a hyperbolic two-dimensional flow as perturbation of the axisymmetric case. They found that even though the cross-section was not circular, the area in the axisymmetric and two-dimensional cases were approximately the same $(Ca_c = 0.145p^{-1/6})$. Thus, it appears that only the area and not the shape is important in the breakup process and suggests the possibility of using the approximation of circular cross-sectional area for other flows.

The analysis corresponding to simple shear flow is substantially more complicated than the case of axisymmetric flow. Hinch and Acrivos (1980) studied the case in which the drop is almost aligned with the streamlines in a shear flow by assuming that the drop cross section remains circular due to the slenderness. They found that the drop axis bends slightly and that the elongated droplet attains a sigmoidal shape, but that the deviation from a straight line was small for most cases of interest. A steady-state analysis of the drop shape indicated that steady shapes were possible for all values of Ca. However, a time dependent simulation indicated that the drops were unstable beyond $Ca_c = 0.0541p^{-2/3}$. In numerical calculations involving jump in the strain rate they found that drops 'broke' (interpreted as failure of the numerical scheme) at the lower strain rate.[29]

Extensions of these analyses yield models which can be incorporated in the context of the flows of Chapters 7 and 8. For example, for the case $p \ll 1$ and $Ca \gg 1$, the dynamics of a nearly axisymmetric drop with pointed ends, characterized by an orientation $\mathbf{m}$ ($\mathbf{m} = 1$) and a length $l(t)$, is given by (Khakhar and Ottino, 1986c)

$$D \ln l(t)/Dt = \mathbf{L} : \mathbf{mm} - \underline{(\sigma/2(5)^{1/2}\mu_e a)[(l(t)/a)^{1/2}/(1 + 0.8p(l(t)/a)^3)]} \tag{9.3.5a}$$

$$Dm/Dt = (G\mathbf{D} + \mathbf{\Omega})^{\mathrm{T}} \cdot \mathbf{m} - (G\mathbf{D} : \mathbf{mm})\mathbf{m} \tag{9.3.5b}$$

where $G(t) = (1 + 12.5a^3/l(t)^3)/(1 - 2.5a^3/l(t)^3)$. The underlined term in (9.3.5a) acts as a resistance to the deformation [contrast with (9.3.1a,b)]. A very long drop, $(l(t)/a) \to \infty$, $G \to 1$, rotates and stretches as a passive element since the resistance to stretching is unable to counterbalance the effect of the outer flow. Note also that since $G > 1$ the droplet 'feels' a flow which is slightly more extensional than the actual flow.

Other researchers have sought a unification of results and developed a format which is useful for both low and high viscosity drops. Olbricht,

Rallison, and Leal (1982) developed two phenomenological models; one in which the microstructure was characterized as a vector [for example $l(t)\mathbf{m}$, as in (9.3.5a,b)], the other as tensor. The vector model is useful for elongated drops and macromolecules, for example in terms of a dumbbell model; the tensor model for high viscosity drops (see 9.3.4). Both models involve a number of phenomenological coefficients, which can be obtained by comparison of the models to particular cases. For example, in an empirical scheme for breakup of drops, one of the coefficients was backed-out from the small deformation results of Barthès-Biesel and Acrivos (1973) and subsequently used to predict the critical strain rate for breakup in other flows. The vector equation, useful for a variety of axisymmetric elements, such as droplets or macromolecules, is

$$D\mathbf{p}/Dt = \mathbf{p}\cdot(G\mathbf{D}+\boldsymbol{\Omega}) - G(F/(1+F))[(\mathbf{D}:\mathbf{pp})/(\mathbf{p}\cdot\mathbf{p})]\mathbf{p} - [\beta/(1+F)]\mathbf{p} \quad (9.3.6)$$

where $\mathbf{p}$ is the length of the element, and G, F, and β are suitably selected parameters.[30] Such models can be incorporated in the context of the simulations described in Chapters 7 and 8.

Example 9.3.1

In the context of the model of Equation (9.3.5a,b) a drop is said to break when it undergoes infinite extension and surface tension forces are unable to balance the viscous stresses. Consider breakup in flows with $\mathbf{L}:\mathbf{mm}$ constant in time (for example, an orthogonal stagnation flow with the drop axis initially coincident with the maximum direction of stretching). Rearranging Equation (9.3.5a), and defining a characteristic length as $a/p^{1/3}$, we obtain

$$\mathbf{L}:\mathbf{mm}/(\mathbf{D}:\mathbf{D})^{1/2} = \text{efficiency} = (1/2(5)^{1/2}E)[(l_s^{1/2}/(1+0.8l_s^3)]$$

where l_s denotes the steady-state length and $E = p^{1/6}Ca$. A graphical interpretation of the roots l_s is given in Figure E9.3.1. The horizontal line represents the asymptotic value of the efficiency (i.e., corresponding to $D\mathbf{m}/Dt = 0$), which in three-dimensions is $(2/3)^{1/2}$, and the value of the resistance is a function of the drop length for various values of the dimensionless strain rate E. For $E < E_c$ there are two steady states: one stable and the other unstable. For $E > E_c$ there are no steady states and the drop extends indefinitely (see Khakhar and Ottino, 1986c).

Problem 9.3.1

Estimate the order of magnitude of the additional body forces that arise due to the translation and rotation of the moving frame. How significant are they to the analysis based on Equation (9.3.3)?

Problem 9.3.2

Stone, Bentley, and Leal (1986) investigated the time evolution of the drop length, $L(t)/a$, for conditions slightly above the critical capillary number in the flow field of Figure P2.5.3. They plotted their results using a dimensionless time, $GK^{1/2}t$. They found that curves of $L(t)/a$ versus $GK^{1/2}t$ for a wide range of K fall approximately on the same curve. Justify.

Problem 9.3.3

Cast the results of Hinch and Acrivos (1980) and Buckmaster (1973) in the form of Equations (9.3.5a,b).

9.4. Active microstructures as prototypes

The interaction of active microstructures and a chaotic flow field might provide a prototype for a variety of physical problems. We will mention a few possible situations. For example, such a picture might help to understand why small amounts of high polymers (or needle-like particles), when added to laminar or turbulent flows, produce (sometimes) consequences at a macroscopic level which are out of proportion to their concentration.[31] Manifestation of these effects might be the drag reduction produced by polymers in turbulent flows and the increase of viscosity and non-Newtonian effects displayed by polymer solutions injected into porous media. Even though the two problems are very different – one is a low

Figure E9.3.1. Graphical interpretation of breakup of pointed drops. The figure corresponds to the case in which the droplet does not rotate with the flow.

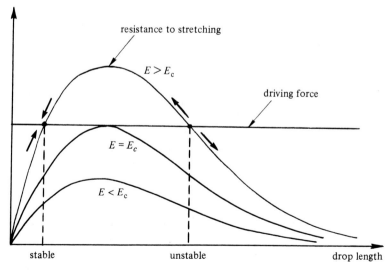

Reynolds number flow, the other is high Reynolds flow – there seems to be agreement that the explanation of the phenomena in the case of polymers is due to flow-induced conformational changes and that much benefit can be obtained by focusing on the behavior of a single micro-structure in a linear, possibly time dependent, flow. Macromolecular stretching can modify the surrounding flow; in turbulence the smallest eddies might be suppressed; in porous media the hydrodynamic interactions might result in non-Newtonian effects.

Even though considerable experimental and theoretical work has been done in both areas, the suggestion here is that microstructures in chaotic flows might have a bearing on these kinds of problems. However, since our primary focus is on a few common aspects of both problems, our review of the literature is rather superficial.[32] A review of experimental studies of flow induced conformation studies is given by Leal (1984). Most of the work focusing on conformational changes has been on well controlled linear flows, for example, opposed jets, two-roll mill, four-roll mill, six-roll mill (see, e.g., Berry and MacKley, 1977); only a few studies have focused on spatially periodic flows (Nollert and Olbricht, 1985). There should be no question that such flows contain some of the relevant aspects of both problems (i.e., porous media, turbulent flows) and that the conceptual and experimental advantages are significant. The flows are relatively simple to characterize completely and measurement of conforma-tional changes by light scattering are relatively simple to perform (as opposed to measurements in an actual turbulent flow or the interior of a real porous media). What is missing, however is the *unsteady* nature of the flow experienced by microstructures in actual turbulent flows and topologically disordered porous media. Our thesis here is that there is some opportunity for an unexplored mid-ground.

One such possible mid-ground is to focus on the behavior of micro-structures in two-dimensional chaotic flows (either experimentally or theoretically). Carefully designed low Reynolds number chaotic flows provide several key ingredients missing in the studies listed above. Let us consider a few:

(i) Chaotic flows provide unsteady (chaotic) Lagrangian histories. The flows are capable of generating regions where the stretching is mild and time-periodic and chaotic regions where, over time, the stretching of material elements is nearly exponential (efficient flows).[33] In the chaotic regions the microstructure is never in steady extension; the maximum directions of stretching constantly change.

(ii) The unsteadiness of the maximum directions of stretching, present in turbulent flows, is also present in chaotic two-dimensional flows, such as the journal bearing flow (Chapter 8); the randomness associated with porous media, usually ascribed to the porous media itself, is present in the partitioned-pipe mixer, which is spatially periodic, or in the eccentric helical annular mixer, which is time periodic.

(iii) Chaotic flows allow the investigation of spatial co-operative effects (e.g., percolation of chaotic regions, percolation of significantly stretched regions, see Figure 7.4.11, color plates).

(iv) Practical mixing problems start from an unmixed state; for example, polymer is injected into a pipe line and the drag reducing effects take place during mixing of the polymer with the flow. This effect cannot be incorporated into the conventional picture; however, chaotic flows can mimic this effect.

The central question is what happens to microstructures, such as those described by Equation (9.3.6), when placed in a chaotic flow (and also, but this is a harder question since it involves hydrodynamic interactions, what happens to the *flow* as a result of the interaction between the microstructure and the flow). The simplest studies can proceed under the assumption that it is possible to decouple the global and local aspects of the problem (i.e., there is no hydrodynamic interactions). Even with this idealization, several questions come to mind. For example, where is the region of significantly stretched elements? The answer to this question might have relevance in several co-operative phenomena. Besides stretching, several other phenomena can be incorporated at the microstructural level; breakup and coagulation/coalescence came to mind. A whole range of phenomena can be anticipated; for example it seems obvious that the process can generate fractal-like structures as obtained in diffusion limited aggregation (see, e.g., Witten and Sander, 1981, 1983; Meakin, 1985).

Example 9.4.1*
Consider stretching and breakup of pointed drops (with volume $(4/3)\pi a^3$) in the flow described in Section 8.2 (partitioned-pipe mixer). The objective is to determine if it is possible to predict (qualitatively) the behavior of the droplets based on the results obtained for the stretching of (passive) material filaments.

In an admittedly oversimplified model (Franjione and Ottino, 1987, unpublished), we assume that the drops are characterized by a vector $\mathbf{p}$ $l = |\mathbf{p}|$ have a viscosity ratio p, and obey an evolution equation of the type (9.3.5), i.e.

$$D\mathbf{p}/Dt = \mathbf{p} \cdot (\nabla \mathbf{v})^* - \{(G-1)[\mathbf{D} : \mathbf{pp}/(\mathbf{p} \cdot \mathbf{p})] + f\}\mathbf{p}$$

where

$$(\nabla v)^* = GD + \Omega,$$

is the velocity gradient experienced by the droplet as it moves in the flow, and G and f are functions defined as

$$G = [1 + 12.5(a/l)^3]/[1 - 2.5(a/l)^3],$$
$$f = [\sigma/2(5)^{1/2}\mu_e a](a/l)^{1/2}/[1 + 0.8p(a/l)^3].$$

Further, we will arbitrarily assume that the drops break when $l/a > 10$.

The droplets, somewhat stretched, are placed at the entrance of the partitioned-pipe mixer and arranged in a square lattice. The initial orientation is purely radial and the initial number of droplets is 3.14×10^4. We will consider two operating conditions, $\beta = 2$ and $\beta = 8$. The values of the parameters for the results presented below are $p = 3 \times 10^{-4}$ and

Figure E9.4.1. Stretching and breakup of microstructures in the flow field of the partitioned-pipe mixer for two operating conditions ((a) $\beta = 2$, (b) $\beta = 8$). The dots represent the region where the microstructures have broken after 10 sections. Note that in (b) a significant amount of breakup occurs near the center of the channel.

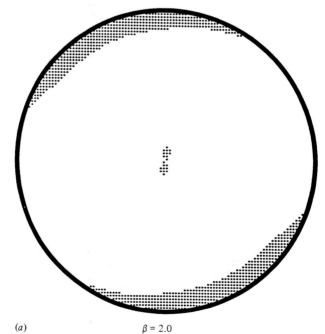

(a) $\beta = 2.0$

$(L/V_z)\sigma/2(5)^{1/2}\mu a = 1$, where L is the length of the plate and V_z the average axial velocity. On physical grounds, we restrict G to be positive. If any of the droplets, as it moves through the mixer, produces $G < 0$ the simulation for that drop is stopped and the drop removed. The central question is to determine the regions of breakup to determine any correlation with the results of Section 8.2 for passive structures (i.e., $G = 1$, $f = 0$).

Figure E9.4.1 shows the results for two different values of the operating parameter β and for a mixer consisting of 10 elements. Figure E9.4.1(a) shows that most of the drop breakup has occurred near the walls, whereas Figure 8.3.10(a1) predicts that most of the stretching (greater than 10^4) occurs in the chaotic regions (compare with Poincaré section) which are quite evenly distributed throughout the flow. The correlation with efficiency is no better; Figure 8.3.10(a2) predicts that some of the most

Figure E9.4.1 *continued*

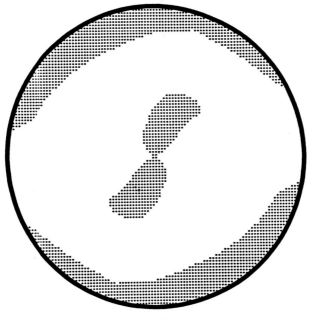

(b) $\beta = 8.0$

efficient regions are the large regular islands where no breakup occurs at all. The results for $\beta = 8$ are no more encouraging. Figure E9.4.1(b) shows that breakup occurs near the walls and in two large central regions, whereas Figure 8.2.10($a1$) and ($a2$) although roughly consistent with each other do not predict such a pattern. It thus appears that these results could hardly have been predicted based on the maps of stretching of Section 8.3. Obviously, extensive studies are needed in order to make qualitative predictions of breakup in this and other complex flows.

*Example 9.4.2**

The chaotic flows of Chapters 7 and 8 can be used to study various aggregation processes. Possibly the simplest example, which can be

Figure E9.4.2. Coagulation in the blinking vortex flow. Figure 9.4.2(a) shows the initial location of particles with capture diameter 0.003 which did not coagulate in the flow field of the blinking vortex flow operating at $\mu = 0.8$ after 25 cycles; the approximate location of the bounding KAM surface is shown by a full line; (b) example of a non-monotonic histogram of cluster order; the operating conditions correspond to $\mu = 0.4$ and a capture diameter 0.001 after 212 cycles of the flow. The initial arrangement was achieved in the following way: 3,600 particles were placed in the region $-1 < x < 1$, $-0.01 < y < 0.01$ and mixed for 5,000 cycles without coagulation; (c) time evolution of cluster order for a well mixed condition ($\mu = 1.0$, $\delta = 0.001$, 5,000 cycles). The symbols represent the computational results (circles, order-1; triangles, order-2; diamond, order-3, squares, the total number of clusters); the curves are the solution predicted by a mean field kinetic model.

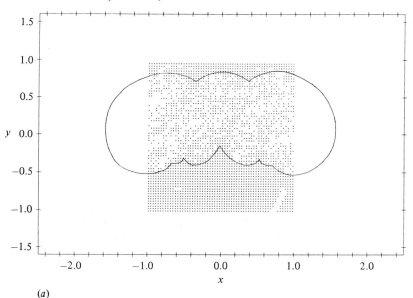

(a)

regarded as model of gradient coagulation (Levich, 1962), corresponds to irreversible binding of particles with capture diameter δ (Muzzio and Ottino, 1988). More complex examples might consider aggregation and breakup processes, growth of fractal-like structures, Brownian diffusion, etc. In this example, when two particles coagulate, the size of the cluster does not change. The number of particles forming part of a cluster is denoted as the *order* of the cluster. Figure 9.4.2(*a*) shows the initial location of particles with capture diameter 0.003 which did not coagulate in the flow field of the blinking vortex flow (Section 7.3) operating at $\mu = 0.8$ after 25 cycles. It is apparent that the regular regions are not nearly as effective as the chaotic regions and that most of the coagulation takes place within the bounding KAM surface. The dynamics of this simple coagulation process is quite complicated and it is possible to find conditions leading to non-monotonic histograms of cluster order; this seems to be due to the presence of relatively large regular and chaotic regions (Figure 9.4.2(*b*) shows a case at a value of μ slightly above the transition to global chaos). Under 'well mixed' conditions, i.e., when the stirring is efficient enough as to destroy large scale concentration variations,

Figure E9.4.2 *continued*

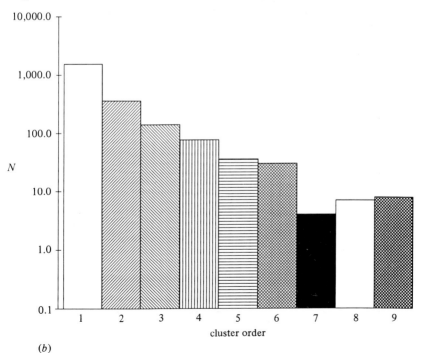

(*b*)

Figure E9.4.2 continued

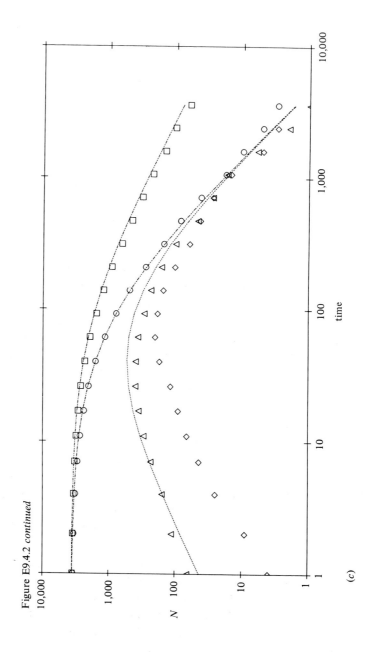

(c)

the kinetics of cluster formation can be described by simple laws, such as Smoluchowski's mean field kinetics (even though there is no diffusion), and the dependency of the kinetic constant upon parameters such as μ and δ, can be predicted based on simple models. Figure 9.4.2(c) shows the type of agreement observed between computations and a kinetic model for clusters of order one (single particles), two (dimers), and the total number of particles.

Bibliography

The content of this chapter is more indicative of future possibilities than of the current state of the art. The common thread is the role played by local and global flows on diffusion and reaction processes and in stretching, breakup, and aggregation processes. For example, the fluid mechanical path is not very important in conversion and selectivity of simple reactions but appears to be very important in complex reactions such as polymerizations; small variations in the coefficients of a dynamical equation appear to determine whether a microstructure breaks or not.

There is a substantial amount of literature on the foundations of both topics and the review presented here is a rather biased account suited to our needs. The sections on diffusion and reaction rely on transport across stretching interfaces. The main ideas and a brief historical review can be found in Ottino (1982). The importance of stretching of material surfaces is addressed in many works; one of the earliest is Batchelor (1952) and various other articles referred to in the chapter. In the context of chemical engineering, the idea of a local flow was used by Fisher (1968) and by Chan and Scriven (1970); in mechanical engineering, the idea has been used in the context of combustion, e.g., Spalding (1978a). The transformations of Section 9.1 can be found in Ranz (1979); Ottino (1982); and Chella and Ottino (1984) (many of the examples are taken from this work). The basic ideas of the lamellar viewpoint are discussed by Ranz (1985).

The review of stretching and breakup does not do justice to the state of the art and the reader should consult the reviews by Rallison (1984); and Acrivos (1983). The most complete summary of experimental work is given by Grace (1982). Probably the most complete experimental studies are those from Leal *et al.* A partial listing includes Bentley and Leal (1986a); and Stone, Bentley, and Leal (1986). Complete details are given by Bentley (1985).

Notes

1 These terms were coined by Aref and Tryggvason (1984).

2 This viewpoint was suggested by Khakar, Chella, and Ottino (1984).

3 This is a good approximation if the distribution of s is narrow. In chaotic flows the distribution can be very broad, especially if there are large regular regions.

4 See Example 9.2.3.

5 If the structure is nearly axisymmetric, as the whorl of Problem 4.2.5, we have, approximately,

$$\left(\frac{\partial}{x\partial x}\left(x\frac{\partial\omega_s}{\partial x}\right), 0, 0\right)$$

where x is the radial distance.

6 In a few simple cases we might be able to solve some problems taking into account the distribution. One case is the vortex wrapping of Problem 4.2.5.

7 Throughout this chapter we will use an approximate form of the continuity equation, namely,

$$\frac{Dc_s}{Dt} = \nabla \cdot (D_s \nabla c_s) + r_s.$$

Similarly, the energy equation is written as

$$\frac{DT}{Dt} = \alpha_T \nabla^2 T + r_T$$

where T is the temperature, α_T is an average thermal diffusivity, and r_T is a generation term that includes energy generation due to chemical reaction, viscous dissipation, and volume change (see Bird, Stewart, and Lightfoot, 1960, Chap. 18). We do not consider multi-component effects.

8 This equation is valid for a material region $S_{\mathbf{X}_m}$ of size d where a linear velocity field approximation is valid. For turbulent flow this scale can be taken as the Kolmogorov scale, $(\mu^3/\rho^2\varepsilon)^{1/4}$, where ε is the viscous dissipation per unit volume. Note, however, that there can be significant stretching and folding below these scales as seen in the examples of Sections 7.1.

9 For example, we know that if the flow is a steady curvilinear flow, the stretching history α decays as $1/t$; in chaotic flows, on the average, α, is nearly constant, etc.

10 A brief list of other works using similar equations are: Saffman, 1963; Tennekes and Lumley, 1980; Foister and van de Ven, 1980; and Batchelor, 1979.

11 The application of a_v to mixing was suggested by Corrsin (1954).

12 See Ottino and Macosko (1980), and references therein, and Townsend's hot spot model, Example 9.1.1.

13 This definition of 'warped time' coincides with that of Ranz (1979) if $t_C = t_{D_0}$.

14 The choice of the scaling time t_C is also important in improving the accuracy of numerical solutions (it should be selected in such a way that each term (e.g., $\partial^2 C_i/\partial x^2$) in Equations (9.1.6a,b) is of the order one. The coefficient (e.g., t_C/t_{D_0}) then reflects the importance of the term). Thus, t_C is chosen to be t_R for 'slow' reactions and t_{D_0} for 'fast' reactions.

15 Also, note that the streams themselves can be partially premixed and that everything can be generalized to the case of three or more streams.

16 Damköhler (1936) defined what is now called the first Damköhler number, Da_I, as (length/velocity)/characteristic reaction time which can be interpreted as the ratio of the characteristic Eulerian time to a characteristic reaction time. Our $Da_I^{(L)}$ is defined in an analogous fashion, the superscript (L) refers to the use of a Lagrangian time-scale.

17 For example, for the reaction $A + vB \rightarrow P$, the conditions for a stationary reacting zone are: $(vc_{A_0}s_{A_0}/c_{B_0}s_{B_0}) = 1$, $(vc_{A_0}/c_{B_0})(D_A/D_B)^{1/2} = 1$.

18 The case of fast reactions has been considered in the context of combustion and has received considerable attention in applied mathematics (see for example Kapila, 1983); the use of fast reactions as tracers is discussed in Ottino (1981).

19 An extreme case is given by polymerizations; see Example 9.2.1. It is possible also to imagine situations where imperfect mixing can lead to 'thermal exlosions' due to imbalance between heat generation and dissipation due to thermal conduction. The presence of a thermal explosion depends strongly upon the Lewis number (Ottino, 1982).

20 This topic dates from the work of Danckwerts (1958) and Zweitering (1959). See references in Chapter 1. A book devoted to these issues is Nauman and Buffham (1983).

21 See Section 8.5.

22 Batchelor and Townsend (1956) obtained, for homogeneous turbulence, $\alpha = 0.43(\varepsilon/\mu)^{1/2}$, $\beta = 0.12(\varepsilon/\mu)^{1/2}$, $\gamma = -0.55(\varepsilon/\mu)^{1/2}$.

23 The tensor $\mathbf{L}(\mathbf{X}, t)$ plays the same role as the stretching function, $\alpha(\mathbf{X}, t)$ in Section 9.1.

24 This formulation is largely due to Taylor (see Chapter 1).

25 As indicated in our presentation in Chapter 1, Taylor (1934) invented the two basic experimental configurations; to mimic a two-dimensional hyperbolic flow he invented the four-roller apparatus, to mimic shear flows he constructed a parallel flow apparatus.

26 A few references discussing various aspects of experimental results, prior to the work of Bentley (1985), are the following: Rumscheidt and Mason (1961), repeated some of the same experiments of Taylor (1934); their results matched the theory for small deformations. Karam and Bellinger (1968), carried out experiments in simple shear flow using a Couette apparatus over a relatively wide range of p and obtained a graph of Ca_c versus p which had a minimum for p in the range 0.2 to 1. They also reported a viscosity ratio of $p \approx 4$ beyond which breakup was not possible in shear flow. A few years later an exhaustive experimental study was reported by Grace (1971, published 1982). In the first half of his paper Grace concentrated on shear and hyperbolic flows (as Taylor had done), but on a much wider range of viscosity ratios (10^{-6} to 950), whereas in the second half of the paper he concentrated on the applicability of the results to static mixers. In addition to Ca_c, the time for breakup and the length and deformation at breakup were also recorded. He also investigated the breakup under conditions of super-critical strain ($Ca > Ca_c$). Grace also found a maximum viscosity ratio, slightly lower than Karam and Bellinger's, $p \approx 3.5$, beyond which drops could not be broken in shear flow. At low viscosity ratios he found $Ca_c \approx p^{-0.55}$ for simple shear flow and $Ca_c \approx p^{-0.16}$ for two-dimensional extensional flow. For drops of higher viscosity ratio, the data are more sparse and indicate an increase of Ca_c with p. There have been relatively few studies focusing on the effect of the history of the imposed flow and the dynamics of the drop. One early study is by Torza, Cox, and Mason (1972). They found that the rate of increase of the strain rate (dS/dt) had a significant impact on breakup. Though data are only reported for relatively low values of dS/dt, they found that the critical strain rate was a function of dS/dt, especially at high dS/dt.

27 See Bentley and Leal (1986a) for details of the apparatus and Bentley and Leal (1986b) for experimental results.

28 It is significant that the breakup mechanism *was not* due to capillary wave instabilities as studied by various researchers (e.g., Mikami, Cox, and Mason, 1975; Khakhar and Ottino, 1987). A long thread is not equivalent to an infinite cylinder and end effects cannot be neglected under most conditions of interest. Obviously, a significant difference between an infinite and a finite cylinder is that an infinite thread is an equilibrium shape whereas a finite cylinder is not; in the case of an infinite cylinder the only question is whether or not it is stable (Stone and Leal, 1989). However, if drops are significantly stretched, $l/a \gg 10^2$, as they are in chaotic flows, the prevalent mode of breakup is by capillary waves.

29 Other results for time dependent flows indicate the severity of the fluid mechanical path. For example, Khakhar and Ottino (1987) considered the number of fragments N produced by bursting of an *infinite* thread by means of a model based on growth of capillary waves. The flow field around the thread was taken to be

$$v_z = \dot{\varepsilon}(t)z, \qquad v_r = -\tfrac{1}{2}\dot{\varepsilon}(t)r.$$

Two cases were compared; one with constant $\dot{\varepsilon}$, the other with $\dot{\varepsilon}(t) = S/(1 + St)$ where S is a constant. The number of fragments goes as $Ca^{2.8}$ for constant $\dot{\varepsilon}$ and as Ca^1 for the case $\dot{\varepsilon}(t) = S/(1 + St)$ (where Ca is the capillary number, $Ca = \mu_c Sd/\sigma$, and d the initial diameter of the thread). This implies that under identical conditions a flow with decaying efficiency will produce significantly fewer fragments than one with constant efficiency. These results appear to have interesting implications in the context of experimental studies of breakup in chaotic flows, such as those of Sections 7.4 and 7.5. Although the experiments are just at the beginning and it is early to draw firm conclusions, it does appear that there is significantly more stretching and breakup within the chaotic regions as compared with the regular regions, where the efficiency of the flow is poor.

30 Choices corresponding to two-dimensional extensional flow and simple shear flow are given by Khakhar and Ottino (1986b).

31 This simple picture implicitly considers the possibility of drag reduction without any wall effects. Such a viewpoint was advocated by de Gennes (1986).

32 For an introduction to the literature of drag reduction see Lumley (1973), Virk (1975), Berman (1978), and many of the articles in Vol. **24**(5), *J. of Rheology*, 1980.

33 Recent studies involving experiments in the journal bearing flow of Section 7.4 show that droplets in chaotic regions can stretch by several orders of magnitude and break primarily by capillary wave instabilities whereas drops placed in large islands, such as those of Figure 7.4.5(*a*), hardly deform at all (Tjahjadi, 1989).

34 Recent theoretical analyses and computations involving fast bimolecular reactions $(A + B \rightarrow P)$ in systems with initially distributed striations of the reactants – such as Figure E9.2.1 – have shown scaling behavior. Owing to the unevenness in the striation distribution the reaction planes move and the average striation thickness increases in time. The remarkable result is that, regardless of the initial distribution, the system evolves in such a way as to produce a self-similar time-dependent striation thickness distribution (F. J. Muzzio and J. M. Ottino, 'Evolution of a lamellar system with diffusion and reaction: a scaling approach', *Phys. Rev. Lett.*, **63**, 47–50, 1989).

Appendix

Cartesian vectors and tensors

The objective of this appendix is to cover some indispensable background with minimal mathematical complexity. The discussion is restricted to Cartesian tensors.[1]

Properties of vector spaces

Vectors, in a general sense,[2] are denoted as **u**, **v**, **w**, etc. We require them to satisfy the following properties:

I Sum

(1) Associative $\mathbf{u} + (\mathbf{v} + \mathbf{w}) = (\mathbf{u} + \mathbf{v}) + \mathbf{w}$
(2) Commutative $\mathbf{u} + \mathbf{v} = \mathbf{v} + \mathbf{u}$
(3) Existence of zero (**0**) $\mathbf{u} + \mathbf{0} = \mathbf{u}$
(4) Existence of negative $\mathbf{u} + (-\mathbf{u}) = \mathbf{0}$.

II Multiplication

(1) Associative $\alpha(\beta\mathbf{u}) = (\alpha\beta)\mathbf{u}$
(2) Unit multipication $1\mathbf{u} = \mathbf{u}$
(3) Distributive $(\alpha + \beta)\mathbf{u} = \alpha\mathbf{u} + \beta\mathbf{u}$
(4) Distributive $\alpha(\mathbf{u} + \mathbf{v}) = \alpha\mathbf{u} + \alpha\mathbf{v}$.

Properties I + II form a vector space.

III Inner product

(1) $\mathbf{u} \cdot \mathbf{v} = \mathbf{v} \cdot \mathbf{u}$
(2) $(\mathbf{u} + \mathbf{v}) \cdot \mathbf{w} = \mathbf{u} \cdot \mathbf{w} + \mathbf{v} \cdot \mathbf{w}$
(3) $\alpha(\mathbf{u} \cdot \mathbf{v}) = \alpha\mathbf{u} \cdot \mathbf{v}$
(4) $\mathbf{u} \cdot \mathbf{u} > 0$ if $\mathbf{u} \neq \mathbf{0}$ (magnitude of **u**, $|\mathbf{u}| = +(\mathbf{u} \cdot \mathbf{u})^{1/2}$, denoted u).
(5) $\mathbf{u} \cdot \mathbf{0} = 0$.

Properties I + II + III form an *Inner Product Space*. No reference is necessary to basis, dimension, etc. Properties I, II, and III, are quite general and with proper definitions of '+', addition, and '·', multiplication, they might be obeyed by a suitable class of functions. In particular, they are obeyed by Cartesian vectors and tensors. Some important properties are a direct consequence of I–III. A space obeying I–III satisfies the inequalities:

$$\text{Cauchy–Schwarz} \qquad |\mathbf{u}\cdot\mathbf{v}| \leqslant |\mathbf{u}|\cdot|\mathbf{v}|$$
$$\text{Triangle} \qquad |\mathbf{u}+\mathbf{v}| \leqslant |\mathbf{u}| + |\mathbf{v}|.$$

Operations

Co-ordinate system, basis, dual basis

From here on the discussion is restricted to three-dimensional space. In general, a point $\mathbf{x}$ in space is located by its co-ordinates, $\mathbf{x} = \mathbf{x}(x^1, x^2, x^3)$. For example, in cylindrical coordinates, $\mathbf{x}$ is specified by r, θ, z. The *natural basis* $\{\mathbf{e}_i\}$ is defined as

$$\mathbf{e}_i - \partial \mathbf{x}/\partial x^i$$

We require the $\{\mathbf{e}_i\}$ to be linearly independent, i.e., $\alpha_i \mathbf{e}_i = 0$ implies that $\alpha_i \equiv 0$.[3] Consider a vector $\mathbf{v}$ located at $\mathbf{x}$, i.e., $\mathbf{v} = \mathbf{v}(\mathbf{x})$. The components defined with respect to the natural basis,

$$v^i(\mathbf{x}) = \mathbf{e}_i \cdot \mathbf{v}(\mathbf{x})$$

are called the *contravariant components*. The dual basis, $\{\mathbf{e}^j\}$, is defined such that

$$\mathbf{e}_i \cdot \mathbf{e}^j = 1 \text{ if } i = j$$
$$\mathbf{e}_i \cdot \mathbf{e}^j = 0 \text{ if } i \neq j$$

The components defined with respect to the *dual basis*

$$v_i(\mathbf{x}) = \mathbf{e}^i \cdot \mathbf{v}(\mathbf{x})$$

are called the *covariant components*. A coordinate system is called *orthogonal* if the natural basis is such that

$$\mathbf{e}_i \cdot \mathbf{e}_j \neq 0 \text{ if } i = j$$
$$\mathbf{e}_i \cdot \mathbf{e}_j = 0 \text{ if } i \neq j$$

For an orthogonal co-ordinate system the vectors

$$\mathbf{e}^{\langle i \rangle} = \mathbf{e}_i/|\mathbf{e}_i| = \mathbf{e}_i/e_i$$

form an *orthonormal* basis, i.e.,

$$\mathbf{e}^{\langle i \rangle} \cdot \mathbf{e}^{\langle j \rangle} = \delta_{ij}$$

where δ_{ij} is the Kroneker delta ($\delta_{ij} = 1$ if $i = j$ and $\delta_{ij} = 0$ if $i \neq j$). The components with respect to the orthonormal basis $\{\mathbf{e}^{\langle i \rangle}\}$ are called *physical components.*[4] For future use we record the magnitudes e_i in two co-ordinate systems

Cylindrical: $e_r = 1, \ e_\theta = r, \ e_z = 1$

Spherical: $e_r = 1, \ e_\theta = r, \ e_\phi = r \sin \theta.$

See Figure A.1.

Figure A.1

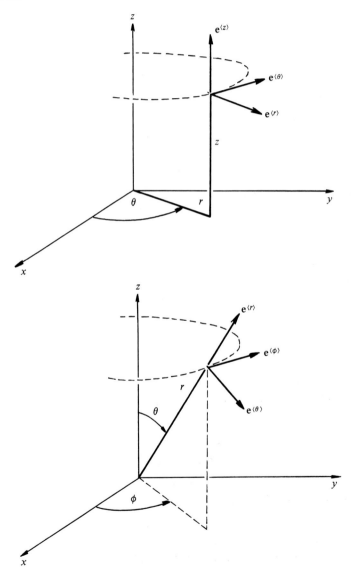

Cross product – definition

The cross product, denoted $\times$, is defined in terms $\{e_i\}$

$$\mathbf{e}_1 \times \mathbf{e}_2 = \mathbf{e}_3$$
$$\mathbf{e}_2 \times \mathbf{e}_1 = -\mathbf{e}_3$$
$$\mathbf{e}_1 \times \mathbf{e}_3 = -\mathbf{e}_2$$

and so forth. In general, $\mathbf{e}_i \times \mathbf{e}_j = \varepsilon_{ijk}\mathbf{e}_k$ (ε_{ijk} is called the permutation symbol) and is such that

$$\varepsilon_{ijk} \equiv \begin{cases} +1, & \text{if } ijk = 123, 231, \text{ or } 312 \\ -1, & \text{if } ijk = 321, 132, \text{ or } 213 \\ 0, & \text{if any two indices are equal} \end{cases}$$

Problem

Verify that $\delta_{mn}\delta_{mn} = 3$.

Problem

Verify that $\varepsilon_{ijk}\varepsilon_{irs} = \delta_{jr}\delta_{ks} - \delta_{js}\delta_{kr}$.

Problem

Using the results given above prove

$$\mathbf{u} \times (\mathbf{v} \times \mathbf{w}) = \mathbf{v}(\mathbf{u}\cdot\mathbf{w}) - \mathbf{w}(\mathbf{u}\cdot\mathbf{v}).$$

This identity provides a means for proving the Cauchy–Schwarz inequality.

Vector representation in terms of bases

Any vector $\mathbf{v}$ admits an infinite number of possible representations. Thus $\mathbf{v}$ can be expressed in terms of the bases $\{\mathbf{e}_i\}$ and $\{\mathbf{e}_i'\}$ as

$$\mathbf{v} = v_i\mathbf{e}_i = v_i'\mathbf{e}_i'.$$

The v_i are the components of $\mathbf{v}$ with respect to $\{\mathbf{e}_i\}$, and the v_i' are the components with respect to $\{\mathbf{e}_i'\}$. Thus, one of the possible ways of identifying $\mathbf{v}$ is by means of its components,

$$[\mathbf{v}] \equiv (v_1, v_2, v_3)$$

where the brackets, [ ], denote matrix representation.[5] It is important not to confuse the vector itself with its component representation.

How to obtain components

Using the inner product and the representation of $\mathbf{v}$ (using any subindex, but never repeating them) we write:

$$\mathbf{v}\cdot\mathbf{e}_j = v_i\mathbf{e}_i\cdot\mathbf{e}_j = v_i(\mathbf{e}_i\cdot\mathbf{e}_j) = v_i\delta_{ij} = v_j$$

so $\mathbf{v} = (\mathbf{v}\cdot\mathbf{e}_j)\mathbf{e}_j$, which is a representation of the vector in terms of itself.[6]

Remark: If the basis is not orthogonal we need the dual basis and $\mathbf{v}$ is expressed as

$$\mathbf{v} = (\mathbf{v} \cdot \mathbf{e}^i)\mathbf{e}_i = (\mathbf{v} \cdot \mathbf{e}_i)\mathbf{e}^i.$$

Addition and subtraction and multiplication by scalar in component form

$$\mathbf{v} + \mathbf{u} \equiv (v_i + u_i)\mathbf{e}_i$$

and

$$\alpha\mathbf{v} = (\alpha v_i)\mathbf{e}_i.$$

Scalar product

$$\mathbf{u} \cdot \mathbf{v} \equiv u_i\mathbf{e}_i \cdot v_j\mathbf{e}_j = u_i v_j\mathbf{e}_i \cdot \mathbf{e}_j = u_i v_j \delta_{ij} = u_i v_i = u_j v_j.$$

Cross product

$$\mathbf{v} \times \mathbf{w} \equiv \mathbf{e}_j v_j \times \mathbf{e}_k w_k = \mathbf{e}_j \times \mathbf{e}_k v_j w_k = \varepsilon_{jki}\mathbf{e}_i v_j w_k = \varepsilon_{ijk}\mathbf{e}_i v_j w_k.$$

The origin of tensors – linear scalar functions and linear vector functions

Linear vector functions provide a convenient and general way of introducing tensors. Before discussing them, let us consider linear scalar functions.

Linear scalar functions (LSF)

Definition: Mapping

$$f(\ ): \mathbb{V} \to \mathbb{R} \text{ (vectors into reals)}$$

such that

$$f(\mathbf{x} + \mathbf{y}) = f(\mathbf{x}) + f(\mathbf{y})$$
$$f(\alpha\mathbf{x}) = \alpha f(\mathbf{x}).$$

Representation theorem
Any LSF has the representation $f(\mathbf{x}) = \mathbf{f} \cdot \mathbf{x}$, where the vector $\mathbf{f}$ is unique and independent of $\mathbf{x}$.

(i) *Uniqueness*: Assume $\mathbf{f}$ and $\mathbf{f}'$ such that $\mathbf{f} \cdot \mathbf{x} = \mathbf{f}' \cdot \mathbf{x} = f(\mathbf{x})$. Then $\mathbf{f} \cdot \mathbf{x} - \mathbf{f}' \cdot \mathbf{x} = 0$ and using property III(2), $(\mathbf{f} - \mathbf{f}') \cdot \mathbf{x} = 0$. Since $\mathbf{x}$ is arbitrary, using property III(5) we obtain $\mathbf{f} = \mathbf{f}'$.
(ii) *Existence*: $f(\mathbf{x}) = f(x_i\mathbf{e}_i)$ and using linearity $f(\mathbf{x}) = x_i f(\mathbf{e}_i) = \mathbf{x} \cdot \mathbf{e}_i f(\mathbf{e}_i) = f(\mathbf{e}_i)\mathbf{e}_i \cdot \mathbf{x}$, so $\mathbf{f} \equiv f(\mathbf{e}_i)\mathbf{e}_i$.

Problem
Show that $f(\cdot) = |\ |$, where the bars denote magnitude, is not a LSF.

Linear vector functions (LVF)

Definition: Mapping $t(\cdot)$ such that $\mathbb{V} \to \mathbb{V}$ (vectors into vectors)

$$t(x + y) = t(x) + t(y)$$
$$t(\alpha x) = \alpha t(x).$$

Example 1

The multiplication of a vector by scalar is a LVF.

Example 2

Define $(a\ b) \cdot u \equiv a(b \cdot u)$, where the dot can be interpreted as $(a\ b)$ acting on u. Note that the result is parallel to a. The entity $(a\ b)$ is called a dyad. Note also that if the order of a and b is reversed the result of the operation is parallel to b, $(b\ a) \cdot u = b(a \cdot u)$. Similarly, $u \cdot (a\ b) \equiv (u \cdot a)b$.

A note about notation

The dyad $(a\ b)$ is often denoted $(a \otimes b)$. Customarily, within this notation the product of a vector and a tensor, here denoted as $T \cdot v$, is denoted as Tv. The contraction (or inner product) of two tensors or a tensor and dyad, here designated as $T:S$ or $T:nn$, is designated as $T \cdot S$ and $T \cdot n \otimes n$, respectively. The main motivation for this notation, which the reader is urged to master, is that the inner product of both vectors and tensors is designated with the same symbol.

Representation theorem

Any LVF has the representation

$$t(u) = (e_1 t_1) \cdot u + (e_2 t_2) \cdot u + (e_3 t_3) \cdot u$$
$$= (e_1 t_1 + e_2 t_2 + e_3 t_3) \cdot u$$
$$= (e_i t_i) \cdot u = e_i (t_i \cdot u)$$

where the t_i are unique.

(i) *Uniqueness*: Consider two representations

$$t(u) = e_i(t_i \cdot u) = e_i(t_i' \cdot u).$$

Taking the inner product with e_i,

$$e_i \cdot t(u) = (t_i \cdot u) = (t_i' \cdot u)$$

and by the same reasoning as with LSF, $t_i = t_i'$.

(ii) *Existence*: We write

$$t(u) \atop \text{vector} = e_i(t_i \cdot u) = \underset{\text{components}}{t_i(u)} \ \underset{\text{basis}}{e_i}$$

It is enough to prove that $t_i(u)$ is a LSF. Write

$$t(u) = t_i(u)e_i$$
$$t(v) = t_i(v)e_i.$$

Then
$$\mathbf{t}(\mathbf{u}) + \mathbf{t}(\mathbf{v}) = t_i(\mathbf{u})\mathbf{e}_i + t_i(\mathbf{v})\mathbf{e}_i = [t_i(\mathbf{u}) + t_i(\mathbf{v})]\mathbf{e}_i.$$

Also since $\mathbf{t}(\ )$ is a LVF,
$$\mathbf{t}(\mathbf{u} + \mathbf{v}) = t_i(\mathbf{u} + \mathbf{v})\mathbf{e}_i,$$

then
$$t_i(\mathbf{u}) + t_i(\mathbf{v}) = t_i(\mathbf{u} + \mathbf{v}).$$

Also
$$\mathbf{t}(\alpha\mathbf{u}) = (\mathbf{e}_i t_i) \cdot \alpha\mathbf{u} = \mathbf{e}_i(t_i \cdot \alpha\mathbf{u}) = \mathbf{e}_i t_i(\alpha\mathbf{u}) = \alpha(\mathbf{e}_i t_i) \cdot \mathbf{u} = \alpha\mathbf{e}_i(t_i \cdot \mathbf{u}) = \mathbf{e}_i \alpha t_i(\mathbf{u})$$

then
$$t_i(\alpha\mathbf{u}) = \alpha t_i(\mathbf{u}).$$

Tensors

A LVF is the most general linear transformation of vectors into vectors. Symbolically, $\mathbf{t}(\mathbf{u})$ is written as
$$\mathbf{t}(\mathbf{u}) = \mathbf{T} \cdot \mathbf{u},$$

where $\mathbf{T} = \mathbf{e}_1 \mathbf{t}_1 + \mathbf{e}_2 \mathbf{t}_2 + \mathbf{e}_3 \mathbf{t}_3$. This sum of dyadics is called a tensor. (More precisely, a second rank tensor. Unless stated explicitly, whenever we use the word tensor we mean a second rank tensor.) Similarly, we can define linear transformations between tensors, and so on.

Components of a tensor

Any of the $\mathbf{t}_i$, for example, $\mathbf{t}_1$, has component representation
$$\mathbf{t}_1 = t_{1j}\mathbf{e}_j, \text{ where } t_{1j} \text{ means the } j\text{th component of } \mathbf{t}_1$$
[Remark: Note convention here. We could just as well have written t_{j1} to mean the same thing.]

Thus, in general,
$$\mathbf{t}_i = t_{ij}\mathbf{e}_j$$

and
$$\mathbf{T} = \mathbf{e}_i \mathbf{t}_i = \mathbf{e}_i t_{ij}\mathbf{e}_j = t_{ij}\mathbf{e}_i\mathbf{e}_j.$$

Usually, the components t_{ij} are denoted with the same letter as the tensor, i.e.,
$$\mathbf{T} = T_{ij}\mathbf{e}_i\mathbf{e}_j.$$

As with vectors, the array of T_{ij}, usually displayed in matrix form, is one of the infinitely many possible representations of the tensor $\mathbf{T}$. Thus, we can write
$$\mathbf{T} = T_{ij}\mathbf{e}_i\mathbf{e}_j = T'_{ij}\mathbf{e}'_i\mathbf{e}'_j$$

as the representations of $\mathbf{T}$ with respect to the bases $\{\mathbf{e}_i\}$ and $\{\mathbf{e}'_i\}$.

How to obtain components

The ij component of $\mathbf{S}$, denoted $[\mathbf{S}]_{ij}$, is obtained as

$$[\mathbf{S}]_{ij} = \mathbf{e}_i \cdot (\mathbf{S} \cdot \mathbf{e}_j).$$

Note that the vector corresponding to the first subindex is placed on the left. It is easy to see why this works:

$$[\mathbf{S}]_{ij} = \mathbf{e}_i \cdot (S_{kl} \mathbf{e}_k \mathbf{e}_l \cdot \mathbf{e}_j) = S_{kl} \mathbf{e}_i \cdot \mathbf{e}_k \mathbf{e}_l \cdot \mathbf{e}_j = S_{kl} \delta_{ik} \delta_{lj} = S_{il} \delta_{lj} = S_{kj} \delta_{ik} = S_{ij}.$$

Some properties of tensors

Zero tensor (0)

Defined such that $\mathbf{0} \cdot \mathbf{u} = \mathbf{0}$ for all $\mathbf{u}$. The components are

$$[\mathbf{0}]_{ij} = \mathbf{e}_i \cdot (\mathbf{0} \cdot \mathbf{e}_j) = \mathbf{e}_i \cdot \mathbf{0} = 0.$$

Thus the component representation is

$$[\mathbf{0}] = \begin{bmatrix} 0 & 0 & 0 \\ 0 & 0 & 0 \\ 0 & 0 & 0 \end{bmatrix}$$

in any basis.

Unit tensor (1)

Defined such that $\mathbf{1} \cdot \mathbf{u} = \mathbf{u}$ for $\mathbf{u}$. The components are

$$[\mathbf{1}]_{ij} = \mathbf{e}_i \cdot (\mathbf{1} \cdot \mathbf{e}_j) = \mathbf{e}_i \cdot \mathbf{e}_j = \delta_{ij}.$$

Thus the component representation is

$$[\mathbf{1}] = \begin{bmatrix} 1 & 0 & 0 \\ 0 & 1 & 0 \\ 0 & 0 & 1 \end{bmatrix}$$

in any basis. Note also that

$$\mathbf{1} = \mathbf{e}_1 \mathbf{e}_1 + \mathbf{e}_2 \mathbf{e}_2 + \mathbf{e}_3 \mathbf{e}_3.$$

Product by scalar

Defined such that $\alpha \mathbf{T} \cdot \mathbf{u} \equiv \mathbf{T} \cdot \alpha \mathbf{u}$.

Addition (and subtraction)

Defined such that $(\mathbf{T} + \mathbf{S}) \cdot \mathbf{u} \equiv \mathbf{T} \cdot \mathbf{u} + \mathbf{S} \cdot \mathbf{u}$.

Inverse tensor

Given $\mathbf{T}$, there exists $\mathbf{T}^{-1}$, such that if $\det \mathbf{T} \neq 0$, then $\mathbf{T} \cdot \mathbf{T}^{-1} = \mathbf{1}$, $\mathbf{T}^{-1} \cdot \mathbf{T} = \mathbf{1}$. If $\det \mathbf{T} \neq 0$, $\mathbf{T}$ is called non-singular. It can be shown that $\det \mathbf{T}$ does not depend on the basis and that $\mathbf{T}^{-1}$ is unique. The rule of computation of $\mathbf{T}^{-1}$ is similar to that used in matrices.

Transpose
Vector function $\mathbf{t}^T(\cdot)$ such that $\mathbf{t}^T(\mathbf{u})\cdot\mathbf{v} = \mathbf{t}(\mathbf{v})\cdot\mathbf{u}$.

Theorem
$\mathbf{t}^T(\ )$ is a LVF: Note

$$\mathbf{t}^T(\mathbf{u} + \mathbf{v})\cdot\mathbf{w} = \mathbf{t}(\mathbf{w})\cdot(\mathbf{u} + \mathbf{v}) = \mathbf{t}(\mathbf{w})\cdot\mathbf{u} + \mathbf{t}(\mathbf{w})\cdot\mathbf{v}$$
$$= \mathbf{t}^T(\mathbf{u})\cdot\mathbf{w} + \mathbf{t}^T(\mathbf{v})\cdot\mathbf{w} = (\mathbf{t}^T(\mathbf{u}) + \mathbf{t}^T(\mathbf{v}))\cdot\mathbf{w}.$$

Since $\mathbf{w}$ is arbitrary,

$$\mathbf{t}^T(\mathbf{u} + \mathbf{v}) = \mathbf{t}^T(\mathbf{u}) + \mathbf{u}^T(\mathbf{v}).$$

Also

$$\mathbf{t}^T(\mathbf{u})\cdot\alpha\mathbf{w} = \alpha\mathbf{t}^T(\mathbf{u})\cdot\mathbf{w} = \alpha\mathbf{t}(\mathbf{w})\cdot\mathbf{u} = \mathbf{t}(\mathbf{w})\cdot\alpha\mathbf{u} = \mathbf{t}^T(\alpha\mathbf{u})\cdot\mathbf{w}$$

and since $\mathbf{w}$ is arbitrary,

$$\alpha\mathbf{t}^T(\mathbf{u}) = \mathbf{t}^T(\alpha\mathbf{u}).$$

Note: The transpose of $\mathbf{T}$ is denoted $\mathbf{T}^T$. Thus, we write $(\mathbf{T}^T\cdot\mathbf{v})\cdot\mathbf{u} = (\mathbf{T}\cdot\mathbf{u})\cdot\mathbf{v}$.

Component representation of transpose

$$[\mathbf{T}^T]_{ij} = \mathbf{e}_i\cdot(\mathbf{T}^T\cdot\mathbf{e}_j) = (\mathbf{T}^T\cdot\mathbf{e}_j)\cdot\mathbf{e}_i = (\mathbf{T}\cdot\mathbf{e}_i)\cdot\mathbf{e}_j = T_{ji}.$$

Note: If $\mathbf{T} = T_{ij}\mathbf{e}_i\mathbf{e}_j$, then $\mathbf{T}^T = T_{ji}\mathbf{e}_i\mathbf{e}_j$. If $\mathbf{T} = \mathbf{T}^T$, $\mathbf{T}$ is called symmetric. If $\mathbf{T} = -\mathbf{T}^T$, $\mathbf{T}$ is called antisymmetric or skew.

Problem
Show that $\mathbf{T} = (\mathbf{T}^T)^T$.

Problem
Show that any tensor $\mathbf{T}$ can be decomposed, uniquely, into symmetric and antisymmetric parts:

$$\mathbf{T}_S = (\mathbf{T} + \mathbf{T}^T)/2, \text{ symmetric.}$$
$$\mathbf{T}_A = (\mathbf{T} - \mathbf{T}^T)/2, \text{ antisymmetric.}$$

Component representation of a dyad

$$[\mathbf{a}\ \mathbf{b}]_{ij} = \mathbf{e}_i\cdot((a_k\mathbf{e}_k b_l\mathbf{e}_l)\cdot\mathbf{e}_j) = (\mathbf{e}_i\cdot a_k\mathbf{e}_k b_l)\mathbf{e}_l\cdot\mathbf{e}_j$$
$$= a_k b_l(\mathbf{e}_i\cdot\mathbf{e}_k)(\mathbf{e}_l\cdot\mathbf{e}_j) = a_k b_l\delta_{ik}\delta_{lj} = a_i b_l\delta_{lj} = a_i b_j.$$

Note:
$$(\mathbf{a}\ \mathbf{b})^T = (\mathbf{b}\ \mathbf{a}).$$

Useful relations involving unit vectors
(1) $\mathbf{e}_i\mathbf{e}_j\cdot\mathbf{e}_k = \mathbf{e}_i(\mathbf{e}_j\cdot\mathbf{e}_k) = \mathbf{e}_i\delta_{jk}$
(2) $\mathbf{e}_i\mathbf{e}_j\cdot\mathbf{e}_k\mathbf{e}_l = \mathbf{e}_i(\mathbf{e}_j\cdot\mathbf{e}_k)\mathbf{e}_l = \mathbf{e}_i\delta_{kj}\mathbf{e}_l$
(3) $\mathbf{e}_i\mathbf{e}_j\times\mathbf{e}_k = \mathbf{e}_i[\mathbf{e}_j\times\mathbf{e}_k]$
(4) $\mathbf{e}_i\times\mathbf{e}_j\mathbf{e}_k = [\mathbf{e}_i\times\mathbf{e}_j]\mathbf{e}_k$
(5) $\mathbf{e}_i\mathbf{e}_j:\mathbf{e}_k\mathbf{e}_l = (\mathbf{e}_j\cdot\mathbf{e}_k)(\mathbf{e}_i\cdot\mathbf{e}_l) = \delta_{jk}\delta_{il}.$

Warning: definition (5) is not universal. For example, some authors define $\mathbf{e}_i\mathbf{e}_j : \mathbf{e}_k\mathbf{e}_l = (\mathbf{e}_j \cdot \mathbf{e}_l)(\mathbf{e}_i \cdot \mathbf{e}_k)$ rather than $(\mathbf{e}_j \cdot \mathbf{e}_k)(\mathbf{e}_i \cdot \mathbf{e}_l)$.

Remarks:
$$v_k \delta_{jk} = v_j$$
$$T_{kl} \delta_{jk} = T_{jl}$$
$$\delta_{jk} = \delta_{kj}.$$

Operations in terms of components

Addition (and subtraction)
$$\mathbf{T} + \mathbf{S} = T_{ij}\mathbf{e}_i\mathbf{e}_j + S_{ij}\mathbf{e}_i\mathbf{e}_j = (T_{ij} + S_{ij})\mathbf{e}_i\mathbf{e}_j, \qquad \text{which is a tensor.}$$

Multiplication by scalar
$$\alpha\mathbf{S} = \alpha S_{ij}\mathbf{e}_i\mathbf{e}_j, \qquad \text{which is a tensor.}$$

Product of vector by tensors
$$\mathbf{T} \cdot \mathbf{v} = T_{ij}\mathbf{e}_i\mathbf{e}_j \cdot v_k\mathbf{e}_k = (T_{ij}v_k)\mathbf{e}_i\delta_{jk} = (T_{ik}v_k)\mathbf{e}_i, \qquad \text{which is a vector.}$$

Note: This corresponds to usual matrix multiplication ($[3 \times 3][3 \times 1]$).
Similarly, $\mathbf{v} \cdot \mathbf{T}$ is interpreted as:
$$\mathbf{v} \cdot \mathbf{T} = v_k\mathbf{e}_k \cdot T_{ij}\mathbf{e}_i\mathbf{e}_j = (v_k T_{ij})\delta_{ki}\mathbf{e}_j = (v_i T_{ij})\mathbf{e}_j.$$

Note: This corresponds to usual matrix multiplication ($[1 \times 3][3 \times 3]$).

Inner product (or double dot product or contraction) of two tensors
$$\mathbf{T} : \mathbf{S} = T_{ij}\mathbf{e}_i\mathbf{e}_j : S_{kl}\mathbf{e}_k\mathbf{e}_l = T_{ij}S_{kl}\delta_{il}\delta_{jk} = T_{ik}S_{ki} = T_{lj}S_{jl}, \qquad \text{which is a scalar.}$$

Similarly,
$$\mathbf{T} : \mathbf{vw} = T_{ij}v_j w_i$$
$$\mathbf{uv} : \mathbf{wz} = u_i v_j w_j z_i.$$

Problem
Show that if $\mathbf{T} = \mathbf{T}^{\mathsf{T}}$ and $\mathbf{S} = -\mathbf{S}^{\mathsf{T}}$ then $\mathbf{T} : \mathbf{S} = 0$. Thus if ':' is interpreted as an inner product this means that the projection is zero.

Product of two tensors
$$\mathbf{T} \cdot \mathbf{S} = T_{ij}\mathbf{e}_i\mathbf{e}_j \cdot S_{kl}\mathbf{e}_k\mathbf{e}_l = T_{ij}S_{kl}\delta_{jk}\mathbf{e}_i\mathbf{e}_l = T_{ik}S_{kl}\mathbf{e}_i\mathbf{e}_l, \qquad \text{which is a tensor.}$$

Note: This corresponds to usual matrix multiplication ($[3 \times 3][3 \times 3]$).

Change of orthonormal basis

The problem is how to find the components of vectors and tensors when they are seen in different bases. Consider two such bases: $\{\mathbf{e}_i\}$ and $\{\mathbf{e}_i'\}$.
Denote
$$a_j^i = \cos(\text{angle between } \mathbf{e}_j' \text{ and } \mathbf{e}_i) = \mathbf{e}_j' \cdot \mathbf{e}_i$$
and arrange the a_j^i as in Table A.1

Table A.1

	$\mathbf{e}_1$	$\mathbf{e}_2$	$\mathbf{e}_3$
$\mathbf{e}_1'$	a_1^1	a_1^2	a_1^3
$\mathbf{e}_2'$	a_2^1	a_2^2	a_2^3
$\mathbf{e}_3'$	a_3^1	a_3^2	a_3^3

Then, for example, from Table A.1

$$\mathbf{e}_2' = a_2^1\mathbf{e}_1 + a_2^2\mathbf{e}_2 + a_2^3\mathbf{e}_3$$
$$\mathbf{e}_2 = a_1^2\mathbf{e}_1' + a_2^2\mathbf{e}_2' + a_3^2\mathbf{e}_3'.$$

In general,

$$\mathbf{e}_i' = a_i^j\mathbf{e}_j$$
$$\mathbf{e}_i = a_j^i\mathbf{e}_j'.$$

We denote the matrix of the a_j^i as $[\mathbf{Q}]$ (note convention here)

$$[\mathbf{Q}] = \begin{bmatrix} a_1^1 & a_1^2 & a_1^3 \\ a_2^1 & a_2^2 & a_2^3 \\ a_3^1 & a_3^2 & a_3^3 \end{bmatrix}.$$

$\mathbf{Q}$ is an orthogonal matrix. Consider now some of the properties of $\mathbf{Q}$. Note that

$$\mathbf{e}_p' \cdot \mathbf{e}_q' = \delta_{qp}.$$

Replacing the transformation we obtain

$$a_p^j\mathbf{e}_j \cdot a_q^k\mathbf{e}_k = \delta_{qp}$$
$$a_p^j a_q^k \delta_{jk} = \delta_{qp}$$
$$a_p^k a_q^k = \delta_{qp}.$$

Similarly,

$$a_k^p a_k^q = \delta_{qp}.$$

Note that $\mathbf{a} \times \mathbf{b} \cdot \mathbf{c} = \text{Volume of prism} = \begin{vmatrix} a_1 & a_2 & a_3 \\ b_1 & b_2 & b_3 \\ c_1 & c_2 & c_3 \end{vmatrix}.$

Then,

$$\mathbf{e}_1' \times \mathbf{e}_2' \cdot \mathbf{e}_3' = 1.$$

Since

$$\mathbf{e}_1' = a_1^1\mathbf{e}_1 + a_1^2\mathbf{e}_2 + a_1^3\mathbf{e}_3$$
$$\mathbf{e}_2' = a_2^1\mathbf{e}_1 + a_2^2\mathbf{e}_2 + a_2^3\mathbf{e}_3$$
$$\mathbf{e}_3' = a_3^1\mathbf{e}_1 + a_3^2\mathbf{e}_2 + a_3^3\mathbf{e}_3$$

then $\det|a_n^m| = 1.$

Problem

Prove that $a_p^s a_q^s = \delta_{qp}$.

Problem

Prove that $\det|a_n^m| = 1$.

Transformations of Cartesian co-ordinates

Vectors

The same vector **v** can be written in $\{e_i'\}$ and $\{e_i\}$ as

$$v = v_s e_s = v_r' e_r'.$$

Since

$$\mathbf{e}_s = a_r^s \mathbf{e}_r'$$

we write

$$v_s a_r^s \mathbf{e}_r' = v_r' \mathbf{e}_r'$$
$$(v_s a_r^s - v_r')\mathbf{e}_r' = \mathbf{0}.$$

Since all the components of **0** are zero,

$$v_r' = a_r^s v_s, \qquad \text{or in matrix form, } [\mathbf{v}'] = [\mathbf{Q}][\mathbf{v}].$$

Similarly,

$$v_s = a_r^s v_r', \qquad \text{or in matrix form, } [\mathbf{v}] = [\mathbf{Q}^{\mathrm{T}}][\mathbf{v}'].$$

Further properties of Q

Since **v** transforms as $[\mathbf{v}'] = [\mathbf{Q}][\mathbf{v}]$ and the length of **v** is invariant, we obtain

$$\left|(\mathbf{a}')^2\right| = [\mathbf{a}'][\mathbf{a}'] = [\mathbf{Q}][\mathbf{a}][\mathbf{Q}][\mathbf{a}] = [\mathbf{Q}^{\mathrm{T}}][\mathbf{Q}][\mathbf{a}][\mathbf{a}] = [\mathbf{a}][\mathbf{a}] = \left|\mathbf{a}^2\right|$$

then,

$$[\mathbf{Q}^{\mathrm{T}}][\mathbf{Q}] = [\mathbf{1}]$$

or

$$[\mathbf{Q}^{\mathrm{T}}] = [\mathbf{Q}^{-1}].$$

Also,

$$\det([\mathbf{Q}^{\mathrm{T}}][\mathbf{Q}]) = \det[\mathbf{1}] = 1$$
$$\det[\mathbf{Q}^{\mathrm{T}}]\det[\mathbf{Q}] = (\det[\mathbf{Q}])^2 = 1$$
$$\det[\mathbf{Q}] = \pm 1.$$

Q is called proper orthogonal if $\det[\mathbf{Q}] = +1$.

Problem

Prove that [**Q**] preserves angles between vectors.

Tensors

Consider two vectors $\mathbf{v}$ and $\mathbf{u}$ and a tensor $\mathbf{T}$ such that $\mathbf{u} = \mathbf{T} \cdot \mathbf{v}$. The question is: What happens when $\mathbf{T}$ is seen in terms of $\{\mathbf{e}'_i\}$ and $\{\mathbf{e}_i\}$? We know that:

$$u'_i = a^j_i u_j$$

and

$$u_j = T_{jq} v_q.$$

Then

$$u'_i = a^j_i T_{jq} v_q$$

$$u'_i = a^j_i T_{jq} a^q_p v'_p.$$

Also, since

$$u'_i = T'_{ip} v'_p$$

we get

$$T'_{ip} = a^j_i a^q_p T_{jq}$$

or, in matrix form, with the usual rules of multiplicaton,

$$[\mathbf{T}'] = [\mathbf{Q}][\mathbf{T}][\mathbf{Q}^{\mathrm{T}}].$$

Problem

Derive the above result using matrix manipulations. Consider

$$[\mathbf{u}'] = [\mathbf{Q}][\mathbf{u}] \qquad \text{and} \qquad [\mathbf{v}'] = [\mathbf{Q}][\mathbf{v}].$$

Since

$$[\mathbf{Q}^{-1}] = [\mathbf{Q}^{\mathrm{T}}]$$

then

$$[\mathbf{Q}^{\mathrm{T}}][\mathbf{v}'] = [\mathbf{v}].$$

In $\{\mathbf{e}_i\}$,

$$[\mathbf{u}] = [\mathbf{T}][\mathbf{v}]$$

whereas in $\{\mathbf{e}'_i\}$,

$$[\mathbf{u}'] = [\mathbf{T}'][\mathbf{v}'].$$

Combining

$$[\mathbf{u}'] = [\mathbf{Q}][\mathbf{T}][\mathbf{v}]$$

$$[\mathbf{u}'] = [\mathbf{Q}][\mathbf{T}][\mathbf{Q}^{\mathrm{T}}][\mathbf{v}']$$

and

$$[\mathbf{T}'] = [\mathbf{Q}][\mathbf{T}][\mathbf{Q}^{\mathrm{T}}].$$

Scalar functions of tensor arguments

There are two kinds:

(i) Those for which the relationship is *dependent* upon the choice of basis.

(ii) Those for which the relationship is *independent* of the choice of basis. These are called *invariant* or *isotropic* functions.

Examples: $f(\mathbf{T}) = T_{12}$ belongs to class (i); as we shall see $\det(\mathbf{T})$ and $\mathrm{tr}(\mathbf{T})$ belong to class (ii).

Trace

Definition: Mapping, $\mathrm{tr}(\ )$, $\mathbb{T} \to \mathbb{R}$, such that

$$\mathrm{tr}(\mathbf{T} + \mathbf{S}) = \mathrm{tr}(\mathbf{T}) + \mathrm{tr}(\mathbf{S})$$
$$\mathrm{tr}(\alpha\mathbf{T}) = \alpha\,\mathrm{tr}(\mathbf{T})$$

(i.e., linear), and

$$\mathrm{tr}(\mathbf{a}\ \mathbf{b}) \equiv \mathbf{a} \cdot \mathbf{b}.$$

Calculation: $\mathrm{tr}(\mathbf{T}) = \mathrm{tr}(T_{ij}\mathbf{e}_i\mathbf{e}_j) = T_{ij}\,\mathrm{tr}(\mathbf{e}_i\mathbf{e}_j) = T_{ij}\delta_{ij} = T_{ii}$.

Properties:

$$\mathrm{tr}(\mathbf{S}^{\mathrm{T}}) = \mathrm{tr}(\mathbf{S})$$
$$\mathrm{tr}(\mathbf{S} \cdot \mathbf{T}) = \mathrm{tr}(\mathbf{T} \cdot \mathbf{S}).$$

Magnitude of a tensor

$$|\mathbf{T}| = |[\mathrm{tr}(\mathbf{T} \cdot \mathbf{T}^{\mathrm{T}})]^{1/2}|.$$

The Cauchy–Schwarz inequality, for tensors, reads

$$|\mathbf{T} : \mathbf{S}| \leqslant |\mathbf{T}||\mathbf{S}|.$$

Problem

Show that $|\mathbf{T}| = |(\mathbf{T} : \mathbf{T}^{\mathrm{T}})^{1/2}|$. (See earlier comments about notation.)

Problem

Show that $|\mathbf{n}\ \mathbf{n}| = 1$ if $|\mathbf{n}| = 1$.

Problem

Show that $\mathbf{T} : \mathbf{T}^{\mathrm{T}} \geqslant 0$.

Problem

Prove that $\mathrm{tr}(\mathbf{A} \cdot \mathbf{B} \cdot \mathbf{C}) = \mathrm{tr}(\mathbf{B} \cdot \mathbf{C} \cdot \mathbf{A})$. Use this result to prove that the trace is invariant.

Problem

Prove $\mathbf{u}\mathbf{v} : \mathbf{w}\mathbf{z} = \mathbf{u}\mathbf{w} : \mathbf{v}\mathbf{z} = (\mathbf{u} \cdot \mathbf{z})(\mathbf{v} \cdot \mathbf{w})$.

Determinant

The determinant is calculated using the representation of **T** in any basis. The result is independent of the basis (without proof).

$$\det \mathbf{T} = \begin{vmatrix} T_{11} & T_{12} & T_{13} \\ T_{21} & T_{22} & T_{23} \\ T_{31} & T_{32} & T_{33} \end{vmatrix}.$$

Problem
Prove that $\det(\mathbf{A}^{-1}) = [\det(\mathbf{A})]^{-1}$.

Problem
Prove that $\det(\mathbf{T}') = \det(\mathbf{T})$.

Eigenvalues and eigenvectors

Given a tensor **T**, its eigenvalues and eigenvectors are the solutions of:

$\mathbf{T} \cdot \mathbf{n} = \lambda \mathbf{n}$ gives the right eigenvectors $\mathbf{n}_i$ and eigenvalues λ_i ($i = 1$ to 3)

$\mathbf{m} \cdot \mathbf{T} = \beta \mathbf{m}$ gives the left eigenvectors $\mathbf{m}_j$ and eigenvalues β_j ($i = 1$ to 3).

Note that the left eigenvectors are also given by $\mathbf{T}^{\mathrm{T}} \cdot \mathbf{m} = \beta \mathbf{m}$.

Procedure
Note that

$$\mathbf{T} \cdot \mathbf{n} - \lambda \mathbf{1} \cdot \mathbf{n} = \mathbf{0} \qquad \text{or} \qquad (\mathbf{T} - \lambda \mathbf{1}) \cdot \mathbf{n} = \mathbf{0}.$$

The system admits a non-trivial solution if and only if

$$\det(\mathbf{T} - \lambda \mathbf{1}) = 0 \qquad \text{(third-order polynomial)}$$

which generates, at the most, three different eigenvalues (real or complex). The corresponding eigenvectors $\mathbf{n}_i$ are the solutions of

$$(\mathbf{T} - \lambda \mathbf{I}) \cdot \mathbf{n} = \mathbf{0},$$

replacing λ by λ_i. Normally, the eigenvalues are normalized, i.e., they are reported such that

$$|\mathbf{n}_i| = 1.$$

Characteristic equation
$\det(\mathbf{T} - \lambda \mathbf{1}) = 0$ can be written as

$$\lambda^3 - \mathrm{I}_{\mathrm{T}}\lambda^2 + \mathrm{II}_{\mathrm{T}}\lambda - \mathrm{III}_{\mathrm{T}}1 = 0$$

where

$$\mathrm{I}_{\mathrm{T}} = \mathrm{tr}(\mathbf{T})$$
$$\mathrm{II}_{\mathrm{T}} = \tfrac{1}{2}[\mathrm{I}_{\mathrm{T}}^2 - \mathrm{tr}(\mathbf{T}^2)]$$
$$\mathrm{III}_{\mathrm{T}} = \det(\mathbf{T}).$$

I_T, II_T, and III_T are called the principal invariants of the tensor. As we have seen I_T, II_T, and III_T are independent of the choice of basis.

Cayley–Hamilton theorem (without proof)
Any tensor **T** satisfies its own characteristic equation, i.e.,
$$T^3 - I_T T^2 + II_T T - III_T 1 = 0.$$
Note: Besides I_T, II_T, and III_T the other invariants commonly encountered in the literature are the moments
$$I'_T = \text{tr}(T), \qquad II'_T = \text{tr}(T^2), \qquad III'_T = \text{tr}(T^3).$$

Representation theorem for symmetric tensor (without proof)
Any isotropic scalar function of a symmetric tensor argument can be expressed as a function of the invariants of the argument:
$$f(T) = g(I_T, II_T, III_T)$$
or, alternatively,
$$f(T) = g'(I'_T, II'_T, III'_T).$$

Example
Show that if **T** is symmetric, the eigenvectors belonging to different eigenvalues are orthogonal.

Denote n_i, λ_i and n_j, λ_j as two sets of eigenvectors–eigenvalues such that $\lambda_i \neq \lambda_j$, then
$$T \cdot n_i = \lambda_i n_i \qquad \text{and} \qquad T \cdot n_j = \lambda_j n_j$$
then, cross-multiplying,
$$(T \cdot n_i) \cdot n_j = \lambda_i n_i \cdot n_j \qquad \text{and} \qquad (T \cdot n_j) \cdot n_i = \lambda_j n_j \cdot n_i.$$
Using the definition of transpose, since **T** is symmetric,
$$(T \cdot n_i) \cdot n_j = (T \cdot n_j) \cdot n_i = \lambda_i n_i \cdot n_j = \lambda_j n_j \cdot n_i$$
and
$$(\lambda_i - \lambda_j) n_i \cdot n_j = 0.$$
Since, by assumption, $\lambda_i \neq \lambda_j$ then n_i and n_j are orthogonal.

Example
Obtain the representation of the above tensor with respect to its eigenvectors. Construct
$$T_{ij} = (T \cdot n_i) \cdot n_j = \lambda_i n_i \cdot n_j = \lambda_i \delta_{ij}.$$
So the matrix representation is
$$[T] = \begin{bmatrix} \lambda_1 & 0 & 0 \\ 0 & \lambda_2 & 0 \\ 0 & 0 & \lambda_3 \end{bmatrix}.$$

Problem

Using the Cayley–Hamilton theorem prove that $\mathbf{T}' = \mathbf{Q} \cdot \mathbf{T} \cdot \mathbf{Q}^{\mathrm{T}}$ and $\mathbf{T}$ have the same invariants.

Problem

Using the Cayley–Hamilton theorem prove that $\mathbf{T}^n$, $n > 3$, can be written in terms of $\mathbf{1}$, $\mathbf{T}$, and $\mathbf{T}^2$.

Problem

Show that if $\mathbf{T}$ is invariant, then $\mathbf{T}^n$ is invariant.

Problem

Show that $\mathbf{T}$ and $\mathbf{T}^{\mathrm{T}}$ have the same eigenvalues.

Problem

Show that if $\mathbf{T} \cdot \mathbf{n} = \lambda \mathbf{n}$, then $\mathbf{T}^{-1} \cdot \mathbf{n} = \lambda^{-1} \mathbf{n}$.

Problem

Show that if $\mathbf{T} = \mathbf{T}^{\mathrm{T}}$ and the T_{ij} are real then the eigenvalues of $\mathbf{T}$ are real.

Problem

Continuing the problem above, construct the matrix $[\mathbf{L}]$ such that its columns, $\mathbf{n}_i$, are the eigenvectors of $\mathbf{T}$, i.e.,

$$[\mathbf{L}] = [\mathbf{n}_1 \mathbf{n}_2 \mathbf{n}_3]$$

where the $\mathbf{n}_i$ are written as (3×1) vectors. Show that $[\mathbf{L}^{\mathrm{T}}][\mathbf{T}][\mathbf{L}]$ is diagonal.

Problem

Prove that $\mathbf{T}$ and $\mathbf{T}' = \mathbf{Q} \cdot \mathbf{T} \cdot \mathbf{Q}^{\mathrm{T}}$ have the same eigenvalues and eigenvectors.

Problem

Show that if $\mathbf{W} = -\mathbf{W}^{\mathrm{T}}$ then $\lambda = 0$ is an eigenvalue.

Problem

Show that the maximum (or minimum) of $\mathbf{D}:\mathbf{nn}$, with $\mathbf{D}$ given and $|\mathbf{n}| = 1$, is given by the solution of the eigenvalue problem $\mathbf{D} \cdot \mathbf{n} = \lambda \mathbf{n}$. Identify λ with a Lagrange multiplier.

Differentiation of scalars, vectors, and tensors

Gradient of a scalar – definition of ∇f (f(x), scalar function)

$$df = f(\mathbf{x} + d\mathbf{x}) - f(\mathbf{x}) \equiv d\mathbf{x} \cdot \nabla f.$$

Denoting $|dx| = ds$

$$\frac{df}{ds} = \frac{dx}{ds} \cdot \nabla f = \mathbf{n} \cdot \nabla f.$$

∇f is a vector.

Gradient of a vector – definition of $\nabla \mathbf{v}$ ($\mathbf{v}(\mathbf{x})$, vector function)

$$d\mathbf{v} = \mathbf{v}(\mathbf{x} + d\mathbf{x}) - \mathbf{v}(\mathbf{x}) \equiv d\mathbf{x} \cdot \nabla \mathbf{v}.$$

Denoting $|d\mathbf{x}| = ds$

$$\frac{d\mathbf{v}}{ds} = \frac{d\mathbf{x}}{ds} \cdot \nabla \mathbf{v} = \mathbf{n} \cdot \nabla \mathbf{v}.$$

$\nabla \mathbf{v}$ is a tensor (second order).

Gradient of a tensor – definition of $\nabla \mathbf{T}$ ($\mathbf{T}(\mathbf{x})$, scalar function)

$$d\mathbf{T} = \mathbf{T}(\mathbf{x} + d\mathbf{x}) - \mathbf{T}(\mathbf{x}) \equiv d\mathbf{x} \cdot \nabla \mathbf{T}.$$

Denoting $|d\mathbf{x}| = ds$

$$\frac{d\mathbf{T}}{ds} = \frac{d\mathbf{x}}{ds} \cdot \nabla \mathbf{T} = \mathbf{n} \cdot \nabla \mathbf{T}.$$

$\nabla \mathbf{T}$ is a third order tensor.

Note: In every case $\nabla(\ )$ is independent of $d\mathbf{x}$. This is what makes the definition useful.

Example

Find the components of $\nabla(\ )$.

Consider ∇f, $f(\mathbf{x})$. Note that the unit vector can be written as

$$\mathbf{n} = n_k \mathbf{e}_k = (dx_k/ds)\mathbf{e}_k.$$

Then, using the chain rule,

$$\frac{df}{ds} = \frac{\partial f}{\partial x_k} \frac{dx_k}{ds}.$$

∇f is vector, since df is scalar. Then

$$df = dx_k (\nabla f)_k$$

and

$$\frac{df}{ds} = \frac{dx_k}{ds}(\nabla f)_k$$

therefore

$$\left[(\nabla f)_k - \frac{\partial f}{\partial x_k} \right]\left(\frac{dx_k}{ds}\right) = 0$$

and

$$(\nabla f)_k = \frac{\partial f}{\partial x_k}.$$

Thus, the operator $\nabla(\ )$, applied to '·' is:

$$\nabla(\cdot) \equiv \mathbf{e}_i \frac{\partial \cdot}{\partial x_i}$$

The symbol '·' denotes a scalar, or tensor[7] preceded by an operation (no symbol, ('·' or '×'). If '·' is a vector or a tensor it should be inserted in component form.

Divergence of a vector field
Definition: ∇ is followed by '·'. Thus,

$$\nabla \cdot \mathbf{v} = \mathbf{e}_i \frac{\partial}{\partial x_i} \cdot v_j \mathbf{e}_j = \mathbf{e}_i \cdot \mathbf{e}_j \frac{\partial v_j}{\partial x_i} = \delta_{ij} \frac{\partial v_j}{\partial x_i} = \frac{\partial v_i}{\partial x_i}, \qquad \text{a scalar.}$$

If $\nabla \cdot \mathbf{v} = 0$, $\mathbf{v}$ is called solenoidal.

Divergence of a tensor field
Definition: ∇ is followed by '·'. Thus,

$$\nabla \cdot \mathbf{T} = \mathbf{e}_i \frac{\partial}{\partial x_i} \cdot T_{jk} \mathbf{e}_j \mathbf{e}_k = \mathbf{e}_i \cdot \mathbf{e}_j \mathbf{e}_k \frac{\partial T_{jk}}{\partial x_i} = \delta_{ij} \frac{\partial T_{jk}}{\partial x_i} \mathbf{e}_k = \frac{\partial T_{jk}}{\partial x_i} \mathbf{e}_k, \qquad \text{a vector.}$$

Laplacian of a scalar field
Definition: divergence of the gradient

$$\nabla \cdot \nabla f = \mathbf{e}_i \frac{\partial}{\partial x_i} \cdot \mathbf{e}_j \frac{\partial f}{\partial x_j} = \delta_{ij} \frac{\partial}{\partial x_i} \frac{\partial f}{\partial x_j} = \frac{\partial}{\partial x_i} \frac{\partial f}{\partial x_i}, \qquad \text{a scalar.}$$

This is often indicated as $\nabla^2 f$.

Gradient of a vector field

$$\nabla \mathbf{v} = \mathbf{e}_i \frac{\partial}{\partial x_i} v_j \mathbf{e}_j = \frac{\partial v_j}{\partial x_i} \mathbf{e}_i \mathbf{e}_j, \qquad \text{a tensor.}$$

Note that the ij component of $\nabla \mathbf{v}$ is $(\partial v_j / \partial x_i)$.

Laplacian of a vector field

$$\nabla \cdot \nabla \mathbf{v} = \mathbf{e}_i \frac{\partial}{\partial x_i} \cdot \mathbf{e}_j \frac{\partial}{\partial x_j} \mathbf{e}_k v_k = \delta_{ij} \frac{\partial}{\partial x_i} \frac{\partial}{\partial x_j} \mathbf{e}_k v_k = \frac{\partial}{\partial x_i} \frac{\partial}{\partial x_i} \mathbf{e}_k v_k.$$

Note: why not

$$\frac{\partial^2 \mathbf{e}_k v_k}{\partial x_i^2}?$$

Curl of a vector field

$$\nabla \times \mathbf{v} = \mathbf{e}_i \frac{\partial}{\partial x_i} \times \mathbf{e}_j v_j = \varepsilon_{ijk} \mathbf{e}_k \frac{\partial v_j}{\partial x_i}, \qquad \text{a vector}$$

If $\nabla \times \mathbf{v} = \mathbf{0}$, $\mathbf{v}$ is called irrotational.

Other operations, such as $\nabla \times \mathbf{T}$, can be defined similarly.

Problem

Prove the following results. Both f and g are scalar functions, $\mathbf{v}$ and $\mathbf{u}$ are vector functions.

$$\nabla(fg) = f\nabla g + g\nabla f$$
$$\nabla(f + g) = \nabla f + \nabla g$$
$$\nabla \cdot (\mathbf{u} + \mathbf{v}) = \nabla \cdot \mathbf{u} + \nabla \cdot \mathbf{v}$$
$$\nabla \cdot (f\mathbf{v}) = (\nabla f) \cdot \mathbf{v} + f(\nabla \cdot \mathbf{v})$$
$$\nabla \cdot (\mathbf{fv}) = \mathbf{f} \cdot (\nabla \mathbf{v})^{\mathrm{T}} + (\nabla \mathbf{f}) \cdot \mathbf{v}$$

Problem

Prove the following results

$$\nabla \cdot (\mathbf{vw}) = \mathbf{v} \cdot \nabla \mathbf{w} + \mathbf{w}(\nabla \cdot \mathbf{v})$$
$$\nabla \times f\mathbf{v} = \nabla f \times \mathbf{v} + f(\nabla \times \mathbf{v})$$
$$\nabla^2(\nabla \cdot \mathbf{v}) = \nabla \cdot \nabla^2 \mathbf{v}$$
$$\nabla \cdot (f\mathbf{1}) = \nabla f$$
$$\nabla \cdot (\nabla \mathbf{v})^{\mathrm{T}} = \nabla(\nabla \cdot \mathbf{v})$$
$$\nabla(\mathbf{a} \cdot \mathbf{x}) = \mathbf{a}, \text{ where } \mathbf{a} \text{ is a constant vector and } \mathbf{x} = x_i \mathbf{e}_i$$
$$\nabla \times \nabla f \equiv 0$$
$$\nabla \cdot \nabla \times \mathbf{v} \equiv 0$$
$$\nabla \mathbf{v} : \mathbf{1} = \nabla \cdot \mathbf{v}$$
$$\nabla(1/|\mathbf{x}|) = -\mathbf{x}/|\mathbf{x}|^3 \text{ where } \mathbf{x} = x_i \mathbf{e}_i.$$

Integral theorems for vectors and tensors

Divergence theorem or Gauss's theorem

Scalars

$$\int_V \nabla f \, dv = \oint_{\partial V} \mathbf{n} f \, ds$$

Vectors

$$\int_V (\nabla \cdot \mathbf{v}) \, dv = \oint_{\partial V} \mathbf{n} \cdot \mathbf{v} \, ds$$

Tensors

$$\int_V (\nabla \cdot \mathbf{T}) \, dv = \oint_{\partial V} \mathbf{n} \cdot \mathbf{T} \, ds.$$

The unit vector $\mathbf{n}$ is the outward normal to ∂V.

Stokes's theorem

Vectors

$$\int_S [\mathbf{n} \cdot (\nabla \times \mathbf{v}) \, ds] = \oint_C (\mathbf{t} \cdot \mathbf{v}) dc$$

Tensors

$$\int_S [\mathbf{n} \cdot (\nabla \times \mathbf{T})] \, ds = \oint_C (\mathbf{t} \cdot \mathbf{T}) \, dc.$$

The symbols are explained in Figure A.2.

Notes

1 Excellent references for this material in the context of continuum mechanics and fluid mechanics, respectively, are Chadwick (1976) and Aris (1962). Portions of this appendix are based on the appendix of Coleman, Markovitz, and Noll (1966).

2 As we shall see this definition includes both 'vectors' and 'tensors'.

3 Unless explicitly stated we use Einstein's summation convention: repeated indices are summed from 1 to 3, i.e., $\alpha_i \mathbf{e}_i = \alpha_1 \mathbf{e}_1 + \alpha_2 \mathbf{e}_2 + \alpha_3 \mathbf{e}_3$. For example, $T_{ij} v_j v_i$ means the double sum $\sum\sum T_{ij} v_j v_i$ with both indices running from 1 to 3.

4 This applies to components of vectors and tensors. Throughout this work, unless explicitly stated otherwise, we will deal with orthonormal bases.

5 As is usual in fluid mechanics we do not distinguish between column and row vectors to simplify the notation; also, the distinction it is always clear by the context of the operation.

Figure A.2.

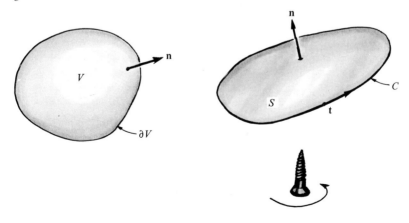

6 The reader might find it instructive to compare this with the expansion of a function in terms of base functions, as in Fourier series for example, and to identify the terms $(\mathbf{v} \cdot \mathbf{e}_j)$ with the Fourier coefficients.

7 In the literature one sometimes encounters operations such as $(\mathbf{v} \cdot \nabla)$. One interpretation is to regard ∇, by itself, as a vector. We prefer, however to regard it as an operator according to the definitions given above.

Frequently used symbols

a	acceleration	**M**	reference state of material filament
a	radius of a spherical droplet	**n**	outside normal to a material
a_v	intermaterial area density		volume
c	concentration	**n**	orientation of an area element,
Ca	capillary number		present state
D	stretching tensor	**N**	reference state of an area element
D	diffusion coefficient	p	pressure
d**a**	differential area element (present	p	viscosity ratio
	state)	**p**	orientation of microstructure with
d**A**	differential area element (reference		length $l = |\mathbf{p}|$
	state)	**p**	a variable in a Hamiltonian system
e	efficiency		(with components p_i)
E	kinetic energy within material	**P**	the fixed point indicated by
	volume V_t		location **P**
f	body force	**q**	heat flux
f	right hand side of a differential	**q**	a variable in a Hamiltonian system
	equation		(with components q_i)
f	typical mapping, as in Sections	**Q**	orthogonal tensor or matrix
	7.2 and 7.3	r	source term in diffusion equation
F	deformation gradient	Re	Reynolds number
G	shear rate	s	striation thickness
$G(\mathbf{X}, t)$	typical Lagrangian function	Sr	Strouhal number
$G(\mathbf{x}, t)$	typical Eulerian function	t	real time
H	curvature	**t**	traction
H	Hamiltonian	T	time period
I	action in Hamiltonian systems	T	temperature
I	intensity of segregation	**T**	stress tensor
I	invariant set in horseshoe map	**v**	velocity
J	Jacobian	V_0	reference material volume
K	flow character in linear flow:	V_t	material volume at time t
	$K = -1$, pure rotational flow,	x	distance normal to striations
	$K = 1$, orthogonal stagnation flow	**x**	position vector (position occupied
l	length of a droplet		by particle **X** at time t)
L_0	reference material length	**X**	particle **X**, usually designated by
L_t	material line at time t		its position at time t
m	orientation of a material filament,		
	present state		

Greek symbols

α	stretching function
α	parameter in tendril–whorl flow
β	parameter in tendril–whorl flow
β	mixing strength in the partitioned pipe mixer
γ	strain
$\dot{\gamma}$	strain rate
Γ	reduced time
$\dot{\varepsilon}$	extensional rate
ε	perturbation parameter
ε	internal energy per unit mass
ε	energy dissipation per unit volume per unit time
η	area stretch
λ	length stretch
μ	viscosity
μ	mixing strength in the blinking vortex flow
ν	kinematical viscosity
ξ	dimensionless space with respect to striation thickness
ρ	density
σ	Liapunov exponent
σ	interfacial tension
τ	warped time
$\boldsymbol{\tau}$	viscous part of the stress tensor
ϕ	potential function
$\boldsymbol{\Phi}_t(\cdot)$	motion or flow
ψ	streamfunction
ω	frequency
ω	vorticity
$\boldsymbol{\Omega}$	vorticity or spin tensor

Subscripts

bot	bottom wall in the cavity flow
c	critical value
e	exterior fluid
i	interior fluid
i	species-i

in	inner cylinder in a journal bearing flow
k	kth term in a vector
n	nth iteration
out	outer cylinder in the journal bearing flow
s	species-s
s	stable manifold
ss	steady state
top	top wall in the cavity flow
u	unstable manifold
0	reference state; initial condition
η	quantities referring to an infinitesimal area element
θ	angle in Hamiltonian systems
λ	quantities referring to an infinitesimal line element
∞	quantities referring to infinite time

Superscripts

(s)	species-s
0	initial position
$'$	quantity referring to the frame F'
∞	quantities referring to flow far away from droplet or microstructure

Special symbols

D	when applied to a vector $\mathbf{f}$, it denotes the matrix $\partial f_i/\partial X_j$ (also $\mathbf{D_X}$)
$D\cdot/Dt$	material derivative
$\{\ \}$	set
$\cup$	union
$\cap$	intersection
$[\mathbf{M}]$	brackets denote matrix representation of tensor $\mathbf{M}$
$\langle\ \rangle$	pointed brackets denote time or area average
$[\,,\,]$	Poisson brackets

References

Abraham, R., and J. E. Marsden (1985) *Foundations of mechanics*, Reading: Benjamin/ Cummings, 1st printing 1978; 5th printing, with corrections.

Abraham, R. H., and C. D. Shaw (1985) *Dynamics – the geometry of behavior: part 1, periodic behavior, part 2, chaotic behavior, and part 3, global behavior*, Santa Cruz: Aerial Press, 3rd printing.

Acrivos, A. (1983) The breakup of small drops and bubbles in shear flows, *Ann. N.Y. Acad Sci.*, **404**, 1–11.

Acrivos, A., and T. S. Lo (1978) Deformation and breakup of a single drop in an extensional flow, *J. Fluid Mech.*, **86**, 641–72.

Allègre, C. J., and D. L. Turcotte (186) Implication of a two-component marble-cake mantle, *Nature*, **323**, 123–7.

Andronov, A. A., E. A. Vitt, and S. E. Khaiken (1966) *Theory of oscillators*, Oxford: Pergamon Press.

Aref, H. (1983) Integrable, chaotic, and turbulent vortex motion in two-dimensional flows, *Ann. Rev. Fluid Mech.*, **15**, 345–89.

Aref, H. (1984) Stirring by chaotic advection. *J. Fluid Mech.*, **143**, 1–21.

Aref, H., and S. Balachandar (1986) Chaotic advection in a Stokes flow, *Phys. Fluids*, **29**, 3515–21.

Aref, H., and N. Pomphrey (1982) Integrable and chaotic motion of four vortices. I. The case of identical vortices, *Proc. Roy. Soc. London.*, **A380**, 359–87.

Aref, H., and G. Tryggvason (1984) Vortex dynamics of passive and active interfaces, *Physica*, **12D**, 59–70.

Aris, R. (1962) *Vectors, tensors, and the basic equations of fluid mechanics*, Englewood Cliffs, New Jersey: Prentice Hall.

Arnold, V. I. (1963) Small denominators and problems of stability in classical and celestial mechanics, *Russian Math. Surveys*, **18**(6), 85–191.

Arnold, V. I. (1980) *Mathematical methods of classical mechanics*, New York, Springer-Verlag (reprint with corrections of 1978 edn).

Arnold, V. I. (1983) *Geometrical methods in the theory of ordinary differential equations*, New York, Springer-Verlag.

Arnold, V. I. (1985) *Ordinary differential equations*, Cambridge, Mass.: MIT University Press, 4th printing, 1985.

Ashurst, W. T., R. Kerstein, R. M. Kerr, and C. H. Gibson (1987) Alignment of vorticity and scalar gradient with strain rate in simulated Navier–Stokes turbulence, *Phys. Fluids*, **30**, 2343–73.

Astarita, G. (1976) Objective and generally applicable criteria for flow classification, *J. non-Newtonian Fluid Mech.*, **6**, 69–76.

Ashurst, W. T., R. M. Kerr, A. Kerstein, and C. H. Gibson (1986) Examination of a three-dimensional turbulent flow field obtained by direct Navier–Stokes simulations, *SANDIA* Rep. (SAND86-8213), March.

Ballal, B. Y., and R. S. Rivlin (1976) Flow of a Newtonian fluid between eccentric rotating cylinders: inertial effects, *Arch. Rat. Mech. Anal.*, **62**, 237–94.

Barthès-Biesel, D., and A. Acrivos (1973) Deformation and burst of a liquid droplet freely suspended in a linear shear field, *J. Fluid Mech.*, **61**, 1–21.

Batchelor, G. K. (1952) The effect of homogeneous turbulence on material lines and surfaces, *Proc. Roy. Soc. Lond.*, **A213**, 349–66.

Batchelor, G. K. (1953) *The theory of homogeneous turbulence*, Cambridge: Cambridge University Press.

Batchelor, G. K. (1959) Small-scale variation of convected quantities like temperature in turbulent fluid. Part 1. General discussion and the case of small conductivity, *J. Fluid Mech.*, **5**, 113–33.

Batchelor, G. K. (1967) *An introduction to fluid mechanics*, Cambridge: Cambridge University Press (reprinted 1979).

Batchelor, G. K. (1979) Mass transfer from a particle suspended in fluid with a steady linear ambient velocity distribution, *J. Fluid Mech.*, **95**, 369–400.

Batchelor, G. K., and A. A. Townsend (1956) Turbulent diffusion, in *Surveys in mechanics*, G. K. Batchelor and R. M. Davies, eds., pp. 352–99. Cambridge: Cambridge University Press.

Bentley, B. J. (1985) Drop deformation and burst in two-dimensional flows, Ph.D. thesis, Dept. of Chemical Engineering, California Institute of Technology, Pasadena.

Bentley, B. J., and L. G. Leal (1986a) A computer-controlled four-roll mill for investigations of particle and drop dynamics in two-dimensional linear show flows, *J. Fluid Mech.*, **167**, 219–40.

Bentley, B. J., and L. G. Leal (1986b). An experimental investigation of drop deformation and breakup in steady, two-dimensional linear flows, *J. Fluid Mech.*, **167**, 241–83.

Berker, R. (1963) Intégration des équations du mouvement d'un fluide visqueux incompressible, in *Handbuck der Physik*, VIII/2, S. Flügge, ed., pp. 1–384, Berlin: Springer-Verlag.

Berman, N. S. (1978) Drag reduction by polymers, *Ann. Rev. Fluid Mech.*, **10**, 47–64.

Berry, M. V., N. L. Balazs, M. Tabor, and A. Voros (1979) Quantum maps, *Ann. Phys.*, **122**, 26–63.

Berry, M. V., and M. R. MacKley (1977) The six roll mill: unfolding an unstable persistently extensional flow, *Phil. Trans. Roy. Soc. Lond.*, **A287**, 1–16.

Bigg, D., and S. Middleman (1974) Laminar mixing in a rectangular cavity, *Ind. Eng. Chem. Fundam.*, **13**, 184–90.

Bird, R. B., W. E. Stewart, and E. N. Lightfoot (1960) *Transport phenomena*, New York: Wiley.

Birkhoff, G. D. (1920) Surface transformations and their dynamical applications, *Acta Mathematica*, **43**, 1–119.

Birkhoff, G. D. (1927) Dynamical systems, *Am. Math. Soc. Pub.*, Vol. IX, Providence: Rhode Island, 1927 (photolithoprinted by Cushing-Malloy, Ann Arbor, Michigan, 1960).

Birkhoff, G. D. (1935) Nouvelles recherches sur les systèmes dynamiques, *Pont. Acad. Sci. Novi Lyncaei*, **1**, 85.

Bowen, R. M. (1976) Theory of mixtures, pp. 1–127 in *Continuum physics*, Vol. 3, A. C. Eringen, ed., London: Academic Press.

Brenner, H. (1984) Antisymmetric stress induced by the rigid-body rotation of dipolar suspension, *Int. J. Eng. Sci.*, **22**, 645–82.

Brothman, A., G. N. Wollan, and S. M. Feldman (1945) New analysis provides formula to solve mixing problems, *Chem. Metall. Eng.*, **52**, 102–106.

Brown, G. L., and A. Roshko (1974) On density effects and large structure in turbulent mixing layers, *J. Fluid Mech.*, **64**, 775–816.

Buckmaster, J. D. (1972) Pointed bubbles in slow viscous flow, *J. Fluid Mech.*, **55**, 385–400.

Buckmaster, J. D. (1973) The bursting of pointed drops in slow viscous flow, *J. Appl. Mech.*, **40**, 18–24.

Burgers, J. M. (1948) A mathematical model illustrating the theory of turbulence, *Adv. Appl. Mech.*, **1**, 171–99.

Cantwell, B. J. (1978) Similarity transformations of the two-dimensional, unsteady, stream-function equations, *J. Fluid Mech.*, **85**, 257–71.

Cantwell, B. J. (1981) Organized motion in turbulent flow, *Ann. Rev. Fluid Mech.*, **13**, 457–515.

Cantwell, B. J., D. Coles, and P. Dimotakis (1978) Structure and entrainment in the plane of symmetry of a turbulent spot, *J. Fluid Mech.*, **87**, 641–72.

Caswell, J. B. (1967) Kinematics and stress on a surface at rest, *Arch. Rat. Mech. Anal.*, **26**, 385–99.

Chadwick, P. (1976) *Continuum mechanics*, London: G. Allen & Unwin.

Chaiken, J., R. Chevray, M. Tabor, and Q. M. Tan (1986) Experimental study of Lagrangian turbulence in Stokes flow, *Proc. Roy. Soc. Lond.*, **A408**, 165–74.

Chan, K. L., and S. Sofia (1987) Validity tests of the mixing-length theory in deep convection, *Science*, **235**, 465–7.

Chan, W. C., and L. E. Scriven (1970) Absorption into irrotational stagnation flow, *Ind. Eng. Chem. Fundam.*, **9**, 114–20.

Chella, R. (1984) Modelling of fluid mechanical and reactive mixing; applications to mixers and reactors with segregated feed, Ph.D. Thesis, Dept. of Chemical Engineering, University of Massachusetts, Amherst.

Chella, R., and J. M. Ottino (1984) Conversion and selectivity modifications due to mixing in unpremixed reactors, *Chem. Eng. Sci.*, **39**, 551–67.

Chella, R., and J. M. Ottino (1985a) Fluid mechanics of mixing in a single screw extruder, *Ind. Eng. Chem. Fundam.*, **24**, 170–80.

Chella, R., and J. M. Ottino (1985b) Stretching in some classes of fluid motions and asymptotic mixing efficiencies as measure of flow classification, *Arch. Rat. Mech. Anal.*, **90**, 15–42.

Chien, W.-L. (1986) Laminar mixing: regular and chaotic mixing in cavity flows, Ph.D. Thesis, Dept. of Chemical Engineering, University of Massachusetts, Amherst.

Chien, W.-L., H. Rising, and J. M. Ottino (1986) Laminar mixing and chaotic mixing in several cavity flows, *J. Fluid Mech.*, **170**, 355–77.

Coleman, B. D., H. Markovitz, and W. Noll (1966) *Viscometric flows of non-Newtonian fluids*, Berlin: Springer-Verlag.

Cooper, A. R. (1966) Diffusive mixing in continuous laminar flow systems. *Chem. Eng. Sci.*, **21**, 1095–1106.

Corcos, G. M., and F. S. Sherman (1984) The mixing layer: deterministic models of a turbulent flow. Part 1. Introduction and the two-dimensional flow, *J. Fluid Mech.*, **193**, 29–65.

Corrsin, S. (1954) A measure of the area of a homogeneous random surface in space, *Quart. Appl. Math.*, **12**, 404–8.

Corrsin, S. (1957) Simple theory of an idealized turbulent mixer, *AIChE J.*, **3**, 329–30.

Corrsin, S. (1972) Simple proof of fluid line growth in stationary homogeneous turbulence, *Phys. Fluids*, **15**, 1370–72.

Corrsin, S., and M. Karweit (1969) Fluid line growth in grid-generated isotropic turbulence, *J. Fluid Mech.*, **39**, 87–96.

Cox, R. G. (1969) The deformation of a drop in a general time-dependent fluid flow, *J. Fluid Mech.*, **37**, 601–23.

Crawford, J. D., and S. Omohundro (1984) On the global structure of period doubling flows, *Physica*, **13D**, 161–80.

Damköhler, G. (1936) Einflüsse der Strömung, Diffusion und des Wärmeüberganges auf die Leistung von Reaktionsöfen. I. Allgemeine Gesichtspunkte für die Übertragung eines chemischen Prozesses aus dem Kleinen ins Grosse, *Z. Elektrochem.*, **42**, 846–62.

Danckwerts, P. V. (1952) The definition and measurement of some characteristics of mixtures, *Appl. Sci. Res.*, **A3**, 279–96.

Danckwerts, P. V. (1953) Continuous flow systems-distribution of residence times, *Chem. Eng. Sci.*, **2**, 1–13.

Danckwerts, P. V. (1958) The effect of incomplete mixing on homogeneous reactions, *Chem. Eng. Sci.*, **8**, 93–9.

Danielson, T. J., Ph.D. Thesis, Dept. of Chemical Engineering, University of Massachusetts, Amherst, in progress.

de Gennes, P. G. (1986) Towards a scaling theory of drag reduction, *Physica*, **140A**, 9–25.

Devaney, R. L. (1986) *An introduction to chaotic dynamical systems*, Menlo Park: Benjamin/Cummings.

Dimotakis, P. E., R. C. Miake-Lye, and D. A. Papantoniou (1983) Structure and dynamics of turbulent round jets, *Phys. Fluids*, **26**, 3185–92.

Doherty, M. F., and J. M. Ottino (1988) Chaos in deterministic systems: strange attractors, turbulence, and applications in chemical engineering, *Chem. Eng. Sci.*, **43**, 139–83.

Dombre, T., U. Frisch, J. M. Greene, M. Hénon, A. Mehr, and A. M. Soward (1986) Chaotic streamlines in the ABC flows, *J. Fluid Mech.*, **167**, 353–91.

Duffing, G. (1924) Beitrag zur Theorie der Flüssigkeitsbewegung zwischen Zapfen und Lager, *Z. angew. Math. Mech. (ZAMM)*, **4**, 297–314.

Eckart, C. (1948) An analysis of the stirring and mixing process in incompressible fluids, *J. Marine Res.*, **VII**, 265–75.

Escande, D. F. (1985) Stochasticity in classical Hamiltonian systems: universal aspects, *Phys. Reports*, **121**, 165–261.

Feigenbaum, M. J. (1980) Universal behavior in non-linear systems, *Los Alamos Science*, **1**, 4–27.

Feingold, M., L. P. Kadanoff, and O. Piro (1988) Passive scalars, three-dimensional volume preserving maps and chaos, *J. Stat. Phys.*, **50**, 529–65.

Fields, S. D., and J. M. Ottino (1987a) Effect of segregation on the course of polymerizations, *AIChE J.*, **33**, 959–75.

Fields, S. D., and J. M. Ottino (1987b) Effect of stretching path on the course of polymerizations: applications to idealized unpremixed reactors, *Chem. Eng. Sci.*, **42**, 467–77.

Fields, S. D., and J. M. Ottino (1987c) Effect of striation thickness distribution on the course of an unpremixed polymerization, *Chem. Eng. Sci.*, **42**, 459–65.

Fields, S. D., E. L. Thomas and J. M. Ottino (1987) Visualization of interfacial urethane polymerization by means of a new microstage reactor, *Polymer*, **27**, 63–72.

Finlayson, B. A. (1972) *Method of weighted residuals and variational principles*, New York: Academic Press.

Fisher, D. A. (1968) A model for fast reactions in turbulently mixed fluids, M.Sci. Thesis, Dept. of Chemical Engineering, University of Minnesota, Minneapolis.

Foister, R. T., and T. G. M. van de Ven (1980) Diffusion of Brownian particles in shear flows, *J. Fluid Mech.*, **96**, 105–32.

Franjione, J. G., Ph.D. Thesis, Dept. of Chemical Engineering, University of Massachusetts, Amherst, in progress.

Franjione, J. G., and J. M. Ottino (1987a) Chaotic mixing in two continuous systems, *Bull. Am. Phys. Soc.*, **32**, 2026 (abstract only).

Franjione, J. G., and J. M. Ottino (1987b) Feasibility of numerical tracking of material lines and surfaces in chaotic flows, *Phys. Fluids*, **30**, 3641–3.

Galloway, D., and U. Frisch (1986) Dynamo action in a family of flows with chaotic streamlines, *Geophys. Astrophys. Fluid Dynam.*, **36**, 53–83.

Galloway, D., and U. Frisch (1987) A note on the stability of space-periodic flows, *J. Fluid Mech.*, **180**, 557–64.

Gibbs, J. W. (1948) The Collected Works of J. W. Gibbs, in *Elementary principles in statistical mechanics*, Vol. II, Part I, pp. 144–151, New Haven: Yale University Press.

Gibson, C. H. (1968) Fine structure of scalar fields mixed by turbulence: I. Zero-gradient points and minimal gradient surfaces, *Phys. Fluids*, **11**, 2305–15.

Giesekus, H. (1962) Strömungen mit konstantem Geschwindigkeitsgradienten und die Bewegung von darin suspendierten Teilchen. Teil II: Ebene Strömungen und ein experimentelle Anordung zu ihrer Realisierung, *Rheol. Acta*, **2**, 113–21.

Gillani, N. W., and W. N. Swanson (1976) Time-dependent laminar incompressible flow through a spherical cavity, *J. Fluid Mech.*, **78**, 99–127.

Goldberg, E. D., I. N. McCave, J. J. O'Brien, and J. H. Steele (eds) (1977) *The Sea, Volume 6: Marine modelling*, New York: Wiley.

Goldstein, H. (1950) *Classical mechanics*, Cambridge, Mass.: Addison-Wesley.

Grace, H. P. (1982) Dispersion phenomena in high viscosity immiscible fluid systems and applications of static mixers as dispersion devices in such systems, *Chem. Eng. Commun.*, **14**, 225–77. Presented originally at the Third Engineering Foundation Conference in Mixing, Andover, N.H., August, 1971.

Greene, J. M. (1986) How a swing behaves, *Physica*, **18D**, 427–47.

Greene, J. M., and J.-S. Kim (1987) The calculation of Lyapunov spectra, *Physica*, **24D**, 213–25.

Guckenheimer, J. (1986) Strange attractors in fluids: another view, *Ann. Rev. Fluid Mech.*, **18**, 15–31.

Guckenheimer, J., and P. Holmes (1983) *Nonlinear oscillations, dynamical systems and bifurcations of vector fields*, New York, Springer-Verlag.

Hama, F. R. (1962) Streaklines in a perturbed shear flow, *Phys. Fluids*, **5**, 644–50.

Hawthorne, W. R., D. S. Wendell, and H. C. Hottel (1948) Mixing and combustion in turbulent gas jets, in *Third Symp. on Combustion and Flame and Explosion Phenomena*, pp. 266–88, Baltimore: Williams & Wikens, 1948.

Helleman, R. H. G. (1980) Self generated chaotic behavior in nonlinear mechanics, Proceedings of the 5th International Summer School on Fundamental Problems in

Statistical Mechanics, published in *Fundamental problems in statistical mechanics V*, E. G. D. Cohen, ed., pp. 165–275, Amsterdam: North-Holland.

Heller, J. P. (1960) An unmixing demonstration, *Am. J. Phys.*, **28**, 348–56.

Hénon, M. (1966) Sur la topologie des lignes de courant dans un cas particulier, *Comptes Rendus Acad. Sci. Paris*, **A262**, 312–14.

Hinch, E. J., and A. Acrivos (1979) Steady long slender drops in two-dimensional straining motion, *J. Fluid Mech.*, **91**, 401–14.

Hinch, E. J., and A. Acrivos (1980) Long slender drops in a simple shear flow, *J. Fluid Mech.*, **98**, 305–28.

Hirsch, M. W., and S. Smale (1974) *Differential equations, dynamical systems, and linear algebra*, New York: Academic Press.

Hoffman, N. R. A., and D. P. McKenzie (1985) The destruction of geochemical heterogeneities by differential fluid motion during mantle convection, *Geophys. J. R. Astr. Soc.*, **82**, 163–206.

Holland, W. R. (1977) Ocean circulation models, in *The Sea, Volume 6: Marine modelling*, E. D. Goldberg, I. N. McCave, J. J. O'Brien, and J. H. Steele, eds., New York: Wiley.

Holloway, G., and S. S. Kirstmannson (1984) Stirring and Transport of tracer fields by geostrophic turbulence, *J. Fluid Mech.*, **141**, 27–50.

Huilgol, R. R. (1975) *Continuum mechanics of viscoelastic liquids*, New York: Halsted Press.

Hyman, D. (1963) Mixing and agitation, *Adv. Chem. Eng.*, **3**, 113–202.

Ioos, G., and D. D. Joseph (1980) *Elementary stability and bifurcation theory*, New York, Springer-Verlag.

Jeffrey, G. B. (1922) The rotation of two cylinders in a viscous fluid, *Proc. Roy. Soc. Lond.*, **A101**, 169–74.

Jones, S. W., and H. Aref (1988) Chaotic advection in pulsed source-sink systems, *Phys. Fluids*, **31**, 469–85.

Jones, S. W., O. M. Thomas, and H. Aref (1990) Chaotic advection by laminar flow in a twisted pipe. *J. Fluid Mech.*, **209**, 335–57.

Kantorovich, L. V., and V. I. Krylov (1964) *Approximate methods of higher analysis*, New York: Wiley.

Kapila, A. K. (1983) *Asymptotic treatment of chemically reacting systems*, London: Pitman.

Karam, H. J., and J. C. Bellinger (1968) Deformation and breakup of liquid droplets in simple shear flow, *Ind. Eng. Chem. Fundam.*, **7**, 576–83.

Kazakia, J. Y., and R. S. Rivlin (1978) Flow of a Newtonian fluid between eccentric rotating cylinders and related problems, *Studies in Appl. Math.*, **58**, 209–47.

Kerr, R. M. (1985) Higher-order derivative correlations and the alignment of small-scale structures in isotropic numerical turbulence, *J. Fluid Mech.*, **153**, 31–58.

Khakhar, D. V. (1986) Fluid mechanics of laminar mixing: dispersion and chaotic flows, Ph.D. Thesis, Dept. of Chemical Engineering, University of Massachusetts, Amherst.

Khakhar, D. V., R. Chella, and J. M. Ottino (1984) Stretching, chaotic motion, and breakup of elongated droplets in time dependent flows, in *Proc. IX Intl. Congress on Rheology, Vol. 2*, B. Mena, A. García-Rejón, and C. Range-Nafaile, eds., pp. 81–8, Mexico City: Universidad Nacional Autonoma de Mexico.

Khakhar, D. V., J. G. Franjione, and J. M. Ottino (1987) A case study of chaotic mixing in deterministic flows: the partitioned-pipe mixer, *Chem. Eng. Sci.*, **42**, 2909–26.

Khakhar, D. V., and J. M. Ottino (1985) Chaotic mixing in two-dimensional flows: stretching of material lines, *Bull. Am. Phys. Soc.*, **30**, 1702 (abstract only).

Khakhar, D. V., and J. M. Ottino (1986a) Fluid mixing (stretching) by time-periodic sequences of weak flows, *Phys. Fluids*, **29**, 3503–5.

Khakhar, D. V., and J. M. Ottino (1986b) A note on the linear vector model of Olbricht, Rallison, and Leal as applied to the breakup of slender axisymmetric drops, *J. non-Newt. Fluid Mech.*, **21**, 127–31.

Khakhar, D. V., and J. M. Ottino (1986c) Deformation and breakup of slender drops in linear flows, *J. Fluid Mech.*, **166**, 265–85.

Khakhar, D. V., and J. M. Ottino (1987) Breakup of liquid threads in linear flows, *Int. J. Multiphase Flow*, **13**, 71–86.

Khakhar, D. V., H. Rising, and J. M. Ottino (1986) An analysis of chaotic mixing in two chaotic flows, *J. Fluid Mech.*, **172**, 419–51.

Kolmogorov, A. N. (1954a) On conservation of conditionally periodic motions under small perturbations of the Hamiltonian, *Dokl. Akad. Nauk. SSSR*, **98**, 527–30.

Kolmogorov, A. N. (1954b) The general theory of dynamical systems and classical mechanics, in *Proceedings of the 1954 Congress in Mathematics*, pp. 315–33, Amsterdam: North-Holland. Translated as Appendix (pp. 741–57) in R. H. Abraham and J. E. Marsden (1978) *Foundations of mechanics*, Benjamin/Cummings: Reading, Massachusetts.

Koochesfahani, M. M., and P. E. Dimotakis (1985) Laser-induced fluorescence measurements of mixed fluid concentration in a liquid plane shear layer, *AIAA J.*, **23**, 1700–7.

Koochesfahani, M. M., and P. E. Dimotakis (1986) Mixing and chemical reactions in a turbulent mixing layer, *J. Fluid Mech.*, **170**, 83–112.

Kusch, H. A., Ph.D. Thesis, Dept. Chemical Engineering, University of Massachusetts, Amherst, in progress.

Lamb, H. (1932) *Hydrodynamics*, 6th edition, Cambridge: Cambridge University Press (reprinted New York: Dover, 1945).

Landford, O. E. (1982) The strange attractor theory of turbulence, *Ann. Rev. Fluid Mech.*, **14**, 347–64.

Leal, L. G. (1984) Birefringence studies of flow-induced conformation changes in polymer solutions, in *Proc. IX Intl, Congress on Rheology, vol. 1*, B. Mena, A.García-Rejón, and C. Rangel-Nafaile, eds, pp. 191–209, Mexico City: Universidad Autonoma de Mexico.

Lee, L. J., J. M. Ottino, W. E. Ranz, and C. W. Macosko (1980) Impingement mixing in reaction injection molding, *Polym. Eng. Sci.*, **20**, 868–74.

Leonard, A., V. Rom-Kedar, and S. Wiggins (1987) Fluid mixing and dynamical systems, *Nucl. Phys. B. (Proc. Suppl.)*, **2**, 179–90.

Leong, C. W., (1990) Chaotic mixing of viscous fluids in time-periodic cavity flows, Ph.D. Thesis, Dept. of Chemical Engineering, University of Massachusetts, Amherst.

Levich, V. G. (1962) *Physicochemical hydrodynamics*, Englewood Cliffs, N.J.: Prentice-Hall.

Lichtenberg, A. J., and M. A. Lieberman (1983) *Regular and stochastic motion*, New York, Springer-Verlag.

Lighthill, M. J. (1963) Introduction. Boundary layer theory, in *Laminar boundary layers*, L. Rosenhead, ed., Chap. II, pp. 46–113, London: Oxford University Press.

Lorenz, E. N. (1963) Deterministic nonperiodic flow, *J. Atmos. Sci.*, **20**, 130–41.

Lumley, J. L. (1973) Drag reduction in turbulent flow by polymer additives, *J. Polym. Sci., Macromolec. Rev.*, **7**, 263–90.

Lundgren, T. S. (1982) Strained spiral vortex model for turbulent fine structure, *Phys. Fluids*, **25**, 2193–2203.

Lundgren, T. S. (1985) The concentration spectrum of the product of a fast bimolecular reaction, *Chem. Eng. Sci.*, **40**, 1641–52.

MacKay, R. S., J. D. Meiss, and I. C. Percival (1984) Transport in Hamiltonian systems, *Physica*, **13D**, 55–81.

MacKay, R. S., and. J. D. Meiss (1987) Hamiltonian dynamical systems: a reprint selection, Philadelphia: Taylor and Francis.

Malvern, L. E. (1969) *Introduction to the mechanics of a continuous medium*, Englewood Cliffs, N.J.: Prentice-Hall.

Marble, F. E., and J. E. Broadwell (1977) The coherent flame model for turbulent chemical reactions, *Project SQUID, Tech. Rep. TRW-9-PU*.

May, R. M. (1976) Models with very complicated dynamics, *Nature*, **261**, 459–67.

McCabe, W. L., and J. C. Smith (1956) *Unit operations of chemical engineering*, New York: McGraw-Hill, 2nd edition.

McKenzie, D. P. (1983) The earth's mantle, *Scientific American*, **249**, 66–78.

McKenzie, D. P., J. M. Roberts, and N. O. Weiss (1974) Convection in the earth's mantle: towards a numerical simulation, *J. Fluid Mech.*, **62**, 465–538.

McMurtry, P. A., W.-H. Jou, J. J., Riley, and R. W. Metcalfe (1986) Direct numerical simulations of a reacting mixing layer with chemical heat release, *AIAA Journal*, **24**, 962–70.

Meakin, P. (1985) Computer simulation of growth and aggregation processes, in *On growth and form: fractal and non-fractal patterns in physics*, H. E. Stanley and N. Ostrowsky, eds., pp. 111–135, Dordrecht: Nijhoff.

Melnikov, V. K (1963) On the stability of the center for time periodic perturbations, *Trans. Moscow Math. Soc.*, **12**, 1–57.

Middleman, S. (1977) *Fundamentals of polymer processing*, New York: McGraw-Hill.

Mikami, T., R. Cox, and R. G. Mason (1985) Breakup of extending liquid threads, *Int. J. Multiphase Flow*, **2**, 113–18.

Milne-Thomson, L. M. (1955) *Theoretical hydrodynamics*, New York: Macmillan, 3rd edition.

Minorsky, N. (1962) *Nonlinear oscillations*, Princeton: Van Nostrand.

Moffat, H. K. (1969) The degree of knottedness of tangled vortex lines, *J. Fluid Mech.*, **35**, 117–129.

Mohr, W. D., R. L. Saxton, and C. H. Jepson (1957) Mixing in laminar flow systems, *Ind. Eng. Chem.*, **49**, 1855–7.

Moser, J. (1962) On invariant curves of area-preserving mappings of an annulus, *Nachr. Akad. Wiss. Göttingen Math. Phys.* **Kl. II**, 1–20.

Moser, J. (1973) *Stable and random motion in dynamical systems*, Princeton: Princeton University Press.

Muzzio, F. J. and J. M. Ottino (1988) Coagulation in chaotic flows, *Phys. Rev. A*, **38**, 2516–24.

Naumann, E. B., and B. A. Buffham (183) *Mixing in continuous flows systems*, New York: Wiley.

Noll, W. (1962) Motions with constant stretch history, *Arch. Rat. Mech. Anal.*, **11**, 97–105.

Nollert, M. U., and W. L. Olbricht (1985) Macromolecular deformation in periodic extensional flows, *Rheol. Acta*, **24**, 3–14.

Olbricht, W. L., J. M. Rallison, and L. G. Leal (1982) Strong flow criteria based on microstructure deformation, *J. non-Newt. Fluid Mech.*, **10**, 291–318.

Onsager, L. (1949) Statistical hydrodynamics, *Nuovo Cimento*, **VI** ser. IX, 279–87.

Oseen, C. W. (1931) *Vehr. des 3. int. Kongr. für techn. Mechanik* [Stockholm 1930] Bd. 1, pp. 3–22, Stockholm.

Ottino, J. M. (1981) Efficiency of mixing from data on fast reactions in multi-jet reactors and stirred tanks, *AIChE J.*, **27**, 184–92.

Ottino, J. M. (1982) Description of mixing with diffusion and reacton in terms of the concept of material surfaces, *J. Fluid Mech.*, **114**, 83–103.

Ottino, J. M., and C. W. Macosko (1980) An efficiency parameter for batch mixing of viscous fluids, *Chem. Eng. Sci.*, **35**, 1454–7.

Ottino, J. M., W. E. Ranz, and C. W. Macosko (1979) A lamellar model for analysis of liquid-liquid mixing, *Chem. Eng. Sci.*, **34**, 877–90.

Ottino, J. M., W. E. Ranz, and C. W. Macosko (1981) A framework for the mechanical mixing of fluids, *AIChE J.*, **27**, 565–77.

Ottino, J. M., C. W. Leong, H. Rising, and P. D. Swanson (1988) Morphological structures produced by mixing in chaotic flows, *Nature*, **333**, 419–25.

Patterson, A. R. (1983) *A first course in fluid mechanics*, Cambridge: Cambridge University Press.

Palis, J. (1969) On Morse–Smale dynamical systems, *Topology*, **8**, 385–405.

Pan, F., and A. Acrivos (1967) Steady flows in rectangular cavities, *J. Fluid Mech.*, **28**, 643–55.

Peixoto, M. M. (1962) Structural stability on two-dimensional manifolds, *Topology*, **2**, 101–20.

Percival, I., and D. Richards (1982) *Introduction to dynamics*, Cambridge: Cambridge University Press.

Perry, A. E., and M. S. Chong (1986) A series-expansion study of the Navier–Stokes equations with applications to three-dimensional separation patterns, *J. Fluid Mech.*, **173**, 2907–23.

Perry, A. E., and M. S. Chong (1987) A description of eddying motions and flow patterns using critical-point concepts, *Ann. Rev. Fluid Mech.*, **19**, 125–55.

Perry, A. E., and B. D. Fairlie (1974) Critical points in flow patterns, *Adv. in Geophys.*, **18B**, 299–315.

Poincaré, H. (1892, 1893, 1899) *Les méthodes nouvelles de la méchanique céleste*, 3 vols., Paris: Gauthier-Villars.

Prandtl, L. (1925) Bericht über Untersuchungen zur ausgebildeten Turbulenz, *Z. Angew. Math. Mech. (ZAMM)*, **5**, 136–139.

Prandtl, L., and O. G. Tietjens (1934) *Fundamentals of hydro- and aero-mechanics* (Engineering Societies Monographs), 1st edn, N.Y.: McGraw-Hill (reprinted by Dover, 1957).

Rallison, J. M. (1984) The deformation of small viscous drops and bubbles in shear flows, *Ann. Rev. Fluid Mech.*, **16**, 45–66.

Ralph, M. E. (1986) Oscillatory flows in wavy-walled tubes, *J. Fluid Mech.*, **168**, 515–40.

Ranz, W. E. (1979) Application of a stretch model to mixing, diffusion, and reaction in laminar and turbulent flows, *AIChE J.*, **25**, 41–7.

Ranz, W. E. (1985) Fluid mechanical mixing-lamellar description, Ch. 1, pp. 1–28 in *Mixing of liquids by mechanical agitation*, J. J. Ulbrecht and G. K. Patterson, eds. New York: Gordon and Breach.

Reichl, L. E. (1980) *A modern course in statistical physics*, Austin: University of Texas Press.

Rhines, P. B. (1977) The dynamics of unsteady currents, in *The Sea, Volume 6, Marine modelling*, E. D. Goldberg, I. N. McCave, J. J. O'Brien, and J. H. Steele, eds., New York: Wiley.

Rhines, P. B. (1979) Geostrophic turbulence, *Ann. Rev. Fluid Mech.*, **11**, 401–41.

Rhines, P. B. (1983) Vorticity dynamics of the oceanic general circulation, *Ann. Rev. Fluid Mech.*, **18**, 433–97.

Riley, J. J., R. W. Metcalfe, and S. A. Orzag (1986) Direct simulations of chemically reacting turbulent mixing layers, *Phys. Fluids*, **29**, 406–22.

Rising H. (1989) Applications of chaos and dynamical systems approaches to mixing in fluid, Ph.D. Thesis. Dept of Mathematics and Statistics, University of Massachusetts, Amherst.

Rising, H., and J. M. Ottino (1985) Use of horseshoe functions in fluid mixing systems, *Bull. Am. Phys. Soc.*, **30**, 1699 (abstract only).

Roberts, F. A. (1985) Effects of a periodic disturbance on structure and mixing in turbulent shear layers and wakes, Ph.D Thesis, Graduate Aeronautical Laboratories, California Institute of Technology.

Rom-Kedar, V., A. Leonard, and S. Wiggins (1990) An analytical study of transport, mixing, and chaos in an unsteady vortical flow, *J. Fluid Mech*, **214**, 347–94.

Rosensweig, R. E. (1985) *Ferrohydrodynamics*, New York: Cambridge University Press.

Roshko, A. (1976) Structure of turbulent shear flows: a new look, *AIAA J.*, **14**, 1349–57.

Roughton, F. J. W., and B. Chance (1963) in *Techniques of organic chemistry*, Vol. 8, A. Weissberger, ed., Chap. XIV, New York: Interscience.

Rumscheidt, F. D., and S. G. Mason (1961) Particle motions in sheared suspensions. XII. Deformation and burst of fluid drops in shear and hyperbolic flows, *J. Colloid Sci.*, **16**, 238–61.

Rüssmann, H. (1970) Über invariante Kurven differenzierbarer Abbildungen eines Kreisringes, *Nachr. Akad. Wiss. Göttingen Math. Phys.*, **Kl**, 67–105.

Ryu, H.-W., H.-N. Chang, and D.-I. Lee (1986) Creeping flows in rectangular cavities with translating top and bottom walls: numerical study and flow visualization, *Korean J. Chem. Eng.*, **32**, 177–85.

Saffman, P. G. (1963) On the fine-scale structure of vector fields convected by a turbulent fluid, *J. Fluid Mech.*, **16**, 545–72.

Salam, F. M. A., J. E. Marsden, and P. P. Varaiya (1983) Chaos and Arnold diffusion in dynamical systems, *IEEE Trans. on Circuits and Systems*, **CAS-30**, 697–708.

Sax, J. E. (1985) Transport of small molecules in polymer blends: transport-morphology relationships, Ph.D. Thesis, Dept. of Chemical Engineering, University of Massachusetts, Amherst.

Sax, J. E., and J. M. Ottino (1985) Influence of morphology on the transport properties of polybutadiene/polystyrene blends: experimental results, *Polymer*, **26**, 1073–80.

Schlichting, H. (1955) *Boundary layer theory*, New York: McGraw-Hill.

Schowalter, W. R. (1979) *Mechanics of non-Newtonian fluids*, Oxford: Pergamon Press.

Schuster, H. G. (1984) *Deterministic chaos: an introduction*, Weinheim: Physik-Verlag.

Serrin, J. (1959) The mathematical principles of classical fluid mechanics, in *Handbuch der Physik*, VIII/1, S. Flügge, ed., pp. 125–263, Berlin: Springer-Verlag.

Serrin, J. (1977) Lectures for 8.430, Mathematical theory of fluid dynamics, University of Minnesota, Minneapolis (unpublished).

Sevick, E. M., P. A. Monson, and J. M. Ottino (1988) Morphology and transport using the Ising lattice as a morphological description. *Chem. Eng. Sci.*, to appear.

Shinnar, R. (1961) On the behaviour of liquid dispersions in mixing vessels, *J. Fluid Mech.*, **10**, 259–75.

Smale, S. (1967) Differentiable dynamical systems, *Bull. Am. Math. Soc.*, **73**, 747–817.

Snyder, W. T., and G. A. Goldstein (1965) An analysis of fully developed flow in an eccentric annulus, *AIChE J.*, 642–467.

Sobey, I. J. (1985) Dispersion caused by separation during oscillatory flow through a furrowed channel, *Chem. Eng. Sci.*, **40**, 2129–34.

Solomon, T. H. and J. P. Gollub (1988) Chaotic particle transport in time dependent Rayleigh-Bénard convection, *Phys. Rev. A*, to appear.

Spalding, D. B. (1977) The ESCIMO theory of turbulent combustion, *Imperial College Mech. Eng. Dept. Rep. No.* HTS/76/13.

Spalding, D. B. (1978a) The influences of laminar transport and chemical kinetics on the time-mean reaction rate in a turbulent flame, in *Proc. 17th Symp. (Int.) on Combustion*, pp. 431–40, The Combustion Institute.

Spalding, D. B. (1978b) Chemical reactions in turbulent fluids, in *Physico-Chemical Hydrodynamics; Proc. Levich 60th birthday Conference*, pp. 321–38, London: Advance Publications, May.

Spencer, R. S., and R. M. Wiley (1951) The mixing of very viscous liquids, *J. Colloid. Sci.*, **6**, 133–45.

Stone, H., B. J. Bentley, and L. G. Leal (1986) An experimental study of transient effects in the breakup of viscous drops. *J. Fluid Mech.*, **173**, 131–58.

Stone, H. A., and L. G. Leal (1989) Relaxation and breakup of an initialy extended drop in an otherwise quiescent fluid. *J. Fluid Mech.* **198**, 399–427.

Stuart, J. T. (1967) On finite amplitude oscillation in laminar mixing layers, *J. Fluid Mech.*, **29**, 417–40.

Swanson, P. D. and J. M. Ottino (1990) A comparative, compuntational and experimental study of chaotic mixing of viscous fluids, *J. Fluid Mech.* **213**, 227–249.

Swanson, P. D., and J. M. Ottino (1985) Chaotic mixing of viscous liquids between eccentric cylinders, *Bull. Am. Phys. Soc.*, **30**, 1702 (abstract only).

Swinney, H. L. (1985) Observations of complex dynamics and chaos, in *Fundamental problems in statistical mechanics VI*, E. G. D. Cohen, ed., pp. 253–89, Amsterdam: Elsevier (the appendix of this paper was published in 1983 in *Physica*, **7D**, 3–15).

Synge, J. L. (1960) Classical dynamics, in *Handbuck der Physik*, III/1, S. Flügge, ed., pp. 1–225, Berlin: Springer-Verlag.

Tadmor, Z., and C. G. Gogos (1979) *Principles of polymer processing*, New York: Wiley Interscience.

Tanner, R. I. (1976) A test particle approach to flow classification for viscoelastic fluids, *AIChE J.*, **22**, 910–14.

Tanner, R. I., and R. R. Huilgol (1975) On a classification scheme for flow fields, *Rheol. Acta*, **14**, 959–62.

Taylor, G. I. (1934) The formation of emulsions in definable fields of flow, *Proc. Roy. Soc., Lond.*, **A146**, 501–23.

Taylor, G. I. (1964) Conical free surfaces and fluid interfaces, in *Proc. 11th Int. Cong. Appl. Mech.*, pp. 790–6, Munich.

Tennekes, H., and J. L. Lumley (1980) *A first course in turbulence*, Cambridge, Mass.: MIT Press, 6th printing.

Thirring, W. (1978) *A course in mathematical physics, I: Classical dynamical systems*, New York, Springer-Verlag.

Torza, S., R. Cox, and S. G. Mason (1972) Particle motions in sheared suspensions. XXVII. Transient and steady deformation and burst of liquid drops, *J. Colloid. Sci.*, **38**, 395–411.

Townsend, A. A. (1951) The diffusion of heat spots in isotropic turbulence, *Proc. Roy.*

Soc., Lond., **A209**, 418–30.

Tritton, D. J. (1977) *Physical fluid dynamics*, New York: Van Nostrand Reinhold (reprinted 1982).

Truesdell, C. A. (1954) *The kinematics of vorticity*, Bloomington: Indiana University Press.

Truesdell, C. A. (1977) *A first course in rational continuum mechanics, Part I: Fundamental concepts*, New York: Academic Press.

Truesdell, C. A., and R. Toupin (1960) The classical field theories, in *Handbuch der Physik*, III/1, S. Flügge, ed., pp. 226–793, Berlin: Springer-Verlag.

Ulbrecht, J. J., and G. K. Patterson, eds. (1985) *Mixing of liquids by mechanical agitation*, New York: Gordon and Breach.

van Dyke, M. (1982) *An album of fluid motion*, Stanford: The Parabolic Press.

Veronis, G. (1973) Large scale ocean circulation, *Adv. Appl. Mech.*, **13**, 1–92.

Veronis, G. (1977) The use of tracer in circulation studies, in *The Sea, Volume 6, Marine modelling*, E. D. Goldberg, I. N. McCave, J. J. O'Brien, and J. H. Steele, eds., pp. 169–88, New York: Wiley.

Virk, P. S. (1975) Drag reduction fundamentals, *AIChE J.*, **21**, 625–56.

von Mises, R., and K. O. Friedrichs (1971) *Fluid mechanics*, New York, Springer-Verlag.

Wallerstein, G. (1988) Mixing in stars, *Science*, **240**, 1743–50.

Walters, P. (1982) *An introduction to ergodic theory*, New York, Springer-Verlag.

Wannier, G. H. (1950) A contribution to the hydrodynamics of lubrication, *Quart. Appl. Math.*, **VIII**, 1–32.

Welander, P. (1955) Studies on the general development of motion in a two-dimensional, ideal fluid, *Tellus*, **7**, 141–56.

Wickert, P. D., C. W., Macosko, and W. E. Ranz (1987) Small-scale mixing phenomena during reaction injection moulding, *Polymer*, **28**, 1105–10.

Wiggins, S. (1988a) On the detection and dynamical consequences of orbits homoclinic to hyperbolic periodic orbits and normally hyperbolic invariant tori in a class of ordinary differential equations, *SIAM J. Appl. Math.*, to appear

Wiggins, S. (1988b) *Global bifurcations and chaos-analytical methods*, New York: Springer-Verlag.

Witten, T. A., and L. M. Sander (1981) Diffusion-limited aggregation, a kinetic critical phenomenon, *Phys. Rev. Lett.*, **47**, 1400–3.

Witten, T. A., and L. M. Sander (1983) Diffusion-limited aggregation, *Phys. Rev.*, **B27**, 5686–97.

Zweitering, Th. N. (1959) The degree of mixing in continuous flow systems, *Chem. Eng. Sci.*, **11**, 1–15.

Author index

Italic numbers refer to figures.

Subject index

Note: A subscript following a page number indicates an endnote; for example 16_2, indicates the second footnote of page 16. An italic number indicates a figure or table.